HEMATOPOIESIS

Biochemical, Cellular, Molecular, and Genomic Perspectives

Modern Trends in Molecular Biology

HEMATOPOIESIS

Biochemical, Cellular, Molecular, and Genomic Perspectives

Edited by

Ravi Kumar Gutti, PhD

Book Series Edited by

Devarajan Thangadurai, PhD

APPLE ACADEMIC PRESS

First edition published 2024

Apple Academic Press Inc.
1265 Goldenrod Circle, NE,
Palm Bay, FL 32905 USA

760 Laurentian Drive, Unit 19,
Burlington, ON L7N 0A4, CANADA

CRC Press
2385 NW Executive Center Drive,
Suite 320, Boca Raton FL 33431

4 Park Square, Milton Park,
Abingdon, Oxon, OX14 4RN UK

Library and Archives Canada Cataloguing in Publication

Title: Hematopoiesis : biochemical, cellular, molecular, and genomic perspectives / edited by Ravi Kumar Gutti, PhD ; book series edited by Devarajan Thangadurai, PhD.
Other titles: Hematopoiesis (Gutti)
Names: Gutti, Ravi Kumar, editor.
Series: Modern trends in molecular biology.
Description: First edition. | Series statement: Modern trends in molecular biology | Includes bibliographical references and index.
Identifiers: Canadiana (print) 20230497837 | Canadiana (ebook) 20230497861 | ISBN 9781774914724 (hardcover) | ISBN 9781774914731 (softcover) | ISBN 9781003413059 (ebook)
Subjects: LCSH: Hematopoiesis. | LCSH: Hematopoietic stem cells.
Classification: LCC QP92 .H46 2023 | DDC 573.1/55—dc23

Library of Congress Cataloging-in-Publication Data

CIP data on file with US Library of Congress

ISBN: 978-1-77491-472-4 (hbk)
ISBN: 978-1-77491-473-1 (pbk)
ISBN: 978-1-00341-305-9 (ebk)

Modern Trends in Molecular Biology Book Series

ABOUT THE SERIES

Series Editor: Devarajan Thangadurai, PhD
Professor, Karnatak University, Dharwad, India
Email: drthangaduraid@gmail.com / drthanga.kud@gmail.com

Advancements in molecular biology have stimulated a great impact on all life sciences investigation to improve the trends in global biodiversity condition. This dynamic force involves the development of highly sophisticated experimental methods in molecular biology with broad interdisciplinary applications, the quickly expanding information flow of scientific discoveries and technical novelties across scientific community, and the development of precise computational tools with constantly updated biological databases for, respectively, analyzing gene expression level, cytogenetic profile, storing data for genotypes, and other molecular features. This has transformed the approach of scientific research allowing revolutionized discoveries in the area of molecular biology. Advances in high-throughput tools, recombinant technology, and NGS made it feasible to unveil genome complexity and to explain the detailed mechanisms for transmission of genetic information. Moreover disease analysis, treatment, and prevention; manipulation of animals and plants for improved phenotypic traits; and synthesis of new peptides and proteins are some of the applications of molecular biology that are frequently being adopted.

There are several approaches from which knowledge in the field of molecular biology is increasingly growing, and it could be challenging for researchers coming into life sciences field. Accordingly there is a need for a platform for preparing molecular biologists to keep themselves abreast of modern trends in molecular biology. Thus, this book series, Modern Trends in Molecular Biology includes the most of the dynamic pieces of research work being done across the globe on numerous themes associated with molecular biology. This book series attempts to enlighten readers regarding novel concepts and findings in the molecular biology field. The scope of this book series includes recent findings and advancements in the areas of transcriptomics, proteomics, genomics, micro RNA, molecular markers, DNA,

RNA and protein isolation, CRISPR, gene editing molecular methods, gene cloning, construction of genome and cDNA library, DNA methylation role for gene expression analysis, AFLP, RAPD, RFLP, VNTRs, genome sequencing tools, phylogenetic analyses, protein modeling, restriction enzymes, metagenomics, metatranscriptomics, metaproteomics, DNA repair, replication and recombination, DNA forensics, pharmacogenomics, functional genomics, structural genomics, and computational analyses for proteins and genes.

This book series, Modern Trends in Molecular Biology, aims to relay state-of-the-art knowledge in field of molecular biology. This book series will provide important resources for college and university students, lecturers and professors, and researchers and scientists in almost all life sciences related disciplines. The book series will offer useful practical and theoretical basis with which to understand the knowledge gaps in relation to a wide range of scientific investigations exploiting molecular biology methods. Overall, presented in concise, organized, and simple format, this series allows smart access to highly considerate topics.

This book series will offer peer-reviewed resources on a variety of topics covering modern aspects of molecular biology, with the goal of helping to keep researchers, scientists, and students abreast of recent advances in the field. Topics will include:

- Molecular Biology: Fundamentals and Applications
- Molecular Parasitology
- Molecular Entomology
- Molecular Virology
- Molecular Phycology
- Molecular Ecology
- Molecular Nematology
- Molecular Physiology
- Molecular Epidemiology
- Molecular Breeding
- Molecular Agriculture
- Molecular Aquaculture
- Molecular Pathology
- Molecular Medicine
- Molecular Genetics
- Molecular Markers
- Molecular Diagnostics
- Molecular Evolution
- Molecular Immunology

- Molecular Microbiology
- Molecular Pharmacology
- Molecular Enzymology
- Molecular Phylogenetics

BOOKS IN THE SERIES

Hematopoiesis: Biochemical, Cellular, Molecular, and Genomic Perspectives
Editor: *Ravi Kumar Gutti, PhD*

Cervical Cancer: Signaling Pathways, Molecular Mechanisms, Natural Products, and Therapeutic Approaches
Authors: *Pratibha Pandey, PhD, Fahad Khan, PhD, Saher Islam, PhD, and Devarajan Thangadurai, PhD*

CRISPR-Based Gene Editing: A Solution for Global Food Security
Editors: *Shakeel Ahmad, PhD, Asif Ali Khan, PhD, and Muhammad Arshad Javed, PhD*

Other Books from AAP by Devarajan Thangadurai and Colleagues

Algal Farming Systems: From Production to Application for a Sustainable Future
Editors: *Jeyabalan Sangeetha, PhD, and Devarajan Thangadurai, PhD*

Algal Genetic Resources: Cosmeceuticals, Nutraceuticals, and Pharmaceuticals from Algae
Editors: *Jeyabalan Sangeetha, PhD, and Devarajan Thangadurai, PhD*

Algal Metabolites: Biotechnological, Commercial, and Industrial Applications
Editors: *Jeyabalan Sangeetha, PhD, Devarajan Thangadurai, PhD, Saher Islam, PhD, and Ravichandra Hospet, PhD*

Beneficial Microbes for Sustainable Agriculture and Environmental Management
Editors: *Jeyabalan Sangeetha, PhD, Devarajan Thangadurai, PhD, and Saher Islam, PhD*

Biodiversity and Conservation: Characterization and Utilization of Plants, Microbes, and Natural Resources for Sustainable Development and Ecosystem Management
Editors: *Jeyabalan Sangeetha, PhD, Devarajan Thangadurai, PhD, Hong Ching Goh, PhD, and Saher Islam, PhD*

Biogenic Nanomaterials: Structural Properties and Functional Applications
Editors: *Devarajan Thangadurai, PhD, Saher Islam, PhD, Jeyabalan Sangeetha, PhD, and Natália Cruz-Martins, PhD*

Biotechnology and Bioinformatics: Advances and Applications for Bioenergy, Bioremediation, and Biopharmaceutical Research
Editors: *Devarajan Thangadurai, PhD, and Jeyabalan Sangeetha, PhD*

Biotechnology of Microorganisms: Diversity, Improvement, and Application of Microbes for Food Processing, Healthcare, Environmental Safety, and Agriculture
Editors: *Jeyabalan Sangeetha, PhD, Devarajan Thangadurai, PhD, Somboon Tanasupawat, PhD, and Pradnya Pralhad Kanekar, PhD*

Cervical Cancer: Signaling Pathways, Molecular Mechanisms, Natural Products, and Therapeutic Approaches
Authors: *Pratibha Pandey, PhD, Fahad Khan, PhD, Saher Islam, PhD, and Devarajan Thangadurai, PhD*

Environmental Biotechnology: Biodegradation, Bioremediation, and Bioconversion of Xenobiotics for Sustainable Development
Editors: *Jeyabalan Sangeetha, PhD, Devarajan Thangadurai, PhD, Muniswamy David, PhD, and Mohd Azmuddin Abdullah, PhD*

Environmental Phycology: Climate Change Mitigation and Hazardous Pollutant Cleanup using Algal Resources
Editors: *Jeyabalan Sangeetha, PhD, and Devarajan Thangadurai, PhD*

Food Microbial and Molecular Biology: From Fundamentals to Applications
Editors: *Saher Islam, PhD, Devarajan Thangadurai, PhD, Jeyabalan Sangeetha, PhD, and Zaira Zaman Chowdhury, PhD*

Fundamentals of Molecular Mycology
Authors: *Devarajan Thangadurai, PhD, Jeyabalan Sangeetha, PhD, and Muniswamy David, PhD*

Genomics and Proteomics: Principles, Technologies, and Applications
Editors: *Devarajan Thangadurai, PhD, and Jeyabalan Sangeetha, PhD*

Global Food Safety: Microbial Interventions and Molecular Advancements
Editors: *Saher Islam, PhD, Devarajan Thangadurai, PhD, Jeyabalan Sangeetha, PhD, and Natália Cruz-Martins, PhD*

Industrial Biotechnology: Sustainable Production and Bioresource Utilization
Editors: *Devarajan Thangadurai, PhD, and Jeyabalan Sangeetha, PhD*

Microalgal Biotechnology: Bioprospecting Microalgae for Functional Metabolites towards Commercial and Sustainable Applications
Editors: *Jeyabalan Sangeetha, PhD, Svetlana Codreanu, PhD, and Devarajan Thangadurai, PhD*

Microbial Fertilizer Technology for Sustainable Crop Production
Editors: *Jeyabalan Sangeetha, PhD, Abdel Rahman Mohammad Al-Tawaha, PhD, and Devarajan Thangadurai, PhD*

Mycorrhizal Technology: Managing Plant Stress and Mitigating Climate Change using Mycorrhizae
Editors: *Jeyabalan Sangeetha, PhD, Abdel Rahman Mohammad Al-Tawaha, PhD, and Devarajan Thangadurai, PhD*

Organic Farming for Sustainable Development
Editors: *Jeyabalan Sangeetha, PhD, Kasem Soytong, PhD, Devarajan Thangadurai, PhD, and Abdel Rahman Mohammad Al-Tawaha, PhD*

Phycobiotechnology: Biodiversity and Biotechnology of Algae and Algal Products for Food, Feed, and Fuel
Editors: *Jeyabalan Sangeetha, PhD, Devarajan Thangadurai, PhD, Sanniyasi Elumalai, PhD, and Shivasharana Chandrabanda Thimmappa, PhD*

Seaweed Biotechnology: Biodiversity and Biotechnology of Seaweeds and Their Applications
Editors: *Jeyabalan Sangeetha, PhD, and Devarajan Thangadurai, PhD*

About the Editor

Ravi Kumar Gutti, PhD
Professor, Department of Biochemistry, School of Life Sciences, University of Hyderabad, India

Ravi Kumar Gutti, PhD, is a Professor in the Department of Biochemistry of the School of Life Sciences at the University of Hyderabad, India. He received his PhD in Molecular Biology and Biotechnology from IARI, New Delhi. He has extensive research and teaching experience from Harvard Medical School (USA), the National Institutes of Health (USA), Moffitt Cancer and Research Center (USA), and RWTH Aachen (Germany). His research is focused on understanding the molecular mechanisms in low platelet diseases. He has guided several PhD and master's students. He is the recipient of the Chancellors Award from the University of Hyderabad for making significant contributions to teaching and research at the university. His contributions to research have been well received by several national and international organizations, including the International Society for Stem Cell Research (USA), Korean Society of Bone Marrow Transplantation (Korea), FIRM-Tissue and Cell Engineering Society (UK), International Union of Biochemistry and Molecular Biology (Canada), Japanese Society of Hematology (Japan), Korean Society of Laboratory Medicine (Korea), Singapore Stem Cell Society (Singapore), Pan Pacific Symposium on Stem Cells and Cancer Research (Taiwan), and Korean Society of Hematology (Korea) with various awards. National funding bodies have recognized the potential of his research in the form of the DBT-Innovative Young Biotechnologist Award and ICMR-Venkoba Rao Award in Physiology. His research is well-funded by DBT, DST, ICMR, CSIR, and SERB grants of the Government of India and international collaboration under Indo-German DAAD Program. He was recently elected as a "Fellow of the Telangana Academy of Sciences (FTAS)" for 2019 in Medical, Health, and Pharmaceutical Sciences. He has authored several publications in reputed international journals and presented his research at several international scientific meetings.

Contents

Contributors

Obul Reddy Bandapalli
Division of Molecular Genetic Epidemiology, German Cancer Research Center (DKFZ), Heidelberg, Germany

Aparajita Chakraborty
Department of Biotechnology, St. Xavier's College, Calcutta, West Bengal, India

Lakshmi Manasa Chaturvedula
Independent Researcher, South Plainfield, NJ, USA

Anuradha Venkatakrishnan Chimata
Department of Biology, University of Dayton, Dayton, USA

Anpreet Choudhary
School of Bioengineering and Biosciences, Lovely Professional University, Phagwara, Punjab, India

Swati Dahariya
Department of Biochemistry, School of Life Sciences, University of Hyderabad, Hyderabad, Telangana, India

Doli Das
Department of Molecular and Human Genetics, Institute of Science, Banaras Hindu University, Varanasi, Uttar Pradesh, India

Raghunadharao Digumarthi
Oncology Services, KIMS ICON, Vizag Steel General Hospital, Visakhapatnam, Andhra Pradesh, India

Dushyant Kumar Gautam
Department of Biochemistry, School of Life Sciences, University of Hyderabad, Hyderabad, Telangana, India

Ravi Kumar Gutti
Department of Biochemistry, School of Life Sciences, University of Hyderabad, Hyderabad, Telangana, India

Arunasree M. Kalle
Department of Animal Biology, School of Life Sciences, University of Hyderabad, Hyderabad, Telangana, India

Raghunath Manchala
Endocrinology and Metabolism Division, National Institute of Nutrition (NIN), Uppal Road, Hyderabad, Telangana, India

Debasmita Naik
Department of Animal Biology, School of Life Sciences, University of Hyderabad, Hyderabad, Telangana, India

Indira Paddibhatla
Department of Biochemistry, School of Life Sciences, University of Hyderabad, Hyderabad, Telangana, India

Akhilesh Prajapati
Department of Biotechnology, School of Science, GSFC University, Vadodara, Gujarat, India

Khushbu Priya
Department of Molecular and Human Genetics, Institute of Science, Banaras Hindu University, Varanasi, Uttar Pradesh, India

Sanjeev Raghuwanshi
Department of Pharmacology and Regenerative Medicine, College of Medicine, University of Illinois, Chicago, USA

Geeta Rai
Department of Molecular and Human Genetics, Institute of Science, Banaras Hindu University, Varanasi, Uttar Pradesh, India

K. Rajender Rao
Molecular Genetics, National Center for Laboratory Animal Sciences, National Institute of Nutrition, Hyderabad, Telangana, India

Keyur Raval
Department of Chemical Engineering, National Institute of Technology Karnataka, Surathkal, Mangalore, Karnataka, India

Ritu Raval
Department of Biotechnology, Manipal Institute of Technology, MAHE, Manipal, Karnataka, India

Rasmita Samal
Department of Life Sciences, Central University of Karnataka, Gulbarga, Karnataka, India

Rekha Sharma
School of Bioengineering and Biosciences, Lovely Professional University, Phagwara, Punjab, India

Anula Divyash Singh
Apollo Hospitals Educational and Research Foundation, Hyderabad, Telangana, India

Suvir Singh
Clinical Hematology and Bone Marrow Transplantation, Christian Medical College, Ludhiana, Punjab, India

Arun Kumar Subramani
Department of Chemical Engineering, National Institute of Technology Karnataka, Surathkal, Mangalore, Karnataka, India

Prashant Suravajhala
Amrita School of Biotechnology, Amrita Vishwa Vidyapeetham, Clappana, Kerala, India

Hiral Thacker
Department of Molecular and Human Genetics, Institute of Science, Banaras Hindu University, Varanasi, Uttar Pradesh, India

Chaturvedula Tripura
CSIR–Center for Cellular and Molecular Biology Annexe-II, Uppal Road, Hyderabad, Telangana, India

Aditya Velidandi
Department of Biotechnology, National Institute of Technology, Warangal, Telangana, India

Satti Vishnupriya
Department of Genetics, Osmania University, Hyderabad, Telangana, India; #102, Apoorva Towers, Rd. No. 2, Banjara Hills, Hyderabad 500034, India

Sugunakar Vuree
MNR Foundation for Research and Innovation, MNR Medical College and Hospital, MNR University, Fasalwadi Village, Sangareddy (District), Telangana 502294, India

Samiksha Wasnik
Division of Regenerative Medicine, Department of Medicine, Loma Linda University, Loma Linda, CA, USA

Abbreviations

ADA	adenosine deaminase
AGM	aorta-gonad-mesonephric
AID	autoimmune and immune-related disorders
AL	acute leukemia
ALCL	anaplastic large cell lymphoma
ALL	acute lymphoblastic leukemia
ALPM	anterior lateral-plate mesoderm
AMKL	acute megakaryoblastic leukemia
AML	acute myeloid leukemia
APC	antigen-presentation cell
APL	acute promyelocytic leukemia
ASIR	age-standardized incidence rate
ATF2	activating transcription factor 2
ATO	arsenic trioxide
B- ALL	B-cell acute lymphoblastic leukemia
BALRs	B-all associated long RNAs
BCP	B-cell precursor
BCR	breakpoint cluster region
BM	bone marrow
BMF	bone marrow failure
CBP	CREB-binding protein
CCAT1	colon cancer-associated transcript-1
cDNA	complementary DNA
ceRNA	competitive endogenous RNA
CGD	chronic granulomatous disease
CHT	caudal hematopoietic tissue
CL	chronic leukemia
CLL	chronic lymphocytic leukemia
CLP	common lymphoid precursor
CML	chronic myelogenous leukemia
CMPs	common myeloid progenitors
CNS	central nervous system
COS	chito oligosaccharides
CPAT	coding potential assessment tool

CPC	coding potential calculator
CR	complete remission
CRc	cytogenetic complete remission
CRISPR	clustered regularly interspaced short palindromic repeats
CRITICA	coding region identification tool invoking comparative analysis
CRm	molecular complete remission
CRm	morphological complete remission
CRMRD	complete remission without measurable residual disease
CRNDE	colorectal neoplasia differentially expressed
Cryo-EM	cryo-electron microscopy
CS	chondroitin sulfate
CSC	cancer stem cells
CSPG4	chondroitin sulfate proteoglycan 4
CTCs	circulating tumor cells
DA	degrees of acetylation
DBA	diamond-Blackfan anemia
DC	dendritic cells
DC	dyskeratosis congenita
DD	degree of deacetylation
DEP	dielectrophoretic
DFS	disease-free survival
DLD	deterministic lateral displacement
DP	degree of polymerization
DRB	5,6-dichloro-1-β-D- ribofuranosylbenzimidazole
DS	dermatan sulfate
DSIF	DRB sensitivity-inducing factor
EBP1	ErbB3-binding protein 1
ECM	extracellular matrix
EC-MV	endothelial cell-derived microvesicles
ELN	European LeukemiaNet
EMS	ethyl methane sulfonate
ESL1	E-selectin ligand-1
EVs	extracellular vesicles
FA	Fanconi anemia
FACS	fluorescence-activated cell sorting
FGF	fibroblast growth factor
FLT3	FMS-like tyrosine kinase 3
GAGs	glycosaminoglycans
GC	glucocorticoid

GEP	gene expression profiling
GMCSFR	granulocyte colony-stimulating factor
GMP	granulocyte monocyte precursor
GO	gene ontology
GPC-1	glypican-1
GPC-3	glypican 3
GPI	glycosyl-phosphatidylinositol
GPS	gene panel sequencing
HA	hyaluronic acid
HAT1	histone acetyltransferase 1
HATs	histone acetyltransferases
HBPs	heparin-binding proteins
HC	hematopoietic cancers
HDAC	histone deacetylase
HE	hereditary elliptocytosis
HMDB	human metabolome database
HMs	hematologic malignancies
HMTs	histone methyl-transferases
HPC	hematopoietic progenitor cell
HS	heparan sulfate
HS	hereditary spherocytosis
HSC	hematopoeitic stem cell
HSPC	hematopoietic stem and progenitor cells
HSPG	heparin/heparan sulfate proteoglycans
HSSTs	HS sulfotransferases
ICAM	intercellular adhesion molecule
ICM	intermediate cell mass
ICMR	Indian Council of Medical Research
IDH1	isocitrate dehydrogenase one
IFN γ	interferon γ
IL	interleukin
IWG	International Working Group
JAK	Janus kinase
KEGG	Kyoto Encyclopedia of Genes and Genomes
KO	knockout
KS	keratan sulfate
KSPG	keratan sulfate proteoglycans
LAD	leukocyte adhesion deficiency
LAIP	leukemia-associated immunophenotype
LCM	laser capture microdissection

LMPP	lymphoid-primed multipotential progenitor
LMV	leukocyte-derived microvesicles
lncRNA	long noncoding RNA
LPS	lipopolysaccharide
LSCs	leukemia stem cells
LT	long-term
LT-HSCs	long term HSCs
LUNAR1	leukemia-induced noncoding activator RNA
MALAT1	metastasis-associated lung adenocarcinoma transcript 1
mCR	morphological CR
mDCs	monocytes DCs
MDS	myelodysplastic syndrome
MEG3	maternally expressed gene 3
MEP	megakaryocyte erythroid precursor
MFE	minimum free energy
MIAT	myocardial infarction-associated transcript
MLL-r	mixed-lineage leukemia-rearranged
MPFC	multi-parameter flow cytometry
MPL	myeloproliferative leukemia
MPN	myeloproliferative neoplasm
MPP	multipotent progenitor
MRD	measurable residual disease/minimal residual disease
mRNA	messenger RNA
MS	multiple sclerosis
MVs	circulating microvesicles
MW	molecular weight
NCCs	neural crest cells
ncRNA	noncoding RNA
NET	neutrophil extracellular trap
NG2	neuron-glial antigen 2
NGS	next-generation sequencing
NK	natural killer
NLR	nod-like receptor
NPM1	nucleophosmin
NTA	nanoparticle tracking analysis
ORF	open reading frame
OS	overall survival
PAMPs	pathogen-associated molecular patterns
PCR	polymerase chain reaction

pDCs	plasmacytoid DCs
PEG	polyethylene glycol
PGs	proteoglycans
PHD	plant homeodomain
PIDs	primary immunodeficiencies
PKD	protein kinase D
PLM	posterior lateral-plate mesoderm
PMN	polymorphonuclear
PMV	platelet-derived microvesicles
PORTRAIT	prediction of transcriptomic ncRNA by ab initio methods
PRC2	polycomb repressive complex 2
PRP	platelet-rich plasma
PRR	pattern recognition receptors
PS	phosphatidylserine
PSGL-1	P-selectin glycoprotein
PTMs	posttranslational modifications
RACE	rapid amplification of cDNA ends
RBC-MV	RBC-derived microvesicles
RBCs	red blood cells
RD	reticular dysgenesis
rDNA	ribosomal DNA
RFS	relapse-free survival
RNA-seq	RNA sequencing
ROS	reactive oxygen species
RPS19	ribosomal protein 19
rRNA	ribosomal RNA
RRS	ribosome release score
RT-PCR	reverse transcriptase polymerase chain reaction
RT-qPCR	reverse transcription-quantitative polymerase chain reaction
SAGE	serial analysis of gene expression
SBDS	Shwachman-Bodian-diamond syndrome
SCA	single-cell analysis
SChLAP1	SWI/SNF complex antagonist associated with prostate cancer 1
SCID	severe combined immunodeficiency
scRNA-seq	single-cell RNA sequencing
SDF-1	stromal-cell derived factor 1
SDS	Shwachman-diamond syndrome

SEC	size-exclusion chromatography
SEER	surveillance, epidemiology, and end results
SIP1	sphingosine one-phosphate
SLRPs	small leucine-rich proteoglycans
SMC1A	structural maintenance of chromosome 1 alpha
SNPs	single-nucleotide polymorphisms
SP cells	side population cells
ST	short-term
STAT	signal transducer and activator of transcription
ST-HSCs	short-term HSCs
TA	transit-amplifying cells
TALEN	transcription activator-like effector nuclease
T-ALL	T-cell acute lymphoblastic leukemia
t-AML	therapy related to AML
TCR	T cell receptor
TdT	terminal deoxynucleotidyl transferase
TEM	transmission electron microscopy
TF	tissue factor
TFPI	tissue-factor pathway inhibitor
TFs	transcription factors
TGF-β	transforming-growth factor-β
TLRs	toll-like receptors
TNF-α	tumor necrosis factor α
UAS	upstream activation sequence
UCA1	urothelial Carcinoma-Associated 1
VAF	variant allelic frequency
VCAM	vascular cell adhesion molecule
VDA	ventral wall of the dorsal aorta
vWF	Von Willebrand factor
WAS	Wiskott-Aldrich syndrome
WASp	WAS protein
WES	whole exome sequencing
WGS	whole genome sequencing
WT1	Wilms' tumor 1

Foreword

With great pleasure and enthusiasm, I introduce the book "Hematopoiesis: Biochemical, Cellular, Molecular and Genomic Perspectives," edited by Dr. Ravi Kumar Gutti.

Hematopoiesis has always been recognized as an emerging and fascinating field. Being a wide array network, this highly complex process is regulated and fine-tuned by several transcription factors that work cooperatively with essential signal transduction pathways to maintain intricate balance. This book comprises and sheds light on different factors that indulge and regulate hematopoietic events. The uniqueness of this book encompasses the myriad of information on the biology of normal and aberrant hematopoiesis, epigenetic factors, and on the research tools to assist in the deeper study of aspects relating to the status of blood cells in the cancerous scenario. In the various chapters of this book, the role and involvement of chitosans, proteoglycans, and glycoproteins in hematopoietic hemostasis are also discussed, as well as the recent advances in single-cell analysis for blood cell evaluation. Some specifics of leukemia/lymphoma governed by various non-conventional molecules are also part of this book.

Several experts in the field have contributed to this book, providing an accurate overview of the functional importance of hematopoiesis from different perspectives and the surrounding molecular mechanisms that influence the process of hematopoiesis in one way or another. In addition to the extremely pertinent and up-to-date information provided in the various chapters, elegant pictures, tables, and schemes support the text, attracting the readers' attention while helping to complement the information and the interpretation of the contents. Thus, each chapter constitutes an enigmatic tool for readers/scholars themselves, being highly recommended to readers and will certainly enrich the course of projects in the area, contributing to the growth and progress of this attractive area of research.

—Natália Cruz-Martins, PhD
Polytechnic and University Higher Education Cooperative (CESPU),
Faculty of Medicine, University of Porto (FMUP),
Porto, Portugal

Preface

I realize that this book will prove to be beneficial for its targeted audience. Each chapter of this book is meticulously written and presented by the respective authors. I found an immense interest while going through each of them. One of the distinguishing qualities of this book is that it avoids a consensus textbook structural pattern, wherein chapters are hierarchized in a specific way to be comprehended. One can start at any point in chapters based on their scientific preferences. The book tends to contain the widespread concerns of hematopoietic challenges in different emerging ways that are also clinically addressed. These subjects, in the form of chapters, are extensively revised and updated with advanced materials, including figures and tables.

The book comprises 11 chapters. Chapter 1 discusses the novel mechanistic approach to zebrafish hematopoiesis. Chapter 2 utilizes the available tools and techniques to isolate and study single blood cells in different model systems. Chapter 3 provides the synergistic effect of immunomodulators with their derivatives under inflammation conditions. Chapter 4 unravels the therapeutic application of proteoglycans and glycosaminoglycans (GAGs) influenced by HSCs functionality. Chapter 5 focuses on the epigenetic role of HATS and HDACs in hematopoiesis. Chapter 6 represents the in-silico work to predict the role of human long noncoding RNA (lncRNA) in hematopoiesis. Chapter 7 targets the possible clinical outcomes of aberrant hematopoiesis, including therapeutics. Chapter 8 describes the biology of blood-derived microvesicles in hematopoiesis and their potential clinical application for diagnostic and drug delivery approaches. Chapter 9 delineates the cellular crosstalk of molecular players during deregulated hematopoiesis. Chapter 10 depicts the status of long noncoding RNA in diverse leukemia with their therapeutic approach. Finally, Chapter 11 projects the role of Minimum Residual Disease along with prognostic markers in acute myeloid leukemia (AML).

The challenge that I faced was comprehending and compiling all the chapters with a focus to reawake the readers' interest. I anticipate this book to be of prolific use to its readers. Lastly, I want to thank all the authors for their substantial effort in making this endeavor possible. I am grateful to the publisher for bringing out this work as a compact book.

—Ravi Kumar Gutti, PhD
University of Hyderabad, Telangana, India

CHAPTER 1

Unrevealing the Mechanism of Zebrafish Hematopoiesis: A Novel Approach

APARAJITA CHAKRABORTY

Department of Biotechnology, St. Xavier's College (Autonomous), Kolkata, West Bengal, India

ABSTRACT

Hematopoiesis gives rise to different blood cells, such as megakaryocytes, erythrocytes, neutrophils, monocyte/macrophages, or lymphocytes. Among vertebrates, zebrafish is an important model for studying hematopoiesis owing to their tendency to produce many mating embryo pairs and their external pregnancy, allowing *in vivo* expression of embryogenesis. There are two types of hematopoiesis coming into action: primitive and definitive hematopoiesis. Primitive hematopoiesis involves players, such as the transcription factors of bHLH stem cell leukemia (scl/tal I) along with Scl expression patterns, GATA binding protein 2a (gata2a), and ETS fli1a family members, flib1 and protein related ets (etrsp/etv2). They may be necessary for the formation of hematopoietic and endothelial growth. Interaction between Gata1 and Pu1 proteins is an essential feature for the regulation of hematopoiesis and erythroid specification. Definitive hematopoiesis involves the Notch-Runx1 signature as an essential component of hematopoietic stem cell (HSC) formation. Several other players, such as IDH1, RPS19, and p53 have been highlighted for their role in hematopoietic development. The chapter provides further insights into the epigenetic, metabolomic, and proteomic studies, which may help to unleash the detailed molecular mechanisms of zebrafish hematopoiesis and mutations responsible for various hematological disorders in the upcoming days.

Hematopoiesis: Biochemical, Cellular, Molecular, and Genomic Perspectives. Ravi Kumar Gutti (Ed.)

1.1 HEMATOPOIESIS: AN IMPORTANT PHENOMENON IN VERTEBRATES

Hematopoiesis refers to the process of blood cell components, which occurs during fetal development to produce and replenish the blood system. Such a process involves the integration of the internal controls of the master transitional, transcriptional, and the signature of molecules from neighboring tissues. The wave of blood in vertebrates incorporates two waves of hematopoiesis, the first wave and the descriptive wave. The first wave activates an erythroid progenitor that supplies erythrocytes and macrophages during embryonic development (Gore et al., 2018). The purpose of it is to produce red blood cells needed to perform tissue oxygenation as the embryo grows rapidly. In mammals and avians, these erythroid progenitor cells begin to emerge from the blood vessels in the extra-embryonic yolk sac at the beginning of development. Descriptive hematopoiesis, in contrast, occurs late in development, especially in a wide range of species. It incorporates powerful hematopoietic stem cells (HSCs) and raises the entire bloodstream of the adult body. Various factors play a role in the renewal of HSC, among which the two signatures are the primary players. The Wingless/Integrated (Wnt) family of molecules is known to be important in the functioning of HSC, according to many recent studies. It highlights the positive role of Wnt in HSCs during development and rebirth. At the same time, there were some studies that contradicted the importance of Wnt methods in normal lymphopoiesis and hematopoiesis (Gore et al., 2018; De Pater & Trompouki, 2018). Another important feature is the Notch method which controls cell specification and pattern formation.

As evidenced by various data to assess the significance of the Notch-dependent regulation of hematopoietic growth, it has also been clarified that such a mechanism promotes the expansion/regeneration of HSC in both mice and individuals in adult hematopoiesis.

1.2 ZEBRAFISH: A GOOD MODEL FOR HEMATOPOIESIS

Among vertebrates, zebrafish emerged as a useful model for studying hematopoiesis. Zebrafish are capable of producing a large number of pairs of mating embryos and are also sexually mature at three months of age. Researchers found an external pregnancy that allows *in vivo* expression of embryogenesis, thereby making them a simpler model for developing developmental scientists. They have many advantages, such as the ease of handling embryos and the ability to create large genetic and chemical screens. In addition, the systems that control zebrafish hematopoiesis appear to be maintained as mammals, thus making them a model for the appropriate clinical program. The genetic features of Gata1 and Pu1 and their physical association are responsible for the regulation of classical hematopoiesis, while Runx1, a member of the runt family of materials, regulates explicit hematopoiesis. Runx1 leads to cmyb-induced growth of the Myb family of proto-oncogenes, and its importance in clear hematopoiesis determined from the death of mice in Myb knockout due to liver failure of the fetal fetus (Mardis et al., 2009; Gore et al., 2018).

1.2.1 PRIMITIVE HEMATOPOIESIS

In ancient hematopoiesis, the first wave of blood cell formation produces erythrocytes preparing for respiratory tissue during rapid embryonic growth. The macrophages that phagocytose bacteria and apoptotic cells emerge as a product of natural development. Microglia that differentiate from old macrophages are important regulators of the central nervous system for neural development and function (Gore et al., 2018).

The site of ancient zebrafish hematopoiesis is an island called intermediate cell mass (ICM), which is formed at the trunk midline from two international lines of the posterior lateral-plate mesoderm (PLM). The first molecular marker of ancient hematopoiesis is transcription factors of bHLH stem cell leukemia (scl/tal I) which are expressed early. The genetic sites are refined at 11 hpf when the end of the mesoderm cell is continuously blocked in hematopoietic lines.

The Scl expression patterns, only LIM domain 2 (lmo2) and vasculogenic genes include GATA binding protein 2a (Gata2a) and ETS fli1a family members, flib1 and protein related ets (etrsp/etv2). Scl also binds its partner Lmo2 downstream bHLH-PAS transcript factor cloche to zebrafish, which is necessary for the formation of hematopoietic and endothelial growth.

In addition to PLM, a second, higher hematopoietic site is dominated by zebrafish. Also known as the rostral arteries (RBI), this site is derived from the anterior lateral-plate mesoderm (ALPM) and produces old macrophages and neutrophil granulocytes (Amsterdam et al., 2004; Bertrand et al., 2008; Gore et al., 2018).

In the case of recurrent hematopoiesis, the relationship between Gata1 and Pu1 in the erythroid specification as compared to the end of the myeloid cell is well established. Gata1 suppresses myeloid differentiation and is an important regulator of erythroid cell fate. Chemical data suggests a physical interaction of Gata1 and Pu1 proteins inhibiting each other's genetic recordings. Gore et al. (2018) stated that in addition to Pu1, the function of Gata1 in controlling the erythroid cell end is modified by its combination with many other factors such as Scl, Friend of Gata (Fog1) and others such as Kruppel (Klfs). As per genetic evidence, Pu.1 measurement also contributes to macrophage cell destruction compared to neutrophilic granulocyte cells. This is because anterior lateral plate mesoderm (ALPM) cells have higher levels of Pu.1, selectively forming macrophages of the protein binding to RNA Elavl1a to promote the elimination of zebrafish erythroid cells by stabilizing Gata1 mRNA. Various substances such as transcript intermediate factor 1γ (tuf1γ), stat1b, and Hox genes, and their cofactors Pbx2/4 and Meis1 may play an important role in the early regulation of erythropoiesis upstream of the Gata1 river; as it affects the loss and analysis of the activity in zebrafish.

1.2.2 DEFINITIVE HEMATOPOIESIS

HSCs mainly thrive in the blood and bone marrow in the case of mammals and other vertebrates (Jing & Zon, 2011). But apart from an anatomical conflict, the stationary and genetic systems that regulate the speculation of hematopoietic machinery are highly conserved between zebrafish and mammals. The transition to clear hematopoiesis occurs in two consecutive events. Firstly, a transient wave passes from the posterior blood vessel (PBI) located at the ventro-caudally end of the yolk extension that forms gata1 cells and imo2-producing cells, also known as erythro-myeloid hematopoietic progenitors (EMPS). They may differ between myeloid and erythroid but not the lymphoid list (Bertrand et al., 2010; Da Pater & Trompouki, 2018). Figure 1 lists the two states of hematopoiesis and the blood cells which are being acted on.

During the second event around 28–30 hpf true multipotent regenerative HSPCs are expressed from endothelial cells in the ventral wall of the dorsal aorta (VDA). It is also known as aorta-gonad-mesonephric (AGM) due to its malignant counterparts.

Such specified HSCs are isolated from the hemogenic endothelium and then incorporated into the axial arteries, which will continue to age as progenitors of various cell types. Runx1 was found to be an important symbol of HSCs with cmyb running downstream. The loss of cmyb activity leads to the perfect expression of HSCs but fails to emerge from the dorsal aorta, while other mutating organisms lose the adult hematopoietic line by about 20dpf but are active until 3 months of age. The Notch-Runx1 signature method is an important component required in building HSCs.

The Notch specification and HSC subtitles are controlled by non-canonical Wnt signing. The ligand Wnt16 attracts ligands deltaC and deltaD in somites around the dorsal aorta, leading to the introduction of runx1- and cmyb-expressing HSCs. A knock on the signal modulating the growth factor β (Tgfβ) arises not only from the duct aorta endothelium but from the area around the AGM leads to a significant loss of Runx1 and other HSP signals. This may lead to HSC rupture impairment without affecting the arterial endothelium. Ciau-Uitz et al. (2014) and Gore et al. (2018) describe the positive interaction between a specific signal and the runx1 expression required for the formation of HSCs in clear hematopoiesis.

There are additional features that work with Notch-Runx 1 for HSPCs produced by the AGM. Various mutations in Cbfβ, a non-DNA binding beta-subunit for the Core Binding Factor, have revealed that the Runx1 partner is an important component of HSP growth. Various features, such as scla/sclb are found to work sequentially during the construction of the HSPC. Examination of the thymus-filled thrombocytes reveals that lower HSPCs produce itga2b, and that thrombocytes produce higher CD41-GFP (+) levels. Thrombocyte formation is shown to be dependent on the Scl. The tissues surrounding the endothelium are known to play important roles in HPC expression. Researchers even found that neural crest cells (NCCs) interact with the dorsal aorta endothelium during HSPC speculation. But whenever NCC contact with the dorsal aorta was damaged or blocked, HSPC data was lost. Even Pdfgra-dependent signaling from neural crest migration has been found to be necessary in this process. This is because Pdfgra-deficient embryos have shown a lack of cmyb and runx1 positive HSPCs, key hematopoiesis, and mature T-lymphocytes in -rag2 Thymus (Paik & Zon, 2010).

1.3 ROLE OF EPIGENETIC REGULATION IN THE HEMATOPOIETIC SYSTEM

Studies have also shown the role of various epigenetic regimens such as histone modification, DNA methylation, small RNAs, and chromatin re-uptake structures in hematopoiesis. Zebrafish serve as an excellent system for testing chromatin sub-channels and also the effects of chromatin-converting drugs *in vivo*. The ATAC-seq method allows for the identification of endothelial enhancers; thus, the role of cohesin has been demonstrated in the reorganization of genomic structures during the maternal transition to zygotic transcription in early embryos (Orkin & Zon, 2008). The de novo DNA methyltransferase family contains dnmt3aa, dnmt3ab, dnmt3ba, dnmt3bb.1, dnmt3bb.2, and dnmt3bb.3. Recently, the expression of dnmt3bb.1 was found to be important in the retention of HSP records. It was evident from the loss of cmyb definition by 72hpf in zebrafish dnmt3bb.1 mutants, followed by HSPC apoptosis and loss of hematopoietic lines below myeloid and lymphoid markers such as l-plastin and rag1, respectively. Conditions such as defective epigenetic regulation have therefore been shown to lead to hematopoietic damage such as leukemias and lymphomas (Dooley et al., 2005).

1.4 REVEALING NEW PLAYERS IN HEMATOPOIESIS

The larvae of zebrafish are able to survive without red blood cells (RBCs) via the distribution of oxygen in the first week, thus allowing for the identification of potential mutations in mice. Such mutations are useful in describing the genetic mechanisms that regulate HSCs and the different gene lines. In the light-adjusted months of unused 1-y (if11 or trim33) embryos were found to possess erythropoiesis defects as they did not express Gata1 factor, an important factor in the process.

Identification of blood mutants in zebrafish embryos has helped reveal the genetic pathways which regulate HSCs or different hematopoietic lineages. For instance in moonshine mutants, deficient for *transcriptional intermediary factor 1-γ* (*Tif1γ* or *trim33*) had severely impaired erythropoiesis owing to their inability to express Gata1, a crucial factor of erythropoiesis. Mutation of cdc73 gene restored hematopoiesis in *tif1γ*$^{-/-}$ embryos, whereas knocking down PAF subunit expression or components of 5,6-dichloro-1-β-D-ribofuranosylbenzimidazole (DRB) sensitivity-inducing factor (DSIF) complex even restored RBCs to wild-type levels in the respective embryos (Jing & Zon, 2011; De Pater & Trompouki, 2018).

Tif1γ has even been found to play an important role in the search for extracellular features such as proper b (p-TEFb) extraction and thus facilitate the transcription of complex chromatin (FACT) complexes. An unexpected role of tif1γ has been elucidated as well during the division of erythroid cells.

Various chemical screens have identified the presence of prostaglandin E2 as a compound needed to increase HSC production. Detailed applications of this prostaglandin are still under investigation to increase HSC (Hoggatt et al., 2009).

Advanced cognitive processes have revealed pathways for thymus development and revealed that HSPCs are enlarged and interact with endothelial cells in caudal hematopoietic tissue (CHT). The role of multiple signature mechanisms and cell communication for the emergence of HSPC has been highlighted. For example, the active signaling provided by neutrophils is required for HSC formation. Such distinct structures reveal the role of macrophages in the promotion of HSP and direct hematopoiesis.

1.4.1 ROLE OF RPS19 AND p53 IN HEMATOPOIETIC DEVELOPMENT

The function of selected genes during hematopoietic growth and various genetic interactions affecting disease phenotypes can be assessed in zebrafish embryos using mRNA microinjection (overexpression) and knockdown oligomers of morpholino antisense. Recent reports have suggested that the depletion of ribosomal protein 19 (RPS19) in zebrafish results in a significant loss of RBCs and DBA-like behavioral abnormalities. The maturation and proliferation of erythroid lines have been found to be significantly inhibited. The loss of RPS19 activated a wide range of p53 pathways and fetal apoptosis, including erythroid cells. Changes in other ribosomal proteins have led to similar activation of the p53 family protein leading to the conclusion that the p53 protein family may be a victim of DBA treatment modalities (Danilova et al., 2008; De Pater & Eirini, 2018).

1.4.2 PROTEOMIC AND METABOLOMIC STUDIES IN USE

In the current era of genome-wide technology, genetic studies are complemented by proteomic studies. The translation effects may sometimes vary widely. Although proteomics and metabolomic studies have not been extensively studied in zebrafish, particularly in the hematopoietic system, there are some studies that have shown differences in the levels of protein expressed by many genes using proteomic analysis either in whole zebrafish embryos or even in individual cells during regeneration (Mardis et al., 2009).

De Pater & Trompouki (2018) stated that metabolomics has been shown to be useful in the study of neurological damage caused by chemical disturbances at zebrafish embryonic levels. The sensitivity of these methods is likely to overcome the problem of static and sub-population with the growing advances in mass-spectrometry analysis.

1.4.3 ROLE OF IDH1: ANOTHER PLAYER IN HEMATOPOIESIS

Recent advances in the analysis have discovered the role of zebrafish isocitrate dehydrogenase one (IDH1) in the regulation of myelopoiesis and clear hematopoiesis. The homodimeric IDH1 NAD phosphate (+/-) associated with nursing protein depends on the chemical process of isocitrate to supply α-ketoglutarate. This is additionally a mammalian-preserved method|biological process, which incorporates isocitrate and isopropylmalate dehydrogenases signature sequences and glycine-rich residues, seven compound residues among the isocitrate-Mg^{2+} binding terminal peroxisomal process (Gross et al., 2010; Shi et al., 2015). The removal of Zidh1 exhibited reductions in runx-1 and c-myb expression among the DA wall. The Loss of Zidh1 activity has been found to extend cell proliferation through pu1 expression (the 1st myeloid progenitor) and to scale back those with vertical and mpeg1 expression (granulocytes and macrophages). However, there was no increase in caspase-mediated necrobiosis of vertebrate-producing cells with mpeg1. This suggested that the suppression/removal of zidh1 and inhibition of myeloid factor division displayed various effects related to a rise in pu1 and a decrease in vertical and mpeg1 expression in zebrafish larvae. Overall, it provides a mechanical link between IDH1 mutation, its loss of performance, and differentiation block in Acute myeloid leukemia (AML) (Bertrand et al., 2010; Shi et al., 2015).

Secondly, the important role of Zidh1 in the regulation of myelopoiesis and definitive hematopoiesis has been demonstrated. It has been found that zidh1 knockdown showed no effect on the expression of target genes below Wnt, Notch or transforming-growth factor-β (TGF-β), or macromolecule expression of β-catenin (Wnt) or phosphor-Smad2 or Notch1 intracellular domain (Notch). This suggests that zidh1 may need a effect on HSCs on ascending control signals. The removal of Zidh1 didn't have an impact on the expression of Scl and lmo2 in ICM, which raises the absence of organic process transporters (Bertrand et al., 2008).

Zebrafish have become a helpful and vital organism for the study of organic process disorders and thus the event of little molecular-based therapies. Zebrafish hematopoiesis is awfully nearly like different vertebrates. Most zebrafish genetically changed orthologs have succeeded in the repetition of human phenotypes. Different sorts of zebrafish cancer models were successfully used to develop epigenetic concepts in additionally to learning concerned biological diseases (Konantz et al., 2019). Several human oncogenes were found to cause blood-like disorders like human cancer, once severely stressed by zebrafish.

An inaccessible overexpression of AML1-ETO fusion macromolecule in fish finishes up in the process of defective granulopoiesis, a makeup conjointly detected in human acute leukemia. Epigenetic inhibitors targeting protein deacetylase or deoxyribonucleic acid methyltransferase activity crop myeloid cell proliferation were found in AML1-ETO and NUP98-HOXA9 embryos and thus highlight the potential role of epigenetics in these diseases. Myc oncogene and Notch intracellular domain overexpression in zebrafish victimization liquid body substance cell promoters finish up in T-cell leukemia (T-ALL). Cells with a GFP label are simple to be differentiated from T-ALL zebrafish models via FACS techniques. Transplantation of well-received isolated zebrafish cells triggers full-blown symptoms of cancer that replicate human-like cancer-like behavior (Mardis et al., 2009; Gross et al., 2010; Konantz et al., 2019).

Several aspects of human disease, including cancer, have been explored with the advantage that the leukemic cells labeled EGFP are detected directly. In humans, the cellular clarification for the progression of T-cell lymphoblastic cancer (T-LBL) and distribution of T-All was unclear. Thus various molecular mechanisms concerned with disease progression were explored through genetic testing. Experiments showed that caspase-mediated necrobiosis regulator BCL12, sphingosine one-phosphate receptor (SIP1), and living adhesion molecule 1 (ICAM1) are involved in T-LBL targeting inhibition of caspase-mediated necrobiosis and inhibition of growth cells in vasculature invasion due to cell proliferation-cell adhesion. Akt regeneration was found to beat the block in harmful implants and promote the proliferation of T-LBL cells (Konantz et al., 2019; Cairns et al., 2011). Numerous studies have shown that the mixture of BCL2 and Akt inhibitors could promote lymphocyte death and stop the proliferation of growth cells. Thus these targeted tags are typically utilized in determinative T-LBL and T-ALL treatment choices.

In the event of leukemia overexpression, the oncogenic AE macromolecule in zebrafish larvae contributed to the first killing of zebrafish and was responsible for re-regulating blood cell ancestor cells to grow into leucocyte cells. Thus the reduction of Scl was found to be necessary in neutering the very best of the cell (Bai et al., 2010; Konantz et al., 2019).

The expression, human IDH1-R132H has shown that the mechanisms concerned in human cancer is maintained by zebrafish embryos. Human AML with IDH1 possessing two mutations has been found to possess impaired TET2 activity and exhibits deoxyribonucleic acid hypermethylation.

A rise in 2-HGlevel, a decrease in 5-hmC, and a rise in myelopoiesis, and responses were reserved by AGI-5198, a selected matter of the human IDH1-R132H protein. The direct site-assisted explanation for zebrafish idh1 equally created an associate in Nursing 2-HG increase and a myeloid increase.

An increase in 2-HG caused myeloid enlargement in zebrafish larvae. Direct treatment of 2-HG zebrafish embryos was ineffective, which determined their invasive nature (Mardis et al., 2009; Shi et al., 2015) across the cell membrane. The co-event leading to leukemogenesis isn't yet clear. But, additionally to blocking the myeloid division as a result of the loss of zidh1 function, concomitant genetic mutations, including internal recurrence of the Fms-like tyrosine kinase 3 (FLT3) system, could even be the underlying causes.

The difficulty is set by the effect of IDH1-R132H mutations on the clinical outcome between glioma and AML11,4. When human IDH1-R132H and FLT3-internal tandem duplication plasmids are grouped together in low numbers of zebrafish embryos (which don't cause myeloid proliferation themselves), they have greatly increased myeloid proliferation. It supports the blending of IDH and -FLT3 in AML leukemogenesis. The role of zidh2 during embryonic development has been of great interest (Danilova et al., 2008; Bertrand et al., 2009).

It has been found that zidh1 and zidh2 show a pattern of exposure to a special location at the embryonic stage and repetition. This was indicated by the incomplete redemption of the pu.1 enlargement of the embryo of zidh1 by zidh2 mRNA and lack of effects on this hematopoiesis explicitly. However, the interaction of both zidh1 and zidh2 has caused a uniform increase in pre-myeloid onset and a decrease in late myeloid/macrophage differentiation, suggesting that genes may have a uniform role in regulating myeloid differentiation, but only zidh1 has some effect clear hematopoiesis. IDH1 mutations and 2-HG increases are associated with TET2 suppression that converts methylcytosine into hydroxymethylcytosine. Thus, the worldwide exposure to methylation in developing larvae in IDH1-R132H expression will provide a mechanical link to myeloid proliferation and prove a legitimate point that pathogenetic mechanisms leading to leukemogenesis are maintained in zebrafish larvae (Paschka et al., 2010). Additionally, the interactions between IDH1-R132H and other frequent changes in AML that were technological challenges and difficult for mouse models were easily tested within the zebrafish model. Deciphering such an important point could even be easier in our understanding of complex leukemogenesis in AML (Gross et al., 2010). Many sorts of anemia (erythrocyte depletion) were mimicked in zebrafish. Thus, hereditary elliptocytosis (HE) and hereditary spherocytosis (HS), two outstanding sorts of hemolytic anemia, were observed to be caused due to a deficiency of a molecule mentioned as molecule four.1 (P4.1). As noted in HE patients, P4.1 deficiency has semiconductor units to varied morphological deviations from elliptical vegetative cell morphology, reducing cell deformity, and disrupting the bone network (Shi et al., 2015; Konantz et al., 2019). Another creature mentioned as riesling was glorious as a result of the HS model. It carries a mutation in sptb, leading to somatic cell morphology thanks to a discontinuous membrane network (the same occurs for humans).

A disorder, hypochromic anemia has been defined by bright and tiny erythrocytes, caused by a deficiency of globin or iron. The zebrafish mutant exhibits anemia due to inembryonic globin production impairment.

1.5 ROLE OF MUTATIONS AND DIFFERENT HEMATOPOIETIC DISORDERS

A mutation in zebrafish dracula (*drc*), manifested as light-dependent lysis of red blood cells, had been reported. In this mutation protoporphyrin IX had been shown to accumulate in the mutant embryos. Thus a deficiency in the activity of ferrochelatase, a terminal enzyme in the heme biosynthetic pathway, was observed. The drcm248 mutant embryo showed a G→T transversion in the ferrochelatase gene and create a premature stop codon. The mutant phenotype displayed light-dependent hemolysis similar to humans with erythropoietic protoporphyria, a disorder of ferrochelatase (Childs et al., 2000; Ward et al., 2010; Konantz et al., 2019).

Other disorders include myeloid neoplasms which are acute or chronic and myeloproliferative neoplasms (MPNs), myelodysplastic syndromes. Driver mutations in either Janus kinase 2 (JAK2), or myeloproliferative leukemia virus oncogene (MPL) occur in the majority of MPN patients (Rhleng et al., 1996; Konantz et al., 2019).

Leukocyte adhesion deficiency (LAD) syndromes are rarely defined by WBC dependence pathology. LAD-I is defined by abnormal or reduced ins2 expression, LAD-II is set by specific defects at intervals in the fucosylation of selectin ligands, and LAD-III patients suffer from impairment in integrin. As a result of the inconsistent adherence areas, all LAD patients were found to possess exaggerated numbers of current neutrophils. The zebrafish model that mimics the phenotypes determined in LAD patients is genetically modified by rac2, specifically targeted in hematopoietic cells.

Mor2 zebrafish producing modified neutrophils, or rac2 TALEN knockout, were all gifts with defective immune defenses thanks to aberrant blood cell or vegetative cell motility, but at a consistent time, several phenotypes were detected in RAC2 deficiency. Just like bipolarity mutations, stimulation from CHT, wasn't accessible in zebrafish knockout models. Thus it suggests that various rac2 isoforms may contribute to composition expression (Bertrand et al., 2008; Konantz et al., 2019). Myelokathexis could even be a rare malady with perennial infections defined by a reduced varying performance of neutrophils (Sabaawy et al., 2006; Ridges et al., 2012; Konantz et al., 2019). WHIM syndrome manifests an honest vary of symptoms, in conjunction with warts, disorder and myelokathexis infection. In many patients, WHIM arises as a result of a change in operating gains in CXCR4. A zebrafish model was therefore developed at intervals where a reduced version of CXCR4 was utterly expressed in neutrophils. The whole mount in situ conjugation techniques show high similarity to WHIM phenotypes seen in patients. Chronic growth malady (CGD) exhibits a selection of symptoms like uncontrolled inflammation, self-injury, and severe infections caused by structural deficiency of NADPH catalyst in neutrophilic and leukocyte granulocytes (Rissone et al., 2015; Konantz et al., 2019).

Different zebrafish morphants verify the standard of NADPH catalyst activation of active gas (ROS)-mediated species of phagocytosed antibodies. However, no CGD zebrafish model has been developed. Fanconi associateemia (FA) is an associated body recessive disorder associated with BM failure. Various syndromic disorders like bone loss and an exaggerated risk of malformations were seen. Altered genetic modification of the FA's genes created the genetic background of syllable disorder required for the repair of economical compounds (Boisset et al., 2010; Danilova et al., 2008; Konantz et al., 2019). Two completely different species of syllable zebrafish are developed thus far. The loss of function leads to the event of an FA-like phenotype, including hypocellular KM, shortened linear measure, and chromosomal instability, which is almost just like the fancd2 morphant (Rissone et al., 2015; Konantz et al., 2019). A summary of different mutants in different hematological disorders using the zebrafish model has been highlighted in Table 1.1.

Lack of reliable cell signals and antibodies against these markers could be a downside within the world of zebrafish hematopoiesis; however, while not some limitations, however, fish offers an outstanding model system to review hematopoiesis and hemopoietic diseases. In the long run, this model can still facilitate the invention of the latest medication and introduce new molecular mechanisms of hematopoiesis (Jing & Zon, 2011; Rudner et al., 2011; Da Pater & Trompouki, 2018).

TABLE 1.1 A Summary of Hematological Disorders Using Zebrafish Models which Highlight the Mutant Genes and Their Respective Effects on the Hematopoietic Development

Hematological Disorder	Mutant Gene	Effect of Mutant Gene on Hematopoietic Development	References
Myeloproliferative neoplasms	*jak2a*$^{V581F-}$	Increased erythropoiesis	Ma et al. (2009)
5q-syndrome and CML-like disease	*rps14*	Anemia phenotype due to a late-stage erythropoiesis defect.	Ear et al. (2016)
B-cell acute lymphoblastic anemia	*TEL-AML1*	A high lymphoblastic count in the peripheral blood.	Sabaawy et al. (2006)
Primary immunodeficiency, leukocyte adhesion deficiency (LAD)	*rac2*	A defect in the host response system due to a neutrophil/ macrophage mobility.	Rosowski et al. (2016)
Inherited bone marrow failure: Diamond Blackfan anemia	*rps19*	Defect in erythropoiesis with abnormalities in development.	Childs et al. (2000); Danilova et al. (2008); Uechi et al. (2008)
Erythropoietic protoporphyria	*Dracula*	Erythrocytes which turn light-sensitive, the Dracula gene encodes the ferrochelatase enzyme in the heme biosynthetic pathway.	–

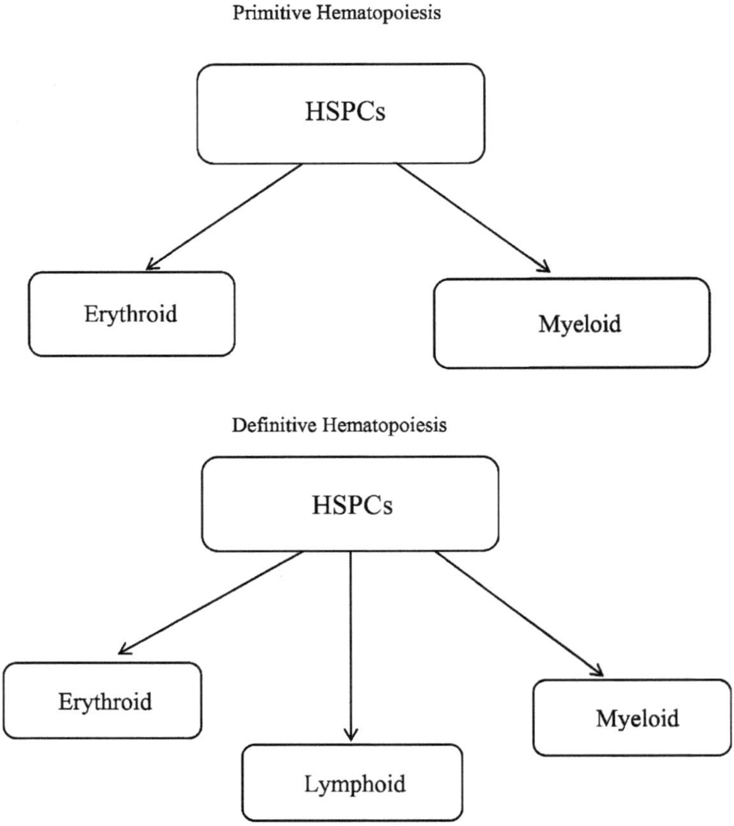

FIGURE 1.1 The two states of hematopoiesis and action on the different blood cells
Source: Adapted from Gore et al., 2018; Jing & Zon, 2011

KEYWORDS

- **definitive**
- **disorders**
- **epigenetics**
- **gata**
- **hematopoiesis**
- **IDH1**
- **interactions**
- **metabolomics**
- **mutations**
- **notch-Runx1 signature**
- **primitive**
- **RPS19**
- **Scl**

REFERENCES

Amsterdam, A., Nissen, R. M., Sun, Z., Swindell, E. C., Farrington, S., & Hopkins, N., (2004). Identification of 315 genes essential for early zebrafish development. *Proceedings of the National Academy of Sciences USA, 101*(35), 12792–12797.

Bai, X., Kim, J., Yang, Z., Jurynec, M. J., Akie, T. E., Lee, J., LeBlanc, J., et al., (2010). TIF1γ controls erythroid cell fate by regulating transcription elongation. *Cell, 142*(1), 133–143.

Bertrand, J. Y., & Traver, D., (2009). Hematopoietic cell development in the zebrafish embryo. *Current Opinion in Hematology, 16*(4), 243–248.

Bertrand, J. Y., Chi, N. C., Santoso, B., Teng, S., Stainier, D. Y., & Traver, D., (2010). Haematopoietic stem cells derive directly from aortic endothelium during development. *Nature, 464*(7285), 108–111.

Bertrand, J. Y., Kim, A. D., Teng, S., & Traver, D., (2008). CD41+ cmyb+ precursors colonize the zebrafish pronephros by a novel migration route to initiate adult hematopoiesis. *Development 135*(10), 1853–1862. https://doi.org/10.1242/dev.015297.

Boisset, J. C., Van, C. W., Andrieu-Soler, C., Galjart, N., Dzierzak, E., & Robin, C., (2010). *In vivo* imaging of haematopoietic cells emerging from the mouse aortic endothelium. *Nature, 464*(7285), 116–120.

Bowman, T. V., & Zon, L. I., (2010). Swimming into the future of drug discovery: *In vivo* chemical screens in zebrafish. *ACS Chemical Biology, 5*(2), 159–161.

Cairns, R. A., Harris, I. S., & Mak, T. W., (2011). Regulation of cancer cell metabolism. *Nature Reviews Cancer, 11*(2), 85–95.

Carradice, D., & Lieschke, G. J., (2008). Zebrafish in hematology: Sushi or science? *Blood, 111*(7), 3331–3342.

Chen, A. T., & Zon, L. I., (2009). Zebrafish blood stem cells. *Journal of Cellular Biochemistry, 108*(1), 35–42.

Childs, S., Weinstein, B. M., Mohideen, M. A. P., Donohue, S., Bonkovsky, H., & Fishman, M. C., (2000). Zebrafish Dracula encodes ferrochelatase and its mutation provides a model for erythropoietic protoporphyria. *Current Biology, 10*(16), 1001–1004.

Ciau-Uitz, A., Monteiro, R., Kirmizitas, A., & Patient, R., (2014). Developmental hematopoiesis: Ontogeny, genetic programming and conservation. *Exp. Hematol., 42*(8), 669–683. https://doi.org/10.1016/j.exphem.2014.06.001.

Danilova, N., Sakamoto, K. M., & Lin, S., (2008). Ribosomal protein S19 deficiency in zebrafish leads to developmental abnormalities and defective erythropoiesis through activation of a p53 protein family. *Blood, 112*(13), 5228–5237.

Davidson, A. J., & Zon, L. I., (2004). The 'definitive' (and 'primitive') guide to zebrafish hematopoiesis. *Oncogene, 23*(43), 7233–7246.

De Domenico, I., Vaughn, M. B., Yoon, D., Kushner, J. P., Ward, D. M., & Kaplan, J., (2007). Zebrafish as a model for defining the functional impact of mammalian ferroportin mutations. *Blood, 110*(10), 3780–3783.

De Pater, E., & Trompouki, E., (2018). Bloody zebrafish: Novel methods in normal and malignant hematopoiesis. *Frontiers in Cell and Developmental Biology, 124*.

Ear, J., Hsueh, J., Nguyen, M., Zhang, Q., Sung, V., Chopra, R., Sakamoto, K. M., & Lin, S., (2016). A zebrafish model of 5q-syndrome using CRISPR/Cas9 targeting RPS14 reveals a p53-independent and p53-dependent mechanism of erythroid failure. *Journal of Genetics and Genomics, 43*(5), 307–318.

Gore, A. V., Pillay, L. M., Venero, G. M., & Weinstein, B. M., (2018). The zebrafish: A fintastic model for hematopoietic development and disease. *Wiley Interdisciplinary Reviews: Developmental Biology, 7*(3), e312.

Gross, S., Cairns, R. A., Minden, M. D., Driggers, E. M., Bittinger, M. A., Jang, H. G., Sasaki, M., et al., (2010). Cancer-associated metabolite 2-hydroxyglutarate accumulates in acute myelogenous leukemia with isocitrate dehydrogenase 1 and 2 mutations. *Journal of Experimental Medicine, 207*(2), 339–344. https://doi.org/10.1084/jem.20092506.

Hoggatt, J., Singh, P., Sampath, J., & Pelus, L. M., (2009). Prostaglandin E2 enhances hematopoietic stem cell homing, survival, and proliferation. *Blood, 113*(22), 5444–5455. https://doi.org/10.1182/blood-2009-01-201335.

Jing, L., & Zon, L. I., (2011). Zebrafish as a model for normal and malignant hematopoiesis. *Disease Models and Mechanisms, 4*(4), 433–438.

Konantz, M., Schürch, C., Hanns, P., Müller, J. S., Sauteur, L., & Lengerke, C., (2019). Modeling hematopoietic disorders in zebrafish. *Disease Models and Mechanisms, 12*(9), dmm040360.

Mardis, E. R., Ding, L., Dooling, D. J., Larson, D. E., McLellan, M. D., Chen, K., Koboldt, D. C., et al., (2009). Recurring mutations found by sequencing an acute myeloid leukemia genome. *New England Journal of Medicine, 361*(11), 1058–1066.

Paik, E. J., & Zon, L. I., (2010). Hematopoietic development in the zebrafish. *Int. J. Dev. Biol., 54*(6, 7), 1127–1137. doi: 10.1387/ijdb.093042ep.

Paschka, P., Schlenk, R. F., Gaidzik, V. I., Habdank, M., Krönke, J., Bullinger, L., Späth, D., et al., (2010). IDH1 and IDH2 mutations are frequent genetic alterations in acute myeloid leukemia and confer adverse prognosis in cytogenetically normal acute myeloid leukemia with NPM1 mutation without FLT3 internal tandem duplication. *Journal of Clinical Oncology, 28*(22), 3636–3643.

Ransom, D. G., Haffter, P., Odenthal, J., Brownlie, A., Vogelsang, E., Kelsh, R. N., Brand, M., et al., (1996). Characterization of zebrafish mutants with defects in embryonic hematopoiesis. *Development, 123*(1), 311–319.

Reynolds, C., Roderick, J. E., LaBelle, J. L., Bird, G., Mathieu, R., Bodaar, K., Colon, D., et al., (2014). Repression of BIM mediates survival signaling by MYC and AKT in high-risk T-cell acute lymphoblastic leukemia. *Leukemia, 28*(9), 1819–1827.

Ridges, S., Heaton, W. L., Joshi, D., Choi, H., Eiring, A., Batchelor, L., Choudhry, P., et al., (2012). Zebrafish screen identifies novel compound with selective toxicity against leukemia. *Blood, 119*(24), 5621–5631.

Rissone, A., Weinacht, K. G., La Marca, G., Bishop, K., Giocaliere, E., Jagadeesh, J., Felgentreff, K., et al., (2015). Reticular dysgenesis–associated AK2 protects hematopoietic stem and progenitor cell development from oxidative stress. *Journal of Experimental Medicine, 212*(8), 1185–1202.

Rosowski, E. E., Deng, Q., Keller, N. P., & Huttenlocher, A., (2016). Rac2 functions in both neutrophils and macrophages to mediate motility and host defense in larval zebrafish. *The Journal of Immunology, 197*(12), 4780–4790.

Rudner, L. A., Brown, K. H., Dobrinski, K. P., Bradley, D. F., Garcia, M. I., Smith, A. C., Downie, J. M., et al., (2011). Shared acquired genomic changes in zebrafish and human T-all. *Oncogene, 30*(41), 4289–4296.

Sabaawy, H. E., Azuma, M., Embree, L. J., Tsai, H. J., Starost, M. F., & Hickstein, D. D., (2006). TEL-AML1 transgenic zebrafish model of precursor B cell acute lymphoblastic leukemia. *Proceedings of the National Academy of Sciences USA, 103*(41), 15166–15171.

Shi, X., He, B. L., Ma, A. C., Guo, Y., Chi, Y., Man, C. H., Zhang, W., et al., (2015). Functions of Idh1 and its mutation in the regulation of developmental hematopoiesis in zebrafish. *Blood, 125*(19), 2974–2984.

Torihara, H., Uechi, T., Chakraborty, A., Shinya, M., Sakai, N., & Kenmochi, N., (2011). Erythropoiesis failure due to RPS19 deficiency is independent of an activated Tp53 response in a zebrafish model of Diamond-Blackfan anaemia. *Br J Haematol., 152*(5), 648–654. doi: 10.1111/j.1365-2141.2010.08535.x.

Ward, P. S., Patel, J., Wise, D. R., Abdel-Wahab, O., Bennett, B. D., Coller, H. A., Cross, J. R., et al., (2010). The common feature of leukemia-associated IDH1 and IDH2 mutations is a neomorphic enzyme activity converting α-ketoglutarate to 2-hydroxyglutarate. *Cancer Cell, 17*(3), 225–234.

CHAPTER 2

Investigative Tools to Study Blood Cells: A Focus on Single Cell Isolation and Analysis

DUSHYANT KUMAR GAUTAM,[1]
ANURADHA VENKATAKRISHNAN CHIMATA,[2]
RAVI KUMAR GUTTI,[1] and INDIRA PADDIBHATLA[1]

[1]*Department of Biochemistry, School of Life Sciences, University of Hyderabad, Hyderabad, Telangana, India*

[2]*Department of Biology, University of Dayton, Dayton, OH, USA*

ABSTRACT

Hematopoiesis is a complex process involving the development and formation of blood cell lineages. Along with a multifactorial network of signaling pathways within the blood cells, the hematopoietic niche also contributes to the overall growth and development of blood cells. Several hereditary, metabolic, pathogenic, autoimmune, and cancer diseases exhibit defective hematopoiesis. A wide spectrum of research tools and methodologies are available for studying the biology of different cells and tissues. While there exists a myriad of information on these investigative techniques, it is essential to understand the potential uses of investigative tools in determining the cellular dynamics of blood cells. Through this chapter, we bring to focus hematopoietic lineages, hallmarks of hematopoietic stem cells (HSCs), and applications of research tools employed to study hematopoiesis with a specific focus on single-cell analysis.

Hematopoiesis: Biochemical, Cellular, Molecular, and Genomic Perspectives. Ravi Kumar Gutti (Ed.)

2.1 INTRODUCTION

Understanding the developmental processes happening within the cells of different tissues and organs is of immense interest to researchers in biomedical sciences (Weatherall, 1998). One such tissue of interest is the connective tissue blood. Exploring disease susceptibility, pathogenesis, and progression is the primary concern of investigators trying to understand human diseases. Technological advancements in science, identification of biomarkers (Marrugo-Ramirez et al., 2018), and decoding the human genome aided in understanding several diseases associated with abnormal numbers of RBCs and WBCs along with bleeding disorders, Iron deficiency and blood cancers, etc. (Adams et al., 2000; Bailey et al., 2014). Chemotherapy has also shown its beneficiary effect in curing blood-related disorders by targeting the essential hematopoietic pathways (Yadav et al., 2021). Blood in the human body has several functions, such as maintaining immunity, coagulation, carrying oxygen, carbon dioxide, and hormones, transporting waste products to the liver and kidneys, and maintaining the body temperature (Ali et al., 2015). Blood cells specifically exhibit an immune defensive role against injury or disease-causing pathogens. Examples of such cells include macrophages, monocytes, neutrophils, and platelets (Manz & Boettcher, 2014). Several studies brought to light the neutrophils' responses and their dependency on interaction with platelets in scenarios such as sepsis (Clark et al., 2007; Ruf & Patscheke, 1995; Singer et al., 2006). During an adaptive immune response, both T and B lymphocytes get activated and proliferate in order to elucidate a response (Heinzel et al., 2018). They also control the levels of cytokines produced, which in turn regulate the magnitude of response. Blood cells perform various functions indigenous to the cell type that is orchestrated by the blood cell markers, signaling pathway molecules, and the soluble factors unique to them. Blood cell homeostasis is dependent on several intrinsic cellular processes, such as sumoylation and ubiquitination, involved in protein regulation. At the molecular level, it is also regulated by epigenetic factors such as (SETDB1, KAT6A, and SIRT1), which are involved in the post-transcriptional modification of histones. Another mechanistic approach intrinsic to the blood cells is the formation of inflammasome required for inflammatory reactions and systemic cell death. In zebrafish, inflammasome is shown to regulate cells intrinsically, hematopoiesis during erythroid versus myeloid lineage specification and also in erythroid differentiation (Chiu et al., 2005; Boulanger et al., 2019; Kalamarz et al., 2012; Kulkarni & Kale, 2020; Paddibhatla et al., 2021). A recent study has also highlighted those extracellular vesicles from the megakaryocytes are activated by TLR2-mediated cytokine signaling to induce platelet production (Undi et al., 2016). Specifically, one study highlights the requirement of TLR4 expression in platelets for neutrophil activation and the formation of extracellular traps, which seems to enhance neutrophil-mediated microbial trapping (Clark et al., 2007). Therefore, blood is the principal tissue that orchestrates metabolic and immune homeostasis, coordinating between different organs and tissues.

In the current chapter, we discuss the characteristics of blood cells along with the relevant tools and techniques utilized to explore hematopoiesis. Furthermore, we analyzed the advantages and disadvantages of single-cell analysis and documented the significant single-cell isolation techniques utilized for the investigation of individual blood cells.

2.2 BLOOD CELL LINEAGES

Hematopoiesis is a meticulous and synchronized process leading to the formation of a functionally and morphologically diverse group of blood cells (Jagannathan-Bogdan & Zon, 2013; Kawahara, 2007). Published and prevailing research pursued extensively in different model organisms contributed to our understanding about the blood cell development and hematopoietic signaling (Gautam et al., 2021). Such observations and conclusions are obtained from both *in vivo* and *in vitro* studies using the heterogenous blood cell populations present within the different anatomical sites that exhibit chronological differences. The bone marrow contains a mixed population of blood cells that include HSCs, blood cell progenitors, blood cell precursors, red blood cells, white blood cells, and platelets (Suda et al., 2011). Each of these hematopoietic progenitors, precursors, and matured cells has a significant and unique role in maintaining hematopoietic homeostasis and blood cell development.

2.2.1 MULTIPOTENT PROGENITORS (MPPs)

Stem cells, in general, have an innate ability to self-renew and to differentiate into any specialized cell in an organism (Shenghui et al., 2009). Their differentiation potential helps to classify them into four different categories, namely totipotent stem cells, pluripotent stem cells, multipotent stem cells, and unipotent stem cells (Singh et al., 2016). Following fertilization event, the cells derived from the zygote up to the 8-cell division stage are considered totipotent (Figure 2.1). These cells have the potency to produce all types of body cells along with extraembryonic tissue (placenta), thus giving rise to the whole organism (Verfaillie et al., 2002; Hood & Rowen, 2013). They are not only different but also create self-copies, thereby maintaining their pool till the embryonic stage (Zakrzewski et al., 2019). Totipotent cells also specialize into pluripotent cells, which further give rise to all three germ layers, yet these pluripotent cells do not show the ability to generate whole organisms (Hanna et al., 2010). Multipotent stem cells differentiate into more than one cell type, but they cannot form the three germ layers. Unipotent stem cells, such as skin and muscle stem cells, show the least differentiation potential, which can produce cell types of one specific lineage only (Biehl & Russell, 2009).

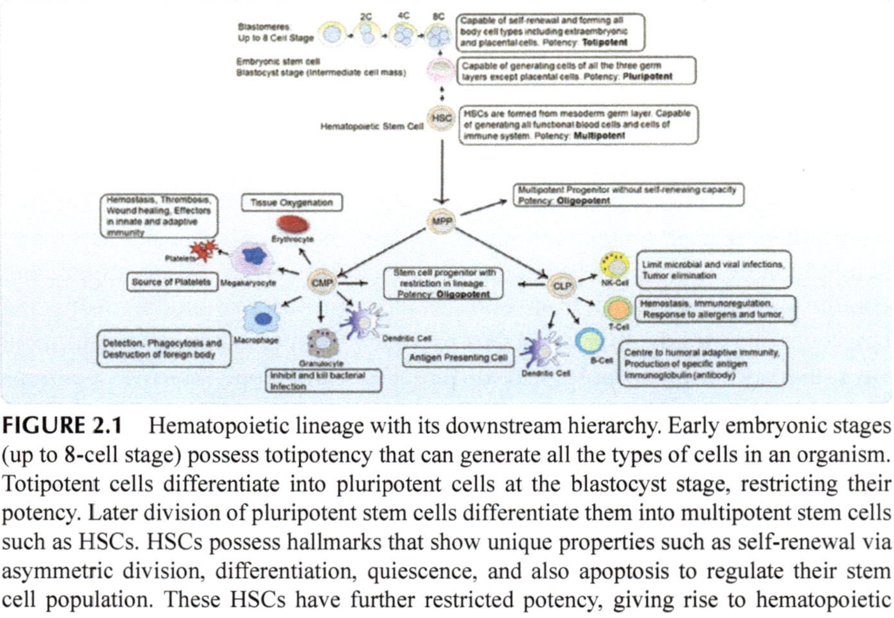

FIGURE 2.1 Hematopoietic lineage with its downstream hierarchy. Early embryonic stages (up to 8-cell stage) possess totipotency that can generate all the types of cells in an organism. Totipotent cells differentiate into pluripotent cells at the blastocyst stage, restricting their potency. Later division of pluripotent stem cells differentiate them into multipotent stem cells such as HSCs. HSCs possess hallmarks that show unique properties such as self-renewal via asymmetric division, differentiation, quiescence, and also apoptosis to regulate their stem cell population. These HSCs have further restricted potency, giving rise to hematopoietic progenitors (MPP, CMP, CLP) that eventually generate all the cells of myeloid and lymphoid lineage (C: cell stage, HSC: hematopoietic stem cell, HPC: hematopoietic progenitor cell, MPP: multipotent progenitor cell, CMP: common myeloid progenitor cell, CLP: common lymphoid progenitor).

HSCs self-renewal ability demarcates the HSC pool into two sub-populations: long-term (LT) and short-term (ST) HSCs. LT-HSCs are known to have self-renewal ability for longer period, to reconstitute the hematopoietic lineage, compared to ST-HSCs (Jurecic, 2019). ST-HSCs self-renewal lasts for about 6 weeks for hematopoietic lineage support. MPPs are the closest progenitor of ST-HSCs, devoid of self-renewal ability, enabling them to maintain hematopoietic system for a transient period (Cheng et al., 2020). MPPs are oligopotent in nature. MPPs exhibit erythroid and myeloid cell lineages with an accumulation of uncommitted cells. MPPs give rise to Common Myeloid Progenitors (CMPs) and common lymphoid progenitors (CLPs) (Figure 2.1), that further differentiate into cells carrying cellular signatures of myeloid and lymphoid lineages, respectively (Seita & Weissman, 2010).

2.2.2 COMMON MYELOID PROGENITORS (CMPs)

CMPs are derived from the bone marrow, and they produce all myeloid cells, such as erythrocytes, megakaryocytes, platelets, macrophages, dendritic cells, and all the granulocytes (Figure 2.1) (Rathinam & Flavell, 2008). These terminally differentiated cells are unipotent in nature. Erythrocytes specifically supply tissue oxygenation for the development of an embryo (Jensen, 2009). Megakaryocytes sheds off platelets into the circulation (Cho, 2015). These platelets function in wound healing and hemostasis, while recent reports have projected their role in immune response also (Wagner & Burger, 2003; Semple et al., 2011). Macrophages specifically evade and phagocytose foreign bodies. Dendritic cells are categorized under both myeloid and lymphoid lineages. Both the CMP and the CLP have the potential to give rise to these dendritic cells (Traver et al., 2000; Manz et al., 2001). These matured dendritic cells are typically antigen-presenting cells (Sun et al., 2013). Granulocytes, i.e., neutrophils, basophils, and eosinophils, help in eliminating the infection (Manz & Boettcher, 2014; Kobayashi et al., 2018). Their population is enhanced during an infection condition helping the body to clear the infection, thereby maintaining immunity and health by releasing growth factors, enzymes, histamine, cytokines, and chemokines at the site of infection.

2.2.3 COMMON LYMPHOID PROGENITOR (CLP)

CLPs are derived from the lymphatic tissue and give rise to all types of lymphocytes. Like CMPs, CLPs are restricted from a wide array of lineage commitments. They are bound to give rise to unipotent immune cells such as Natural Killer cells (NK cells), B-cells, and T-cells (Kondo et al., 1997). NK cells are considered to be part of the innate immune system and are the first effector lymphocytes to encounter microbial and viral pathogens along with hindering the growth of tumors (Vivier et al., 2008; Lodoen & Lanier, 2006). CLPs, from the bone marrow, give rise to T-cells, whose development is initiated in the thymus. T-cells are known to play a central role in the adaptive immune response (Boutros, Heigwer, & Laufer, 2015; Chen et al., 2012). One of the functions of T cells is immune-mediated cell death via CD8+ and CD4+ subtypes of T cells (Klein et al., 2009). B cells are central to the adaptive humoral immune system. They produce antigen-specific immunoglobins (Ig), popularly known as antibodies, to neutralize the effect of invasive pathogens. In addition to their role in humoral immunity, they have also been involved in the initiation of T cells immune response and another immune-related homeostasis. Dendritic cells are also known as accessory cells that perform immune surveillance by presenting the antigens produced both endogenously and exogenously (Figure 2.1).

2.3 HALLMARKS OF HSCs

HSCs are the classical example of multipotent stem cells. They give rise to the progenitors of the two separate hematopoietic branches, i.e., myeloid and lymphoid lineages. It is these lineages that further differentiate forming all the different types of blood cells (Kondo, 2010). HSCs manifest several unique properties that separate them from other hematopoietic progenitors (Domen & Weissman, 1999). HSC hallmarks include self-renewal, differentiation, migration, quiescence, and apoptosis (Figure 2.2).

2.3.1 SELF-RENEWAL

Self-renewal is considered as one of the essential characteristics of HSCs. This ability aids HSCs in reproducing their exact copies with similar potential while restricting their entry into differentiation (McCulloch & Till, 2005). There is a continuous replenishment of matured blood cells through the differentiation of HSCs (Reya et al., 2001). The fate of HSCs is determined by asymmetric and symmetric division enabling the balance between their potential for self-renewal and differentiation (Ito & Ito, 2018).

2.3.2 DIFFERENTIATION

Differentiation of HSCs is driven by a complex network of cytokines and growth factors. These factors are regulated as per cells' requirement or

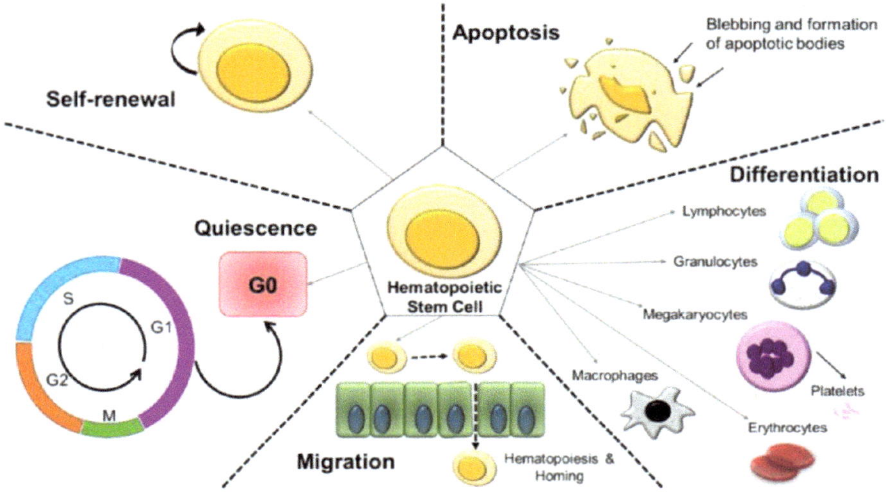

FIGURE 2.2 Hallmarks of HSCs. HSCs have specialized capabilities which enable them to live long and maintain their numbers to establish hematopoietic homeostasis. HSCs strike an intricate balance between self-renewal and differentiation in order to continuously supply mature blood cells. These cells maintain themselves in a quiescent state for a long time and do not cross the Hayflick limit divide for a limited number of times in their lifetime. Apoptosis is another characteristic feature that controls the number of cells in circulation. Migration is a critical property of HSCs that is required for their homing to the site of requirement.

controlled by an extrinsic stimulus (Zhang & Lodish, 2008). Blood cell differentiation is considered as an irreversible dynamic process which once committed cannot be revert to their progenitor state (Wang & Wagers, 2011). Therefore, HSCs need specific signals that help in the generation of hierarchical progenitors and their differentiation into matured cells. Self-renewal is a salient and unique feature of HSCs, while differentiation seems to be an eventual outcome (Seita & Weissman, 2010).

2.3.3 MIGRATION

Migration of the HSCs is essential for the development of hematopoietic tissues. This property confers the regenerative potential of the HSCs (Magnon & Frenette, 2008). Migration of HSCs to both the embryonic and the extra embryonic sites, such as fetal liver, spleen, and finally to bone marrow, is observed at different developmental stages. Until there is a requirement for HSCs to differentiate into matured cells, they are maintained within these anatomical sites (Orkin & Zon, 2008). Circulating HSCs "home" themselves in the bone marrow microenvironment region termed a "niche" (Martinez-Agosto et al., 2007; Morrison & Spradling, 2008). The need for the HSC movement from the bone marrow to anatomical sites via the circulating blood is a tightly regulated and coordinated process. This natural process is exploited by researchers where the mobilization is regulated by certain growth factors (cytokines, G-CSF) and chemotherapeutics leading to the release of HSCs facilitating blood collection by leukapheresis (Magnon & Frenette, 2008).

2.3.4 QUIESCENCE

Hematopoietic cell production is a life-long process served by the HSCs with minimal stem cell exhaustion (Seita & Weissman, 2010). HSCs pursue this strategically by being in the dormant stage. The fate of HSCs' quiescence largely depends upon the intrinsic and extrinsic signals induced in response to stress and blood loss (Wilson & Trumpp, 2006). These signals help in maintaining HSC quiescence by avoiding its entry into the cell cycle. Compared to quiescent cells, the proliferating cells are more prone to genetic mutations, and upon reaching the Hayflick limit, they undergo apoptosis, avoiding defective hematopoiesis. Therefore, blood cell quiescence is an essential property that protects HSCs from developing hematopoietic malignancies.

2.3.5 APOPTOSIS

Apoptosis is a self-destructive, energy-dependent, biochemical event that maintains the hemostasis mechanism during development, aging, and also during immune reactions (Elmore, 2007). The size of the hematopoietic stem cell mass is directly linked to the intricate balance maintained between self-renewal and apoptosis (Seita & Weissman, 2010). Any sort of imbalance between self-renewal and apoptosis exhibits hematopoietic insufficiencies and hematopoietic malignancies (Alenzi et al., 2009). Like other cells, HSCs number are also regulated through programmed cell death (apoptosis).

2.4 TOOLS AND TECHNIQUES TO STUDY HEMATOPOIESIS

The availability of the complete genetic sequences of humans and research model organisms made it easier to study genetic variations and the signaling pathways affected by them (Celniker et al., 2009). Such vast knowledge will benefit scientists with the availability of advanced tools and techniques for the manipulation of known gene sequences that help in identifying the role of genes and proteins. Several genetic tools and techniques are used to imitate disease models and in deciphering the underlying mechanisms contributing to their pathophysiology (Wang, 2021). Along with disease conditions, such models are also used to understand the developmental processes and the "intra" and "inter" cellular interactions in healthy cells and tissues (Turei et al., 2021). We summarize genetic tools that are common and used successfully in various model organisms (Table 2.1). Several genetic tools allow manipulation of gene expression, and these include expression systems that use transcription activators and promoters. These genetic systems helped in organ or tissue-specific expression as observed in *D. melanogaster*, *M. musculus*, and *D. rerio* model. Many such tools are bipartite systems with spatial-temporal control of target gene expression in the tissue of interest. Some examples include the UAS-Gal4 system, Q-system, *Flp* system (Potter et al., 2010; Brand & Perrimon, 1993; Fisher et al., 2017). One of the most widely used tools to drive *in vivo* gene expression is the bipartite UAS-Gal4 system that exploits the specificity of Gal4 (yeast transcription factor) to UAS (Upstream activation sequence) (Potter et al., 2010; Brand & Perrimon, 1993; Fisher et al., 2017). This allows the manipulation of specific enhancers to drive expression of transgenes creating both gain-of-function and loss-of-function (with the use of RNAi lines) genetic models. Other tools such as Gal80 and Flp-DD are also helpful in altering gene expression, conditionally, either by temperature or by chemical control (Seita & Weissman, 2010). Functions of some essential patterning genes like *antennapedia*, *eyeless* and *decapentaplegic* in development have been elucidated using mutant lines and by exploiting the Gal4-UAS system (Capdevila & Guerrero, 1994; Halder, Callaerts, & Gehring, 1995). Another classical route in deciphering gene function is the use of loss-of-function mutants. Some methods of mutagenesis such as ethyl methane sulfonate (EMS), X-rays, *P-elements* produce random mutations, whereas transcription activator-like effector nuclease (TALEN) and CRISPR (clustered regularly interspaced short palindromic repeats)/ Cas9 are examples of site-directed mutagenesis. Large-scale mutagenesis screens helped in identifying the dorso-ventral (DV) axis of the *Drosophila* embryo patterning, revealing the functional significance of 12 genes that include *gastrulation-defective*, *dorsal*, *windbeotel*, *nudel*, *tube*, *pipe*, *snake*, *caster*, *Toll*, *spaetzle*, *pelle* and *cactus* (Nusslein-Volhard, Wieschaus, & Kluding, 1984). Tumor suppressor network members like *warts, lats,* and *hippo* belonging to the Hippo pathway were also identified in such mutagenesis screens (Xu et al., 1995; Udan et al., 2003). Apart from providing resources for future research, these studies establish a basis for identifying genetic networks. Several genes associated with aging and cancer have been identified by using mutants of *C. elegans* and *Drosophila* (Klass, 1983; Minakhina & Steward, 2006). Transposon-mediated mutagenesis made it feasible to insert DNA sequences randomly into the genome of vertebrates and invertebrates (Rushforth et al., 1993; Zwaal et al., 1993; Cooley et al., 1988; Ding et al., 2005). Alternatively, ZFN, TALEN and CRISPR/Cas9 have specific DNA binding regions which identify and bind to the target genes and alter the sequences (Gaj et al., 2013). While the above-discussed genetic tools (Table 2.1) aid in deciphering genetic interactions, there are other techniques (Table 2.2) which are needed to study physiological, morphological and biomolecular changes among others. Morphological changes can be studied using visual observations and several microscopy techniques from bright field microscopy to electron microscopy. Microscopy gives us the opportunity to study the shape of cells, numbers, and even their atypical appearance (Wang & Bodovitz, 2010). Technological advances in hematological research has exploited microscopy and live cell imaging to visualize fluorescently labeled blood cells to study interaction between circulating cells (Morikawa & Takubo, 2017). Live single-cell imaging is another technique used to track gaseous exchange in blood cells (Cole, 2014). Transcript levels are conveniently studied using reporter, real time PCR, *in situ* and microarray while proteins are studied following the traditional methodologies like immunohistochemistry, immunofluorescence, and western blotting. Techniques in large scale leading to the "omics" studies are immensely contributing to our understanding about not only the changes in specific gene and protein levels, but also in studying surface markers and proteins even in single cell level (Wang & Bodovitz, 2010). Techniques like liquid chromatography/mass spectrometry and NMR-based spectroscopy have been used to study for metabolic profiles in different cell populations (Emwas et al., 2019; Kamleh et al., 2009). Deregulated metabolism is observed in several diseases like cancers and neurodegeneration (Houck et al., 2018). Genetic tools have also been instrumental in understanding the development and function of blood cells. The use of temperature sensitive mutants has unraveled the role of conserved transcription factors GATA1, GCM and their differential regulation during blood cell development (Letourneau et al., 2016). *Drosophila* mutants and clonal studies with FLP/FRT system helped in demonstrating the requirement of Notch signaling for proliferation in the hematopoietic organ, the lymph gland (Paddibhatla et al., 2019; Lebestky et al., 2003).

TABLE 2.1 Genetic Tools in Model Organisms*

Tool/ Technique	Type	Mode of Action	Model Organism	References
FlpStop	Gene expression	Conditional gene disruption	*Drosophila melanogaster*	Fisher et al. (2017)
GAL80-DD	Gene expression	Chemical control and knockdown of Gal4-dependent expression systems	*Drosophila melanogaster*	Sethi & Wang (2017)
FLP-DD	Gene expression	Chemical control of gene recombination	*Drosophila melanogaster*	Sethi & Wang (2017)
GAL4/UAS	Gene expression	Knockdown of gene expression, overexpression of gene.	*Drosophila melanogaster*	Brand & Perrimon (1993)
		Overexpression, knockout, mutation.	*Danio rerio*	Asakawa & Kawakami (2008); Asakawa & Kawakami (2009); Scheer & Campos-Ortega (1999)
QF/QUAS	Gene expression	QUAS-dependent gene expression	*Drosophila melanogaster*	Potter et al. (2010)
LexA::VP16, LexA::GAD	Gene expression	Clonal gene expression and repression.	*Drosophila melanogaster*	Lai & Lee (2006)
InSITE	Gene expression	Allows change of genetic effectors	*Drosophila melanogaster*	Gohl et al. (2011)
Tetracycline-inducible systems	Gene expression	Controlled transgene induction.	*Drosophila melanogaster*	Stebbins et al., 2001b)
		Tetracycline-gene overexpression.	*Mus musculus*	Akagi et al. (2001); Gossen et al. (1995)
		Gene overexpression	*Danio rerio*	Akagi et al. (2001); Gossen et al. (1995)
RU486-inducible GAL4 (GeneSwitch)	Gene expression	Tissue-specific and controlled RU486-dependent expression	*Drosophila melanogaster*	Osterwalder et al. (1993)
FINGR system/ ET-FLPx2	Gene expression	Tissue-specific expression or repression of genes	*Drosophila melanogaster*	Bohm et al. (2010)
MARCM	Gene expression	To generate mutations and study the gene of interest	*Drosophila melanogaster*	Lee & Luo (1999, 2001)
Q-system	Gene expression	Transgene expression, lineage tracing, and mosaic analysis.	*Drosophila melanogaster*	Riabinina & Potter (2016)

TABLE 2.1 *(Continued)*

Tool/ Technique	Type	Mode of Action	Model Organism	References
deGradFP	Gene expression	Knockdown of GFP-tagged proteins.	*Drosophila melanogaster*	Caussinus & Affolter (2016)
TARGET	Gene expression	Temporal and regional gene expression targeting.	*Drosophila melanogaster*	McGuire et al. (2004)
CRISPR	Mutagenesis	Gene editing, targeted mutagenesis.	*Drosophila melanogaster*	Bassett et al. (2013)
		Gene editing, targeted mutagenesis.	*Mus musculus*	Shen et al. (2013)
		Gene editing, targeted mutagenesis.	*Danio rerio*	Chang et al. (2013); Hwang et al. (2013)
TALEN	Mutagenesis	Locus-specific gene editing and mutagenesis.	*Drosophila melanogaster*	Katsuyama et al., 2013; Lee et al. (2015)
		Site-specific induction of DNA double-strand breaks, mutagenesis, gene-specific knockout.	*Mus musculus*	Sommer et al. (2015); Sung et al. (2013); Yasue et al. (2015); Zhang et al. (2014)
		Locus-specific variations.	*Danio rerio*	Bedell et al. (2012); Huang et al. (2011); Sander et al. (2011)
PiggyBac	Transposon system	Mutagenesis, gene disruption.	*Drosophila melanogaster*	Kim et al. (2012); Thibault et al. (2004)
		Transgenic expression, insertional mutagenesis.	*Mus musculus*	Ding et al. (2005); Rad et al. (2015)
		Gene transfer and germline transformation.	*Danio rerio*	Lobo et al. (2006)
P-Elements	Transposon system	Gene disruption	*Drosophila melanogaster*	Thibault et al. (2004)
TILLING	Mutagenesis system	Creating genetic lesions, mutations, identifies mutation.	*Drosophila melanogaster*	Winkler et al. (2005)
Sleeping beauty system	Transposon system	Genetic transformation, insertional mutagenesis, gene discovery.	*Mus musculus*	Dupuy et al. (2006, 2009); Ivics et al. (1997); Newberg et al. (2018); Tschida et al. (2014)
		Insertional mutagenesis, transgenic gene expression, Enhancer trapping.	*Danio rerio*	Balciunas et al. (2004); Davidson et al. (2003); Ivics et al. (1997); McGrail et al. (2011)

TABLE 2.1 *(Continued)*

Tool/ Technique	Type	Mode of Action	Model Organism	References
Cre-inducible sleeping beauty system	Transposon system	Cre-mediated recombination and mutagenesis.	*Danio rerio*	Dupuy et al. (2009)
T2/Onc3	Transposon system	Mutagenesis	*Danio rerio*	Dupuy et al. (2009)
GFP, Gal4FF	Gene expression	Site-specific gene expression, knockouts.	*Danio rerio*	Kawakami et al. (2010)
Tol2 system	Transposon system	Transgenic expression.	*Danio rerio*	Clark et al. (2011); Kawakami (2007); Suster et al. (2009)

**The table shows the predominantly used genetic tools to express and manipulate gene expression in vivo model systems. The tools vary from transcriptional control of genes to transposons and mutagenic tools.*

TABLE 2.2 Techniques to Study Morphological and Molecular Changes*

Technique	Type	Purpose	References
Live imaging	Morphological	To investigate the dynamics and response of blood cells to tissue injury.	Babcock et al. (2008)
Single-cell O_2 saturation imaging	Morphological	To track gas handling and exchange by RBCs.	Richardson et al. (2020)
Continuous single-cell imaging.	Morphological	To understand the developmental origin of blood cells.	Eilken et al. (2009)
Label-free spectrally encoded flow cytometry.	Morphological	To assess blood parameters.	Golan et al. (2012)
Flourescent microscopy	Morphological	To determine circulating hemocyte concentration.	–
Confocal microscopy	Morphological	Live imaging, 3D imaging, fluorescent marker imaging.	Bayguinov et al. (2018); Paddock & Eliceiri (2014)
TEM	Morphological	To study structural components of cells using high resolution imaging.	Koster & Klumperman (2003); Cizmar & Yuana (2017)
SEM	Morphological	To study topographical components.	Koster & Klumperman (2003); Lea et al. (1994)

TABLE 2.2 *(Continued)*

Technique	Type	Purpose	References
Western blot	Protein expression	To identify specific proteins segregated based on size.	Mahmood & Yang (2012)
Probe-based hybridization and expression analysis.	Protein/gene expression	To study gene expression.	Watkins et al. (2009)
Tryptophan fluorescence-based WF assay, liquid chromatography-tandem mass spectrometry	Protein/gene expression	To study membrane or surface protein composition.	Bryk & Wisniewski (2017)
Mass spectrometry	Protein/gene expression	To study membrane or surface protein composition.	Ravenhill et al. (2019)
Single cell mass cytometry	Protein/gene expression	To study surface proteins to distinguish distinct cells and understand their interaction.	Balog et al. (2021)
In vivo immunostaining and live imaging.	Protein/gene expression	To study both genetic (transcriptional reporter) and immunological markers.	Csordas et al. (2014)
LC-MS/MS	Metabolites	To screen for biomarkers in diseased background.	Asante et al. (2018, 2019)
NMR spectroscopy-based profiling	Metabolites	Metobolite quantification.	Sengupta & Weljie (2019); Nagana et al. (2018)
Raman spectroscopy	Chemical composition	To measure the chemical composition of biological tissue.	Butler et al. (2016)
FACS	Protein/gene expression	To isolate cells based on cell-specific markers.	Bonner et al. (1972); Herzenberg et al. (1976)
MACS	Protein/gene expression	To isolate cells based on cell-specific markers.	Schmitz et al. (1994)
Image-based single-cell sorting	Protein/gene expression	To isolate cells based on cell-specific markers.	Sesen & Whyte (2020)
In-situ hybridization	Protein/gene expression	To study transcript expression	Tautz & Pfeifle (1989); Choi et al. (2018)

**We have listed the techniques used to study the morphological or molecular changes in the hematopoietic system. Furthermore, we have listed several techniques utilized for the isolation of single cells.*

While these techniques and tools are instrumental in investigating the role of unknown genes and proteins and their interacting molecules, the starting input for the experiments are obtained from a large population of cells from a particular tissue. The output of such assays or experiments is the representative result of the entire population of cells under study. In the past decade, single cell analysis has gained a lot of attention since researchers are having the techniques and tools to study single cells.

2.5 SINGLE CELL ANALYSIS (SCA) OF BLOOD CELLS

Various fields of study including neurobiology, hematopoiesis, computational biology, and systems biology have delineated the functional complexity and the cellular dynamics at the tissue and organ level. The plethora of information obtained from genomics, transcriptomics, metabolomics, and proteomics on heterogenous cell populations provide valuable insights into understanding growth, development, and disease (Gawad et al., 2016; Stegle et al., 2015; Zenobi, 2013; Pham et al., 2021). Yet there remain several unanswered questions regarding the individual cells contributing to the results and conclusions of the heterogenous populations. The past decade of research on SCA benefitted the scientific investigators in understanding the functioning, development, and maintenance of single cells within their heterogenous cellular environment. Evolving technical advancements employed for the analysis of single cells facilitated researchers in unraveling the type of single cells contributing to the overall phenotype obtained from the heterogenous population. The genomic, transcriptomic and metabolomic networking of each blood cell type is significantly different at a specific time point in its development. These differences can be clearly identified and elucidated using the powerful techniques employed for single cell isolation and analysis (Gross et al., 2015; Hu et al., 2016).

2.5.1 ADVANTAGES AND DISADVANTAGES OF SCA

While the study of single cells is highly benefitting the scientific world, there exist multiple advantages in studying single cells along with some drawbacks that need attention (Lahnemann et al., 2020; Wang & Bodovitz, 2010). Years of research brought to focus information on the genetic and epigenetic differences between healthy cells and aberrant cells such as cancerous cells, neurodegenerative cells, etc. (Moosavi & Motevalizadeh, 2016). Such scientific evidence are obtained from studying the heterogenous population of cells supporting the involvement of specific genes and proteins that affect signaling pathways altering the molecular architecture. Yet there exist many unanswered questions regarding the cells within the heterogenous population under scrutiny. Primarily studying single cells present within the heterogenous population is highly beneficial in determining the responses of individual cells to the overall output produced from the heterogenous population. The feasibility of studying genomics (Gawad et al., 2016), transcriptomics (Stegle et al., 2015), proteomics (Pham et al., 2021), and metabolomics (Zenobi, 2013) of the single cells is another advantage. SCA study also provides an opportunity to differentiate the characteristics between the single cells under observation based on the signatures of age exhibiting morphological, physiological, and biochemical differences (Luo et al., 2019). It is beneficial to understand in a given population of similar cells with the significant differences. The other advantage of SCA study is bringing to focus the single cell specific traits. This is achieved by exploiting techniques such as single-cell RNA sequencing (scRNA-seq) which helps in defining the specificity of the cell type and also discovering new cell types (Eberwine et al., 2014). Finally, researchers can utilize intact, live, or fixed cells for the SCA study similar to the heterogenous populations, tissues, and organs. Even though SCA study has gained a lot of significance there exist some drawbacks (Lahnemann et al., 2020). The significant issue faced by researchers studying single cells primarily involves obtaining a sufficient sample size for processing successfully. This is challenging because during the process of isolating single cells, there is a chance for damage (Yuan et al., 2017). Nevertheless, SCA is a beneficial technique, and the output of such SCA leads to a large amount of data. The advantages and disadvantages are reflected in Figure 2.3.

2.5.2 TECHNIQUES INVOLVED IN SINGLE-CELL ISOLATION

Pioneering research using SCA technique was done on mRNA obtained from mouse blastomere. mRNA-Seq-Assay was performed, and the gene expression profile identified the expression of 5270 (75%) more genes in comparison to microarray techniques. This study is a classic example of how the knowledge gained from the analysis of single cell type is profound, different, and greater compared to the representative results obtained from heterogenous cell populations. There are various tools and techniques employed to study single cells (Figure 2.4). We discuss some important isolation techniques that can be employed for studying.

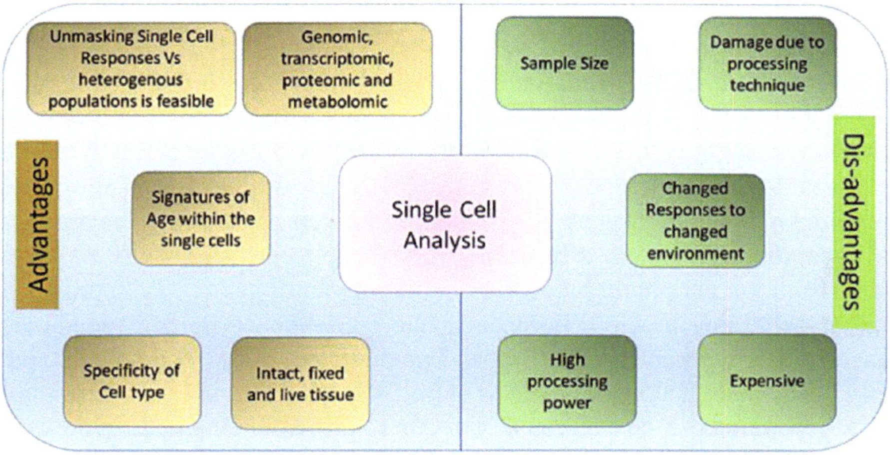

FIGURE 2.3 Advantages and disadvantages of single cell analysis. Single-cell analysis is a high-throughput technique that requires isolation of a specific subset of cell populations, processing, and final investigation to study at gene, transcript, protein, or metabolite level. This process has several pros and cons that are schematically mentioned in this figure.

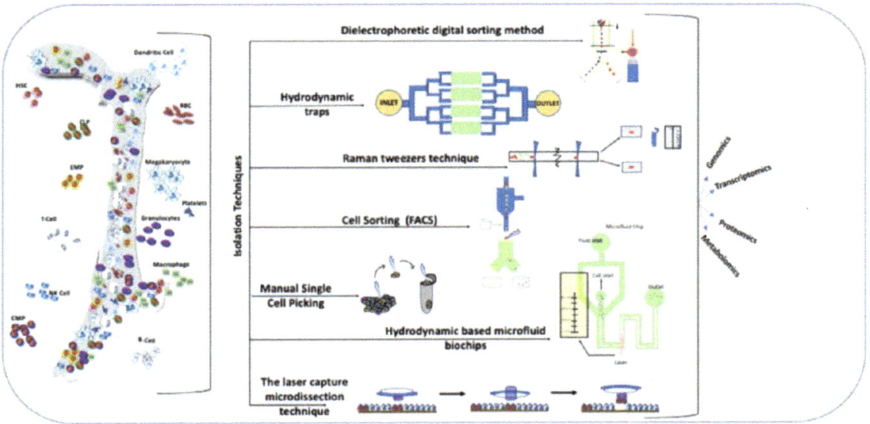

FIGURE 2.4 Single-cell isolation methods. There are several types of blood cell populations in all the niche like the bone marrow in the human body. In the bone marrow, such cells include precursors and progenitors like HSCs, CLP, EMP, and CMP and mature blood lineages like RBCs, platelets, lymphocytes, NK cells, macrophages, and granulocytes. In order to study these cells individually, several isolation techniques like cell sorting (MACS, FACS, DACS), hydrodynamic traps, dielectrophoretic digital sorting method, Raman tweezers, microfluidics, and manual methods are used and are schematically represented in the figure.

2.5.2.1 DIELECTROPHORETIC DIGITAL SORTING METHOD

Dielectrophoretic (DEP) digital sorting method works on the DEP forces, exerted by each cell types. This technique exploits the fact that the dielectric properties, such as membrane permittivity and cytoplasm conductivity are different for each cell type (Valero et al., 2010). Use of this method allows for minimal manipulation as no other labels or markers are used for the purpose of separation (Burt et al., 1990). DEP method helps in successfully isolating cells from the peripheral blood the circulating tumor cells (CTCs) or breast tumor cells, due to the difference in the cells phenotype and membrane capacitance, without any labeling and modification (Han, Yang, & Frazier, 2007; Voldman, 2006; Gascoyne & Shim, 2014). Additionally, dielectrophoretic separation can be combined with other purification methods in order to effectively trap cells of interest (Lee et al., 2020).

2.5.2.2 HYDRODYNAMIC TRAPS

Hydrodynamic trapping is basically a single-cell isolation technique. This method involves obstructions such as mechanical barriers that help in the separation of the target particles from the main inlet. As soon as the target particles are separated, they are retained in the hydrodynamic trapping sites. Later, these sites could be thoroughly screened by many investigations as a part of research interest. These screening procedures include types of walls or pores of different size and pattern that in the trapping site ultimately to capture single cells (Johnson-Chavarria, Tanyeri, & Schroeder, 2011).

2.5.2.3 RAMAN TWEEZERS TECHNIQUE

The Raman tweezers technique is the combination of Raman spectroscopy and the optical tweezers (also known as optical laser trapping). This powerful technique aids cell biologists as a diagnostic tool where the single cells can be trapped and manipulated in a laser beam using a high numerical aperture imaging microscope (Snook et al., 2009). This method has been successfully used for studying single live blood cells to obtain various biochemical parameters. The advantage is that it can be used for live cells. Raman tweezers are also used for trapping the RBCs and WBCs for studying their intrinsic induced effects due to external stimulus (Bankapur et al., 2010).

2.5.2.4 *FLUORESCENCE ACTIVATED CELL SORTING (FACS)*

Fluorescence-activated cell sorting (FACS) is a type of flow cytometry with specialized cell sorting capacity. It is based on light or fluorescent characteristics of each cell. This technique proves to be highly beneficial and sophisticated with user-friendly approach. It has the ability to characterize particular cell type from the heterogeneous population of cells based on specific properties such as cell size, granularity, and fluorescence (Gross et al., 2015). While FACS has been long used for the separation of cells, with the current technical advances, it is now being used for sorting single cells from heterogenous population of cells (Schulz et al., 2012). In a recent study, FACS was combined with 5′-end scRNA-seq and allows even analysis of surface markers simultaneously by combining transcriptomics and surface marker expression (Waghmare & Page-McCaw, 2018).

2.5.2.5 *MANUAL SINGLE CELL PICKING*

Manual cell picking is a method for isolating single cells that is efficient compared to other single-cell isolating techniques. This technique comprises of an inverted microscope connected with movable micro-pipettes through motorized mechanical stage. Single cells are visualized under the microscope and then photographed. This technique possesses an advantage over many techniques in that it can isolate a single cell from live cultures or embryo cells (Hu et al., 2016).

2.5.2.6 *HYDRODYNAMIC-BASED MICROFLUID BIOCHIPS*

Microfluidic generally uses two strategies for trapping cells. The first one is active methods of contactless cell trapping (non-microfluidic) based on an external force field like dielectrophoresis, magnetophoresis, acoustophoresis, optical tweezers, etc., and the second one is passive method of contact cell immobilization based on surface treatment or nonlinear hydrodynamics forces (Microfluidic) or biomimetic or hydrophoretic and deterministic lateral displacement (DLD) (Karimi et al., 2013).

2.5.2.7 *LASER CAPTURE MICRODISSECTION TECHNIQUE*

Laser capture microdissection (LCM) is an advanced technique that has the ability to isolate the single cells or population of cells from the heterogenous population mainly from solid tissue samples or microscope slides. LCM is broadly classified into two categories: Infrared LCM and ultraviolet LCM (Hu et al., 2016). This method captures cells or tissues by using laser and thus allowing for higher precision. The isolated cell(s) can then be analyzed for molecular or biochemical parameters. A recent study also utilized LCM for T-cell clonality analysis of lymphoma samples (Yakirevich et al., 2007).

2.6 SCA OF TRANSCRIPTOMICS, PROTEOMICS, AND METABOLOMICS

2.6.1 GENOMICS

Single-cell genome sequencing aids in understanding the complexity of the genetic heterogenous cell population and thus disease etiology (Gawad et al., 2016). The former approach also helps us in analyzing the heritable germline mutations and somatic mutations in normal vs. malignant cells (Kashima et al., 2020). The advantage of SCA of genome largely depends on the study of single cells isolated from heterogenous population. To begin with, the first task is to get single viable cells from the primary complex source of tissues. Techniques such as laser capture microdissection, microfluidic, and bead-based method are utilized for the isolation of single cells (Emmert-Buck et al., 1996; Navin, 2014). Various approaches have been developed to isolate single viable cells in order to get them suspension. In past decades Whole genome sequencing has made a considerable amount of progress. This technique basically amplifies the genome copy while limiting the background noise such as mutation, chimeras, and genome loss (Blainey, 2013).

2.6.2 TRANSCRIPTOMICS

Single-cell analysis in the past recent years has emerged as one of the potent tools to decipher the understanding of metabolism at the single-cell level. So far in the existing field of single--cell analysis, RNA sequencing (RNA-seq) is the widely accepted technique for transcriptomic studies. RNA-seq works on the principle of reverse transcription of mRNA molecules resulting into the corresponding complementary DNA (cDNA) molecules followed by polymerase chain reaction (PCR) for the exponential amplification and further followed by deep sequencing (Sandberg, 2014). This technique was basically utilized for commercial use for the available high-density DNA microarray chips and for single-cell RNA sequencing (scRNA-seq). Entire transcriptome sequencing at the level of a single cell was first reported by James Eberwine, Iscove, and colleagues.

2.6.3 PROTEOMICS

Proteins are the major macromolecules that constitute the structural framework of the cell, replicate the genetic material, regulate gene expression, support the transport and cargo system within a cell, and catalyze and regulate the cells metabolic reaction (Pham et al., 2021). Human genome, according to an estimate, is said to have 20,000 protein-coding genes while the number of proteins existing is considerably high. The possible explanation for the cause was variations in DNA sequence, alternate mRNA splicing, post-translational modification, and proteolytic cleavage (Aebersold et al., 2018; Uhlen et al., 2015). No, tool till date now exists for whole-proteome analysis at the single-cell level. Some of the widely accepted tool is genetically tagged fluorophore to the protein connected with high resolution-imaging; antibody-tagged strategies coupled with flow cytometry, mass spectrometry, and microfluidics. These techniques have their own sensitivity and resolution that are oriented and applicable to specific set problems (Pham et al., 2021).

2.6.4 METABOLOMICS

The metabolome refers to the global set of molecules such as amino acids, fatty acids, carbohydrates, vitamins, and lipids whose molecular weight is generally <1.5 kDa (German et al., 2005; Wood, 2014; Sengupta & Weljie, 2019; Zhang et al., 2012). They are produced during metabolic reactions in a specific cell, organ, or even an organism. They have a role on the growth and homeostasis of the cell. The largest comprehensive human metabolome database (HMDB) available online (www.hmdb.ca) with approximately 114,000 entries recorded with various data such as clinical data, protein sequence, and disease-responsive compounds. This database continues to grow and facilitate the researchers with updated information (Wishart et al., 2013). Single-cell metabolomics extends its application to various fields, specifically in cancer research. This application can be modulated to identify cancer-circulating cells that lead to metastasis. The advancement in the techniques such as Mass spectrometry, fluorescence biosensors, FRET biosensors, and optical spectrometry has enabled us to estimate the metabolite activities and sensitivities in a single cell (Zenobi, 2013). The fast dynamics of single-cell metabolomics still remain one of the major drawbacks for its estimation. This is due to the high change in the metabolome reactions influenced by the environment within the span of seconds or even milliseconds (Weibel et al., 1974).

2.7 CONCLUSIONS

Hematopoietic development and regulation are crucial in the maintenance of blood cells, immune responses, and development. Defects in hematopoietic stem cells also lead to the transformation of such cells into leukemic progenitors. These cells are also tightly regulated by signals in order to maintain blood cell populations. Therefore, studying such cells for any changes is pivotal to decipher their role in development and disease. In this chapter, we have discussed about the cells at different developmental potentials, and their hallmarks. Furthermore, we have highlighted the important role of genetic tools that can be used in model organisms to decipher their function. Additionally, there are several techniques available to assess such cells and tissues for any molecular or morphological changes. Despite having gold standards that are used in research, there is continuous evolution in such methodologies. One such technique is single-cell analysis. Several recent studies have exploited such developments to understand the role of single cells in diseases and even normal development.

In this chapter, we have provided the advantages, disadvantages, and current challenges of single-cell techniques and analysis with an emphasis on studying the blood cells. The enormous growth in the single cell isolation techniques and analysis over the past few decades has led the scientific community to a big leap. The single-cell isolation techniques research offers an unprecedented challenges and opportunities in studying high throughput single-cell behavior that clearly improves our understanding of cellular heterogeneity. We have listed the investigative tools and techniques that can be employed as an approach to study the outcome of biological samples (Table 2.1). In order to achieve and strengthen our comprehensive understanding about SCA, the SCA is often integrated or combined with multi-omics (e.g., genomics, transcriptomics, proteomics and metabolomics). On a broader aspect cellular heterogeneity in turn proves to be informative in terms of analyzing single cells at the DNA sequence, RNA expression and protein levels. But omics-based SCA methods enable in reshaping and reducing the noise of cellular heterogeneity into a meaningful set of discoveries. Also, recent omics-based SCA methods require specific timepoints to study genes, proteins, and metabolites outside their physiological niche eventually by cell lysis process. Despite the recent advancement made in single-cell isolation techniques, obvious limitation also persists. For the detection of diverse global genomic, transcriptomic, and epigenetic events sufficient amount of DNA or RNA is the prime most obstacles. Comparatively, the proteome analysis turns out to be more challenging than genomics and transcriptomics events for its complexity and lack of high-sensitive techniques and high-affinity probes to detect low levels and post-translational modifications of proteins.

In a nutshell, SCA has emerged to be a new verge in 'Omics.' With the rapid advancement in single-cell isolation or analysis technology, the posed shortcomings will be eradicated with the possible solution in the near future. We hope that the single-cell techniques will emerge as one of the beneficial tools to extricate the unanswered questions in both the field of biological research and clinical diagnostics.

KEYWORDS

- **blood**
- **hematopoiesis**
- **hematopoietic stem cells**
- **single cell**

REFERENCES

Adams, M. D., Celniker, S. E., Holt, R. A., et al., (2000). The genome sequence of *Drosophila melanogaster*. *Science, 287*(5461), 2185–2195.

Aebersold, R., Agar, J. N., Amster, I. J., et al., (2018). How many human proteoforms are there? *Nat. Chem. Biol., 14*(3), 206–214.

Alenzi, F. Q., Alenazi, B. Q., Ahmad, S. Y., Salem, M. L., Al-Jabri, A. A., & Wyse, R. K., (2009). The haemopoietic stem cell: Between apoptosis and self-renewal. *Yale J. Biol. Med., 82*(1), 7–18.

Ali, R. A., Wuescher, L. M., & Worth, R. G., (2015). Platelets: Essential components of the immune system. *Curr. Trends Immunol., 16*, 65–78.

Asante, I., Pei, H., Zhou, E., et al., (2018). Simultaneous quantitation of folates, flavins and B6 metabolites in human plasma by LC-MS/MS assay: Applications in colorectal cancer. *J. Pharm. Biomed. Anal., 158*, 66–73.

Asante, I., Pei, H., Zhou, E., et al., (2019). Exploratory metabolomic study to identify blood-based biomarkers as a potential screen for colorectal cancer. *Mol. Omics, 15*(1), 21–29.

Babcock, D. T., Brock, A. R., Fish, G. S., et al., (2008). Circulating blood cells function as a surveillance system for damaged tissue in *Drosophila* larvae. *Proc. Natl. Acad. Sci. USA, 105*(29), 10017–10022.

Bailey, J. N., Pericak-Vance, M. A., & Haines, J. L., (2014). The impact of the human genome project on complex disease. *Genes (Basel), 5*(3), 518–535.

Balog, J. A., Honti, V., Kurucz, E., et al., (2021). Immunoprofiling of *Drosophila* hemocytes by single-cell mass cytometry. *Genomics Proteomics Bioinformatics, 19*(2), 243–252.

Bankapur, A., Zachariah, E., Chidangil, S., Valiathan, M., & Mathur, D., (2010). Raman tweezers spectroscopy of live, single red and white blood cells. *PLoS One, 5*(4), e10427.

Bassett, A. R., Tibbit, C., Ponting, C. P., & Liu, J. L., (2013). Highly efficient targeted mutagenesis of *Drosophila* with the CRISPR/Cas9 system. *Cell Rep., 4*(1), 220–228.

Bayguinov, P. O., Oakley, D. M., Shih, C. C., Geanon, D. J., Joens, M. S., & Fitzpatrick, J. A. J., (2018). Modern laser scanning confocal microscopy. *Curr. Protoc. Cytom., 85*(1), e39.

Biehl, J. K., & Brenda, R., (2009). Introduction to stem cell therapy. *Journal of Cardiovascular Nursing.*

Blainey, C. P., (2013). The future is now: Single-cell genomics of bacteria and archaea. *FEMS Microbiol. Rev., 37*(3), 407–427.

Bonner, W. A., Hulett, H. R., Sweet, R. G., & Herzenberg, L. A., (1972). Fluorescence activated cell sorting. *Rev. Sci. Instrum., 43*(3), 404–409.

Boulanger, M., Paolillo, R., Piechaczyk, M., & Bossis, G., (2019). The SUMO pathway in hematomalignancies and their response to therapies. *Int. J. Mol. Sci., 20*(16).

Boutros, M., Heigwer, F., & Laufer, C., (2015). Microscopy-based high-content screening. *Cell, 163*(6), 1314–1325.

Brand, A. H., & Perrimon, N., (1993). Targeted gene expression as a means of altering cell fates and generating dominant phenotypes. *Development, 118*(2), 401–415.

Bryk, A. H., & Wisniewski, J. R., (2017). Quantitative analysis of human red blood cell proteome. *J. Proteome Res., 16*(8), 2752–2761.

Burt, J. P., Pethig, R., Gascoyne, P. R., & Becker, F. F., (1990). Dielectrophoretic characterization of friend murine erythroleukaemic cells as a measure of induced differentiation. *Biochim. Biophys. Acta, 1034*(1), 93–101.

Butler, H. J., Ashton, L., Bird, B., et al., (2016). Using Raman spectroscopy to characterize biological materials. *Nat. Protoc., 11*(4), 664–687.

Capdevila, J., & Guerrero, I., (1994). Targeted expression of the signaling molecule decapentaplegic induces pattern duplications and growth alterations in *Drosophila* wings. *EMBO J., 13*(19), 4459–4468.

Celniker, S. E., Dillon, L. A., Gerstein, M. B., et al., (2009). Unlocking the secrets of the genome. *Nature, 459*(7249), 927–930.

Chang, N., Sun, C., Gao, L., et al., (2013). Genome editing with RNA-guided Cas9 nuclease in zebrafish embryos. *Cell Res., 23*(4), 465–472.

Chen, S., Zhao, M., Wu, G., Yao, C., & Zhang, J., (2012). Recent advances in morphological cell image analysis. *Comput. Math. Methods Med., 2012*, 101536.

Cheng, H., Zheng, Z., & Cheng, T., (2020). New paradigms on hematopoietic stem cell differentiation. *Protein Cel., 11*(1), 34–44.
Chiu, H., Brian, C. R., Richard, P. S., Marta, K., Dan, G., & Shubha, G., (2005). dUbc9 negatively regulates the toll-NF-κB pathways in larval hematopoiesis and drosomycin activation in *Drosophila. Developmental Biology.*
Cho, J., (2015). A paradigm shift in platelet transfusion therapy. *Blood, 125*(23), 3523–3525.
Choi, H. M. T., Schwarzkopf, M., Fornace, M. E., et al., (2018). Third-generation in situ hybridization chain reaction: Multiplexed, quantitative, sensitive, versatile, robust. *Development, 145*(12).
Cizmar, P., & Yuana, Y., (2017). Detection and characterization of extracellular vesicles by transmission and cryo-transmission electron microscopy. *Methods Mol. Biol., 1660,* 221–232.
Clark, S. R., Ma, A. C., Tavener, S. A., et al., (2007). Platelet TLR4 activates neutrophil extracellular traps to ensnare bacteria in septic blood. *Nat. Med., 13*(4), 463–469.
Cole, R., (2014). Live-cell imaging. *Cell Adh. Migr., 8*(5), 452–459.
Cooley, L., Kelley, R., & Spradling, A., (1988). Insertional mutagenesis of the *Drosophila* genome with single P elements. *Science, 239*(4844), 1121–1128.
Csordas, G., Varga, G. I., Honti, V., Jankovics, F., Kurucz, E., & Ando, I., (2014). *In vivo* immunostaining of hemocyte compartments in *Drosophila* for live imaging. *PLoS One, 9*(6), e98191.
Ding, S., Wu, X., Li, G., Han, M., Zhuang, Y., & Xu, T., (2005). Efficient transposition of the piggyBac (PB) transposon in mammalian cells and mice. *Cell, 122*(3), 473–483.
Domen, J., & Weissman, I. L., (1999). Self-renewal, differentiation or death: Regulation and manipulation of hematopoietic stem cell fate. *Mol Med Today, 5*(5), 201–208.
Eberwine, J., Sul, J. Y., Bartfai, T., & Kim, J., (2014). The promise of single-cell sequencing. *Nat Methods, 11*(1), 25–27.
Eilken, H. M., Nishikawa, S., & Schroeder, T., (2009). Continuous single-cell imaging of blood generation from haemogenic endothelium. *Nature, 457*(7231), 896–900.
Elmore, S., (2007). Apoptosis: A review of programmed cell death. *Toxicol. Pathol., 35*(4), 495–516.
Emmert-Buck, M. R., Bonner, R. F., Smith, P. D., et al., (1996). Laser capture microdissection. *Science, 274*(5289), 998–1001.
Emwas, A. H., Roy, R., McKay, R. T., et al., (2019). NMR spectroscopy for metabolomics research. *Metabolites, 9*(7).
Fisher, Y. E., Yang, H. H., Isaacman-Beck, J., Xie, M., Gohl, D. M., & Clandinin, T. R., (2017). FlpStop, a tool for conditional gene control in *Drosophila. Elife, 6.*
Gaj, T., Gersbach, C. A., & Barbas, C. F. 3rd., (2013). ZFN, TALEN, and CRISPR/Cas-based methods for genome engineering. *Trends Biotechnol., 31*(7), 397–405.
Gascoyne, P. R., & Shim, S., (2014). Isolation of circulating tumor cells by dielectrophoresis. *Cancers (Basel), 6*(1), 545–579.
Gautam, D. K., Chimata, A. V., Gutti, R. K., & Paddibhatla, I., (2021). Comparative hematopoiesis and signal transduction in model organisms. *J. Cell Physiol., 236*(8), 5592–5619.
Gawad, C., Koh, W., & Quake, S. R., (2016). Single-cell genome sequencing: Current state of the science. *Nat. Rev. Genet., 17*(3), 175–188.
German, J. B., Hammock, B. D., & Watkins, S. M., (2005). Metabolomics: Building on a century of biochemistry to guide human health. *Metabolomics, 1*(1), 3–9.

Golan, L., Yeheskely-Hayon, D., Minai, L., Dann, E. J., & Yelin, D., (2012). Noninvasive imaging of flowing blood cells using label-free spectrally encoded flow cytometry. *Biomed Opt. Express, 3*(6), 1455–1464.

Gross, A., Schoendube, J., Zimmermann, S., Steeb, M., Zengerle, R., & Koltay, P., (2015). Technologies for single-cell isolation. *Int. J. Mol. Sci., 16*(8), 16897–16919.

Halder, G., Callaerts, P., & Gehring, W. J., (1995). Induction of ectopic eyes by targeted expression of the eyeless gene in *Drosophila. Science, 267*(5205), 1788–1792.

Han, A., Yang, L., & Frazier, A. B., (2007). Quantification of the heterogeneity in breast cancer cell lines using whole-cell impedance spectroscopy. *Clin. Cancer Res., 13*(1), 139–143.

Hanna, J. H., Krishanu, S., & Rudolf, J., (2010). Pluripotency and cellular reprogramming: Facts, hypotheses, unresolved issues. *Cell 143*(4), 508–525. doi:10.1016/j.cell.2010.10.008.

Heinzel, S., Marchingo, J. M., Horton, M. B., & Hodgkin, P. D., (2018). The regulation of lymphocyte activation and proliferation. *Curr. Opin. Immunol., 51*, 32–38.

Herzenberg, L. A., Sweet, R. G., & Herzenberg, L. A., (1976). Fluorescence-activated cell sorting. *Sci. Am., 234*(3), 108–117.

Hood, L., & Rowen, L., (2013). The human genome project: Big science transforms biology and medicine. *Genome Med., 5*(9), 79.

Houck, A. L., Seddighi, S., & Driver, J. A., (2018). At the crossroads between neurodegeneration and cancer: A review of overlapping biology and its implications. *Curr. Aging Sci., 11*(2), 77–89.

Hu, P., Zhang, W., Xin, H., & Deng, G., (2016). Single cell isolation and analysis. *Front Cell Dev. Biol., 4*, 116.

Hwang, W. Y., Fu, Y., Reyon, D., et al., (2013). Efficient genome editing in zebrafish using a CRISPR-Cas system. *Nat. Biotechnol., 31*(3), 227–229.

Ito, K., & Ito, K., (2018). Hematopoietic stem cell fate through metabolic control. *Exp. Hematol., 64*, 1–11.

Jagannathan-Bogdan, M., & Zon, L. I., (2013). Hematopoiesis. *Development, 140*(12), 2463–2467.

Jensen, B. F., (2009). The dual roles of red blood cells in tissue oxygen delivery: Oxygen carriers and regulators of local blood flow. *J. Exp. Biol., 212*(Pt 21), 3387–3393.

Johnson-Chavarria, E. M., Tanyeri, M., & Schroeder, C. M., (2011). A microfluidic-based hydrodynamic trap for single particles. *J. Vis. Exp.,* (47).

Jurecic, R., (2019). Hematopoietic stem cell heterogeneity. *Adv Exp. Med. Biol., 1169*, 195–211.

Kalamarz, M. E., Paddibhatla, I., Nadar, C., & Govind, S., (2012). Sumoylation is tumor-suppressive and confers proliferative quiescence to hematopoietic progenitors in *Drosophila melanogaster* larvae. *Biology Open.*

Kamleh, M. A., Dow, J. A., & Watson, D. G., (2009). Applications of mass spectrometry in metabolomic studies of animal model and invertebrate systems. *Brief Funct. Genomic Proteomic, 8*(1), 28–48.

Karimi, A., Yazdi, S., & Ardekani, A. M., (2013). Hydrodynamic mechanisms of cell and particle trapping in microfluidics. *Biomicrofluidics, 7*(2), 21501.

Kashima, Y., Sakamoto, Y., Kaneko, K., Seki, M., Suzuki, Y., & Suzuki, A., (2020). Single-cell sequencing techniques from individual to multiomics analyses. *Exp. Mol. Med., 52*(9), 1419–1427.

Kawahara, R., (2007). Hematopoiesis. *xPharm: The Comprehensive Pharmacology Reference*, 1–5.

Klass, R. M., (1983). A method for the isolation of longevity mutants in the nematode *Caenorhabditis elegans* and initial results. *Mech Ageing Dev., 22*(3, 4), 279–286.
Klein, L., Hinterberger, M., Wirnsberger, G., & Kyewski, B., (2009). Antigen presentation in the thymus for positive selection and central tolerance induction. *Nat. Rev. Immunol., 9*(12), 833–844.
Kobayashi, S. D., Malachowa, N., & DeLeo, F. R., (2018). Neutrophils and bacterial immune evasion. *J. Innate. Immun., 10*(5, 6), 432–441.
Kondo, M., (2010). Lymphoid and myeloid lineage commitment in multipotent hematopoietic progenitors. *Immunol. Rev., 238*(1), 37–46.
Kondo, M., Weissman, I. L., & Akashi, K., (1997). Identification of clonogenic common lymphoid progenitors in mouse bone marrow. *Cell, 91*(5), 661–672.
Koster, A. J., & Klumperman, J., (2003). Electron microscopy in cell biology: Integrating structure and function. *Nat. Rev. Mol. Cell Biol.*, S6–10.
Kulkarni, R., & Kale, V., (2020). Physiological cues involved in the regulation of adhesion mechanisms in hematopoietic stem cell fate decision. *Front Cell Dev. Biol., 8*, 611.
Lahnemann, D., Koster, J., Szczurek, E., et al., (2020). Eleven grand challenges in single-cell data science. *Genome Biol., 21*(1), 31.
Lea, P. J., Temkin, R. J., Freeman, K. B., Mitchell, G. A., & Robinson, B. H., (1994). Variations in mitochondrial ultrastructure and dynamics observed by high resolution scanning electron microscopy (HRSEM). *Microsc. Res. Tech., 27*(4), 269–277.
Lebestky, T., Seung, H. J., & Utpal, B., (2003). A serrate-expressing signaling center controls *Drosophila* hematopoiesis. *Genes and Development.*
Lee, S., Roh, S. M., Lee, E., et al., (2020). Applications of converged various forces for detection of biomolecules and novelty of dielectrophoretic force in the applications. *Sensors (Basel), 20*(11).
Letourneau, M., Lapraz, F., Sharma, A., Vanzo, N., Waltzer, L., & Crozatier, M., (2016). Drosophila hematopoiesis under normal conditions and in response to immune stress. *FEBS Lett., 590*(22), 4034–4051.
Lodoen, M. B., & Lanier, L. L., (2006). Natural killer cells as an initial defense against pathogens. *Curr. Opin. Immunol., 18*(4), 391–398.
Luo, T., Fan, L., Zhu, R., & Sun, D., (2019). Microfluidic single-cell manipulation and analysis: Methods and applications. *Micromachines (Basel), 10*(2).
Magnon, C., & Frenette, P. S., (2008). Hematopoietic stem cell trafficking. In *StemBook.* Cambridge (MA).
Mahmood, T., & Yang, P. C., (2012). Western blot: Technique, theory, and trouble shooting. *N. Am. J. Med. Sci., 4*(9), 429–434.
Manz, M. G., & Boettcher, S., (2014). Emergency granulopoiesis. *Nat. Rev. Immunol., 14*(5), 302–314.
Manz, M. G., Traver, D., Miyamoto, T., Weissman, I. L., & Akashi, K., (2001). Dendritic cell potentials of early lymphoid and myeloid progenitors. *Blood, 97*(11), 3333–3341.
Marrugo-Ramirez, J., Mir, M., & Samitier, J., (2018). Blood-based cancer biomarkers in liquid biopsy: A promising non-invasive alternative to tissue biopsy. *Int. J. Mol. Sci., 19*(10).
Martinez-Agosto, J. A., Mikkola, H. K., Hartenstein, V., & Banerjee, U., (2007). The hematopoietic stem cell and its niche: A comparative view. *Genes Dev., 21*(23), 3044–3060.
McCulloch, E. A., & Till, J. E., (2005). Perspectives on the properties of stem cells. *Nat. Med., 11*(10), 1026–1028.

Minakhina, S., & Ruth, S., (2006). Melanotic mutants in *Drosophila*: Pathways and phenotypes. *Genetics, 174*(1), 253–263.
Moosavi, A., & Motevalizadeh, A. A., (2016). Role of epigenetics in biology and human diseases. *Iran Biomed. J., 20*(5), 246–258.
Morikawa, T., & Takubo, K., (2017). Use of imaging techniques to illuminate dynamics of hematopoietic stem cells and their niches. *Front Cell Dev. Biol., 5*, 62.
Morrison, S. J., & Spradling, A. C., (2008). Stem cells and niches: Mechanisms that promote stem cell maintenance throughout life. *Cell, 132*(4), 598–611.
Nagana, G. G. A., Djukovic, D., Bettcher, L. F., Gu, H., & Raftery, D., (2018). NMR-guided mass spectrometry for absolute quantitation of human blood metabolites. *Anal. Chem., 90*(3), 2001–2009.
Navin, E. N., (2014). Cancer genomics: One cell at a time. *Genome Biol., 15*(8), 452.
Nusslein-Volhard, C., Wieschaus, E., & Kluding, H., (1984). Mutations affecting the pattern of the larval cuticle in *Drosophila melanogaster*: I. Zygotic loci on the second chromosome. *Wilehm Roux Arch. Dev. Biol., 193*(5), 267–282.
Orkin, S. H., & Leonard, Z. I., (2008). Hematopoiesis: An evolving paradigm for stem cell biology. *Cell, 132*(4), 631–644.
Paddibhatla, I., Dushyant, G. K., & Rakesh, M. K., (2019). SETDB1 modulates the differentiation of both the crystal cells and the lamellocytes in *Drosophila. Developmental Biology*.
Paddock, S. W., & Eliceiri, K. W., (2014). Laser scanning confocal microscopy: History, applications, and related optical sectioning techniques. *Methods Mol. Biol., 1075*, 9–47.
Pham, T., Tyagi, A., Wang, Y. S., & Guo, J., (2021). Single-cell proteomic analysis. *Wiley Interdiscip. Rev. Syst. Biol. Med., 13*(1), e1503.
Potter, C. J., Tasic, B., Russler, E. V., Liang, L., & Luo, L., (2010). The Q system: A repressible binary system for transgene expression, lineage tracing, and mosaic analysis. *Cell, 141*(3), 536–548.
Rathinam, C., & Flavell, R. A., (2008). The hematopoiesis paradigm: Clarity or ambiguity? *Blood, 112*(9), 3534–3535.
Ravenhill, B. J., Kanjee, U., Ahouidi, A., et al., (2019). Quantitative comparative analysis of human erythrocyte surface proteins between individuals from two genetically distinct populations. *Commun. Biol., 2*, 350.
Reya, T., Morrison, S. J., Clarke, M. F., & Weissman, I. L., (2001). Stem cells, cancer, and cancer stem cells. *Nature, 414*(6859), 105–111.
Richardson, S. L., Hulikova, A., Proven, M., et al., (2020). Single-cell O_2 exchange imaging shows that cytoplasmic diffusion is a dominant barrier to efficient gas transport in red blood cells. *Proc. Natl. Acad. Sci. USA, 117*(18), 10067–10078.
Ruf, A., & Patscheke, H., (1995). Platelet-induced neutrophil activation: Platelet-expressed fibrinogen induces the oxidative burst in neutrophils by an interaction with CD11C/CD18. *Br. J. Haematol., 90*(4), 791–796.
Rushforth, A. M., Saari, B., & Anderson, P., (1993). Site-selected insertion of the transposon Tc1 into a *Caenorhabditis elegans* myosin light chain gene. *Mol. Cell Biol., 13*(2), 902–910.
Sandberg, R., (2014). Entering the era of single-cell transcriptomics in biology and medicine. *Nat. Methods, 11*(1), 22–24.
Schmitz, B., Radbruch, A., Kummel, T., et al., (1994). Magnetic activated cell sorting (MACS) – a new immunomagnetic method for megakaryocytic cell isolation: Comparison of different separation techniques. *Eur. J. Haematol., 52*(5), 267–275.

Schulz, K. R., Danna, E. A., Krutzik, P. O., & Nolan, G. P., (2012). Single-cell phosphoprotein analysis by flow cytometry. *Curr. Protoc. Immunol., 8*, 1–20.

Seita, J., & Weissman, I. L., (2010). Hematopoietic stem cell: Self-renewal versus differentiation. *Wiley Interdiscip. Rev. Syst. Biol. Med., 2*(6), 640–653.

Semple, J. W., Italiano, J. E. Jr., & Freedman, J., (2011). Platelets and the immune continuum. *Nat. Rev. Immunol., 11*(4), 264–274.

Sengupta, A., & Weljie, A. M., (2019). NMR spectroscopy-based metabolic profiling of biospecimens. *Curr. Protoc. Protein Sci., 98*(1), e98.

Sesen, M., & Whyte, G., (2020). Image-based single cell sorting automation in droplet microfluidics. *Sci. Rep., 10*(1), 8736.

Shen, B., Zhang, J., Wu, H., et al., (2013). Generation of gene-modified mice via Cas9/RNA-mediated gene targeting. *Cell Res., 23*(5), 720–723.

Shenghui, H., Daisuke, N., & Sean, M. J., (2009). Mechanisms of stem cell self-renewal. *Annual Review of Cell and Developmental Biology*.

Singer, G., Urakami, H., Specian, R. D., Stokes, K. Y., & Granger, D. N., (2006). Platelet recruitment in the murine hepatic microvasculature during experimental sepsis: Role of neutrophils. *Microcirculation, 13*(2), 89–97.

Singh, V. K., Abhishek, S., Manisha, K., Neeraj, K., & Ramesh, C., (2016). Describing the stem cell potency: The various methods of functional assessment and in silico diagnostics. *Front Cell Dev Biol. 134*, 4. doi:10.3389/fcell.2016.00134.

Snook, R. D., Harvey, T. J., Correia, F. E., & Gardner, P., (2009). Raman tweezers and their application to the study of singly trapped eukaryotic cells. *Integr. Biol. (Camb), 1*(1), 43–52.

Stegle, O., Teichmann, S. A., & Marioni, J. C., (2015). Computational and analytical challenges in single-cell transcriptomics. *Nat. Rev. Genet., 16*(3), 133–145.

Suda, T., Takubo, K., & Semenza, G. L., (2011). Metabolic regulation of hematopoietic stem cells in the hypoxic niche. *Cell Stem Cell., 9*(4), 298–310.

Sun, X., Jones, H. P., Dobbs, N., Bodhankar, S., & Simecka, J. W., (2013). Dendritic cells are the major antigen presenting cells in inflammatory lesions of murine mycoplasma respiratory disease. *PLoS One, 8*(2), e55984.

Tautz, D., & Pfeifle, C., (1989). A non-radioactive in situ hybridization method for the localization of specific RNAs in *Drosophila* embryos reveals translational control of the segmentation gene hunchback. *Chromosoma, 98*(2), 81–85.

Traver, D., Akashi, K., Manz, M., et al., (2000). Development of CD8 alpha-positive dendritic cells from a common myeloid progenitor. *Science, 290*(5499), 2152–2154.

Turei, D., Valdeolivas, A., Gul, L., et al., (2021). Integrated intra– and intercellular signaling knowledge for multicellular omics analysis. *Mol. Syst. Biol., 17*(3), e9923.

Udan, R. S., Kango-Singh, M., Nolo, R., Tao, C., & Halder, G., (2003). Hippo promotes proliferation arrest and apoptosis in the Salvador/warts pathway. *Nat. Cell Biol., 5*(10), 914–920.

Uhlen, M., Fagerberg, L., Hallstrom, B. M., et al., (2015). Proteomics. Tissue-based map of the human proteome. *Science, 347*(6220), 1260419.

Undi, R. B., Sarvothaman, S., Narasaiah, K., Gutti, U., & Gutti, R. K., (2016). Toll-like receptor 2 signaling: Significance in megakaryocyte development through Wnt signaling cross-talk and cytokine induction. *Cytokine, 83*, 245–249.

Valero, A., Braschler, T., Demierre, N., & Renaud, P., (2010). A miniaturized continuous dielectrophoretic cell sorter and its applications. *Biomicrofluidics, 4*(2).

Verfaillie, C. M., Pera, M. F., & Lansdorp, P. M., (2002). Stem cells: Hype and reality. *Hematology Am. Soc. Hematol. Educ. Program*, 369–391.
Vivier, E., Tomasello, E., Baratin, M., Walzer, T., & Ugolini, S., (2008). Functions of natural killer cells. *Nat. Immunol., 9*(5), 503–510.
Voldman, J., (2006). Electrical forces for microscale cell manipulation. *Annu. Rev. Biomed. Eng., 8*, 425–454.
Waghmare, I., & Page-McCaw, A., (2018). Wnt signaling in stem cell maintenance and differentiation in the *Drosophila germarium. Genes (Basel), 9*(3).
Wagner, D. D., & Burger, P. C., (2003). Platelets in inflammation and thrombosis. *Arterioscler Thromb. Vasc. Biol., 23*(12), 2131–2137.
Wang, D., & Bodovitz, S., (2010). Single cell analysis: The new frontier in 'omics.' *Trends Biotechnol., 28*(6), 281–290.
Wang, L. D., & Wagers, A. J., (2011). Dynamic niches in the origination and differentiation of haematopoietic stem cells. *Nat. Rev. Mol. Cell Biol., 12*(10), 643–655.
Wang, Q. D., (2021). Emerging trends in deciphering the pathogenesis of human diseases through genetic analysis. *Genes (Basel), 12*(1).
Watkins, N. A., Gusnanto, A., De Bono, B., et al., (2009). A HaemAtlas: Characterizing gene expression in differentiated human blood cells. *Blood, 113*(19), e1–9.
Weatherall, D., (1998). The future role of molecular and cell biology in medical practice in the tropical countries. *Br. Med. Bull., 54*(2), 489–501.
Weibel, K. E., Mor, J. R., & Fiechter, A., (1974). Rapid sampling of yeast cells and automated assays of adenylate, citrate, pyruvate and glucose-6-phosphate pools. *Anal. Biochem., 58*(1), 208–216.
Wilson, A., & Trumpp, A., (2006). Bone-marrow haematopoietic-stem-cell niches. *Nat. Rev. Immunol., 6*(2), 93–106.
Wishart, D. S., Jewison, T., Guo, A. C., et al., (2013). HMDB 3.0 – the human metabolome database in 2013. *Nucleic Acids Res., 41*, D801–807.
Wood, L. P., (2014). Mass spectrometry strategies for clinical metabolomics and lipidomics in psychiatry, neurology, and neuro-oncology. *Neuropsychopharmacology, 39*(1), 24–33.
Wu, X., Yamamoto, M., Akira, S., & Sun, S. C., (2009). Regulation of hematopoiesis by the K63-specific ubiquitin-conjugating enzyme Ubc13. *Proc. Natl. Acad. Sci. USA, 106*(49), 20836–20841.
Xu, T., Wang, W., Zhang, S., Stewart, R. A., & Yu, W., (1995). Identifying tumor suppressors in genetic mosaics: The *Drosophila* lats gene encodes a putative protein kinase. *Development, 121*(4), 1053–1063.
Yadav, R. K., Gautam, D. K., Muj, C., Gajula, B. M. B., & Paddibhatla, I., (2021). Methotrexate negatively acts on inflammatory responses triggered in *Drosophila* larva with hyperactive JAK/STAT pathway. *Dev. Comp. Immunol., 123*, 104161.
Yakirevich, E., Jackson, C. L., Meitner, P. A., et al., (2007). Analysis of T-cell clonality using laser capture microdissection and high-resolution microcapillary electrophoresis. *J. Mol. Diagn., 9*(4), 490–497.
Yang, L., Hu, M., Lu, Y., Han, S., & Wang, J., (2021). Inflammasomes and the maintenance of hematopoietic homeostasis: New perspectives and opportunities. *Molecules, 26*(2).
Yuan, G. C., Cai, L., Elowitz, M., et al., (2017). Challenges and emerging directions in single-cell analysis. *Genome Biol., 18*(1), 84.
Zakrzewski, W., Dobrzynski, M., Szymonowicz, M., & Rybak, Z., (2019). Stem cells: Past, present, and future. *Stem. Cell Res. Ther., 10*(1), 68.

Zenobi, R., (2013). Single-cell metabolomics: Analytical and biological perspectives. *Science, 342*(6163), 1243259.

Zhang, A., Sun, H., & Wang, X., (2012). Serum metabolomics as a novel diagnostic approach for disease: A systematic review. *Anal. Bioanal. Chem., 404*(4), 1239–1245.

Zhang, C. C., & Lodish, H. F., (2008). Cytokines regulating hematopoietic stem cell function. *Curr. Opin. Hematol., 15*(4), 307–311.

Zwaal, R. R., Broeks, A., Van, M. J., Groenen, J. T., & Plasterk, R. H., (1993). Target-selected gene inactivation in *Caenorhabditis elegans* by using a frozen transposon insertion mutant bank. *Proc. Natl. Acad. Sci. USA, 90*(16), 7431–7435.

CHAPTER 3

Evaluation of Chitosan and Its Derivatives in Immunomodulating Blood Sentinel Cells

ARUN KUMAR SUBRAMANI,[1] KEYUR RAVAL,[1] and RITU RAVAL[2]

[1]*Department of Chemical Engineering, National Institute of Technology Karnataka, Surathkal, Mangalore, Karnataka, India*

[2]*Department of Biotechnology, Manipal Institute of Technology, MAHE, Manipal, Karnataka, India*

ABSTRACT

Opportunistic pathogens can invade and thus cause infections in the body. However, in most instances, the body mounts an immune response to counter the infection. Sentinel cells are the body's second line of defense, which act against invading microbes. An important part of the innate system, these sentinel cells include macrophages, neutrophils, and dendritic cells. The plethora of pattern recognition receptors helps them to combat the invading pathogen from colonizing the body. They do so by inducing inflammation and oxidative stress or by the adaptive immune cells releasing chemokines. However, a paradigm shift of unrestrained inflammation or oxidative stress can also lead to a hypersensitivity reaction, sepsis, or autoimmune disorder. On the other extreme, an under-responsive sentinel cell might lead to frequent infections. Immunomodulators like chitosan can assist the sentinel cells in the titrated priming of the pathogen, thereby reducing chronic inflammation conditions. Chitosan, the amorphous derivative of chitin is the second largest abundant polymer after cellulose. The physicochemical properties of chitosan and COS depend on molecular weight, degree of polymerization

Hematopoiesis: Biochemical, Cellular, Molecular, and Genomic Perspectives. Ravi Kumar Gutti (Ed.)

(DP), and degree of deacetylation (DD). Thus, the modulation of these properties can impart chitosan and its oligosaccharides (COS) for various biomedical applications. In the present chapter, an overview of the innate responses of the sentinel cells and the synergistic roles played by chitosan and its derivatives to regularize various responses.

3.1 INTRODUCTION

The susceptibility of humans to disease-causing pathogens is increasing. As a result, a lot of research has been undertaken to address the issue. The physical and chemical barriers of the innate system represent the first line of defense against any intruding pathogen. Failure to combat the infection results in the activation of the second line of defense constituted by the sentinel cells. These cells are present in blood, skin (langerhans cells), and other organs like the liver (kupffer cells), brain (microglia), lungs (alveolar macrophages), etc. In addition to these hematopoietic origin cells, other cell types, such as fibroblasts and epithelial cells take part in the defense fight. The chapter will discuss the sentinel cells of hematopoietic origin, their role in inflammation, and their immunomodulation with chitosan.

3.2 BLOOD SENTINEL CELLS

3.2.1 MACROPHAGES

Macrophages were first described in 1882 by Elia Metchnikoff as an innate immune cell in starfish' larvae after penetration of tangerine tree thorns and later in *Daphnia magna,* which was infected with the fungal spores. Macrophages hold the task of immune surveillance with phagocytosis as their major tool of attack. In the process, they secrete various cytokines, and chemokines (Shimokado et al., 1985; Nathan, 1987) and summon help from the tissue housed or blood surveilling sentinel and adaptive immune cells. Monocytes, the progenitors of the macrophages originate from the yolk sac and in the maturation, sequel leave the blood stream, house themselves in the tissue and are then known as the macrophages. Macrophages express pattern recognition receptors (PRRs) such as the classical toll-like receptors (NOD)-like receptors, which identify the pathogenic components such as lipopolysaccharides (LPS), genetic materials, proteins, etc. (Epelman et al., 2014). Upon encountering infection or damage, the macrophages are the first cells to respond and release cytokines and chemokines, inducing inflammation (Kasahara & Matsushima, 2001).

3.2.1.1 INFLAMMATION MECHANISM

Toll-like receptors (TLRs) act as pioneering pattern recognition receptors (PRR) for a wide variety of microbes expressing pathogen-associated molecular patterns (PAMPs). Lipopolysaccharide (LPS) of gram-negative bacteria (TLR4), peptidoglycan, lipoarabinomannan of gram-positive bacteria (TLR2), and flagellin (TLR5) are some of the microbial components which activate TLRs. One such PAMP is LPS, which first binds to soluble LPS-binding protein (LBP) in the blood. This complex binds with TLR4 via the CD14 receptors, which are expressed on most of the cell types except for the endothelial cells. The binding of the LPS complex to the TLR activates the signaling pathway of nuclear factor (NF-κB). The activated signal cascade results in the release of pro-inflammatory cytokines; interleukin (IL)-1, IL-12, and tumor necrosis factor α (TNF-α), etc., which form an integral part of the innate response (Takeda et al., 2003; Fujiwara & Kobayashi, 2005). However, if the process of inflammation goes unchecked, it can lead to tissue injury. Hence, to prevent the process to cause injury to self-tissue, anti-inflammatory responses are up regulated. These include the secretion of cytokines like TGF-ß and IL-10 which would allow the damaged tissue s to be repaired (Wynn, 2003; Wang & Arase, 2014).

3.2.2 NEUTROPHILS

Neutrophils, the polymorphonuclear (PMN) leukocytes are a part of the circulating sentinel cells. These cells like their counter tissue stationed macrophages originate in the bone marrow then enter the blood, and finally circulate as differentiated cells. These short-lived cells move through the tight gaps between cells or within narrow pores in the extracellular matrix (ECM) and migrate from the blood into tissue (Hoffmann et al., 2007; Carvalho et al., 2015). In the event of infection, the cytokines released by the macrophages stationed in tissues help recruit the neutrophils from the bloodstream. This diapedesis process involves three steps, involving tethering and rolling, firm adhesion, and transmigration of the neutrophil from the bloodstream into the tissue (Carmona-Rivera & Kaplan, 2016).

In the tethering and rolling process, the endothelial cell membrane expressing P-selectin glycoprotein (PSGL-1), which binds to E-selectin-ligand-1 (ESL1) receptor, expressed by activated neutrophils. These non-covalent interactions trigger intracellular cascades that phosphorylate cytosolic proteins, which leads to the activation of syk (spleen tyrosine kinase). Activated syk triggers cytoskeletal rearrangement in the neutrophil resulting in integrin activation (Yago et al., 2010; Woodfin et al., 2010).

In firm adhesion steps, neutrophil rolling along the endothelium induces conformational changes in integrin adhesion receptors, which bind to different adhesion molecules located on an endothelial membrane such as intercellular adhesion molecule (ICAM), vascular cell adhesion molecule (VCAM). These interactions activate strong adhesion of the neutrophil to the endothelium, which favors transmigration (Phillipson et al., 2006).

In the transmigration process, neutrophils transmigrate into the affected tissues through the junctions between endothelial cells. Once neutrophils migrate through the endothelium, it binds to bacteria and undergoes mechanisms such as phagocytosis, degranulation, and neutrophil extracellular trap (NET) formation. This induces inflammation in response to pathogen-associated molecular patterns (PAMPs), such as lipopolysaccharide (LPS), flagellin, and nucleic acids released into the tissue during infection. During this process, neutrophils damage host cells too (Yang et al., 2006; Carman et al., 2007; Woodfin et al., 2009).

3.2.2.1 MICROBIAL REMOVAL MECHANISM

Three crucial processes are performed by the neutrophils to control the pathogenic proliferation. The sealing of the microbe to an area initiates with NET formation It prevents the spread of infection or damage to other adjacent cells. The molecular mechanisms of NET formation are not yet fully elucidated. Some molecules can activate NET formation, such as receptor-mediated signals or through calcium channel formation through ionophores. In this process, without cell lysis, neutrophils' nuclear membrane will disintegrate the decondensed DNA mixes with the cytoplasmic proteins. This complex when secreted to ECM forms a NET trap. This trap favors the release and development of antimicrobial agents that act on microbes effectively.

The second mechanism undertaken by the neutrophils to combat infection growth is granulation, which is responsible for inflammation. Neutrophils produce four different cytosolic granules. They are primary (azurophilic), secondary (specific), tertiary (gelatinase), and secretory vesicles released (Smith, 1994). The functions of the four different granules are listed in Table 3.1.

The third mechanism enthralled in the microbe killing involves phagocytosis, which is performed by neutrophils. Neutrophils initiate phagocytosis through the stimulation of Toll-like receptors (TLRs) and other pattern-recognition receptors. Like in macrophages, TLR triggers inflammatory gene expression regulated by NF-kB elements, resulting in the production of reactive oxygen species (ROS) for the lethal attack against microbes.

TABLE 3.1 Functions of Different Types of Granules of Neutrophils

Granule Type	Function	References
Primary	• Stimulate TLR dependent expression of monocytes. • Acts on foreign particles and degrade it. • Activation of epithelial cells, endothelial cells, macrophages, lymphocytes, and platelets. • Degrades ECM proteins. • Induces chemotaxis of CD4þ and CD8þ cells.	Chistiakov et al. (2015); Faurschou & Borregaard (2003); Wen et al. (2011); Kunes et al. (2009); Rotondo et al. (2011); Selders et al. (2017)
Secondary	• Wide range of microbicidal activity against pathogens • Recognize microbial presence. • Activates lectin pathways by interacting with pathogens. • Binds with LPS and reduce cytosine production. • Stimulates phagocytosis.	
Tertiary	• Inhibits T-cell proliferation. • Tissue remodeling. • Degrades major structural components of ECM.	
Secondary vesicles	• Stimulates neutrophils apoptosis. • Stimulates pro-inflammatory agent	

3.2.3 DENDRITIC CELLS

Dendritic cells (DC) are the last category of innate immunity sentinel cells. These cells too originate from the bone marrow, mature, and differentiate later where they form a network of interstitial DC. They have the major task of antigen presentation to the T-cells and thereby modulate the adaptive immune response. Generally, DCs are classified into conventional DCs (cDCs), plasmacytoid DCs (pDCs), and DCs derived from monocytes (mDCs) which differentiate in the bone marrow (Onai et al., 2007; Waskow et al., 2008). The role of these cells mainly involves in hypersensitivity, autoimmune disease, and cancer (Clark et al., 2000; Liu & Nussenzweig, 2010; Merad et al., 2013). In the present chapter, we would restrict ourselves to the role of macrophage and neutrophils and their immunomodulation with chitosan.

3.3 ROLE OF SENTINEL CELLS IN ACUTE AND CHRONIC INFLAMMATION

Inflammation is a response by the body to the harmful stimuli from the pathogens, damaged cells, toxic compounds, or irradiation (Medzhitov, 2010), and acts by removing injurious stimuli and thereby initiating the healing process (Nathan, 2010). It is categorized into two types, acute and chronic inflammation. Acute inflammation lasts for hours or days. however, chronic inflammation lasts for days or months, or even years. abnormality in the inflammation process leads to severe organ damage and even death. Macrophages, neutrophils, and dendritic cells, also known as sentinel cells are mainly involved in this process by releasing different chemokines and cytokines.

3.3.1 ACUTE INFLAMMATION

The primary mechanism involves mononuclear phagocyte cells' activation to infection sites resulting in edema and local fluid into the bloodstream from affected tissue. The phagocytic cells engulf the pathogen activating the early response cytokines (IL-1, TNF). These cytokines stimulate specific cell populations that release inflammatory cytokines such as IL-6 and IL-8 and promote leukocyte diapedesis and activation. Inflammation induced by the macrophages and neutrophils are a result of the upregulation of one of the three pathways which involves TLR-mediated pathways, Nod-like receptor (NLR) mediated pathways, and helicase pathways. In TLR-dependent pathways the bacterial membrane materials like LPS, and follows NF-κB transcription factors activation, which transcribes inflammatory genes. In NLR pathways, bacterial peptidoglycan compound meso-diaminopimelic acid will activate the NOD1 and NOD2 receptors, which in turn will activate inflammasome complex formation, which converts inactivated inflammatory cytokines to activated cytokines. In helicase pathways, ds RNA of microbes will activate specific gene cascade, which release inflammatory cytokines.

However, in some situations where the infection is high, the inflammation becomes intense and unregulated. This results in the overrun of leukocyte migration in organs and the development of inflammatory cell influx. These responses lead to septic shock to target organs, including the liver, lung, and kidney leading to organ dysfunction and failure happen (Ptaschinski & Lukacs, 2018).

3.3.2 CHRONIC INFLAMMATION

Chronic inflammation typically lasts for days, months, and years. Bacterial substance like LPS not only activates the release of cytokine and chemokine for inflammation. It also stimulates critical antigen-presentation cell (APC) functions, which in turn activates T and B lymphocyte activation for pathogen clearance. However, uncontrolled immune responses can lead to continual inflammatory cell recruitment and tissue damage. Typically, during the immune process, dendritic cells act as APC and recruit T and B-lymphocytes, which undergo cell-based immune response. The activated T lymphocytes produce Tβ and GATA 3 transcription factors, which in turn stimulate the release of cytokine like interferon γ (IFN γ), IL-1, IL-4, IL-5, IL-13, etc., which are responsible for inflammation. Unregulated inflammation leads to severe disorders such as rheumatoid arthritis, chronic obstructive pulmonary disease, asthma, mycobacterial diseases, multiple sclerosis (MS), and viral hepatitis. Unregulated B-lymphocytes results in allergic conditions, autoimmune diseases, where immune complexes are deposited in the tissues. This can lead to tissue damage due to inappropriate activation of inflammatory effector cell populations, particularly the release of their granular products that can destroy host tissue leading to autoimmune disease conditions such as systemic lupus and RA (Ptaschinski & Lukacs, 2018).

3.4 CHITIN AND ITS DERIVATIVES

3.4.1 CHITIN

Chitin is the second most abundant polysaccharide after cellulose. The presence of chitin is evident in various life kingdoms ranging from exoskeleton of crustaceans such as crab, shrimp. It also shows a minor presence in insects and the cell walls of certain fungi and algae (Tharanathan & Kittur, 2003; Younes & Rinaudo, 2015). Chitin is composed of N-acetyl glucosamine monomeric units bound by β-1-4 glycosidic linkages. Chitin as a biomaterial has applications in the field of healthcare, agriculture, drug delivery, tissue engineering, and cosmetics but suffers from vast applicability due to its crystalline nature (Kumar, 2000; Jang et al., 2004; Ahmed et al., 2020). A brief attribute to its structure and the limitations of chitin will be discussed in subsections.

3.4.1.1 HISTORY

The history of chitin dates back to 1811, Prof. Henri Braconnot, a French scientist who first discovered a biopolymer different from cellulose in mushroom. Emulated by Braconnot's success with mushrooms, Odier identified the presence of the same biopolymer in the cuticles of insects in the year 1823. He later named the polymer as "chitin," derived from the Greek word sense tunic, covering, or envelope. Later in 1859, Prof. C. Rouget found a novel derivative of chitin named "chitosan." In 1878, the structural studies of chitin were undertaken by Ledderhose, and the units deciphered as glucosamine and acetic acid. The structural facts followed by the applications in various medical, environmental, and agricultural fields have led to an array of derivatives (Annu et al., 2017).

3.4.1.2 SOURCES AND STRUCTURE OF CHITIN

Chitin is distributed in both the aquatic as well as the terrestrial ecosystem with the major source of the raw material being the marine organisms Table 3.2. The chitin content in the crustaceans range from 27% in crabs to 70% in lobsters (Annu et al., 2017). On the other hand, in insects, the content ranges from 18% in cockroaches to 64% in butterfly (Nwe et al., 2010). The other life domain, which exhibits the presence of chitin, is the fungal kingdom with the minimum content of 2% in *Saccharomyces cerevisiae* to 42% in *Aspergillus niger* extractions (Synowiecki & Al-Khateeb, 2003).

Chitin is the structural polysaccharide similar to cellulose with the difference lying in the monosaccharide unit of N-acetyl glucosamine instead of glucose in the case of cellulose. In nature, chitin exists in three polymorphic forms: α-chitin, β-chitin, and γ-chitin. The structural difference of these forms is the arrangement of chains. Each chain is linked together by N-H O=C hydrogen bond by amine groups (Blackwell, 1969). In α-chitin, the antiparallel arrangement of chains one above another is observed, whereas β chitin possesses a parallel array of chains. The third polymorph; γ chitin, both parallel and antiparallel chain array exists. In α-chitin, due to the antiparallel arrangements of the chains, the structure is more stable and resistant to hydration. While the β-chitin, which is usually seen in the squid pens, the structure is less crystalline, and is sensitive to hydration. Thus, α-chitin is more abundant in the hard exoskeleton of shrimps, insect cuticle, crab, krill, lobster, and yeast cell wall, which are stiff and rigid. On the other hand γ-chitin is present in the fungal cell walls and has become a recent area of research (Acosta et al., 1993; Ahmed & Ikram, 2015; Jones et al., 2020). The extraction of chitin is generally performed from the crustacean shells due to their large abundance. The process involves demineralization followed by deproteination (Roberts, 2008) and decolorization (Nwe et al., 2010).

TABLE 3.2 Some Examples of Natural Sources of Chitin (Annu et al., 2017)

Terrestrial	*Porcellio*, *Armadillidium vulgare*, *Sipyloidea sipylus*, *Drosophila melanogaster*, *Extatosoma tiaratum*, *Bombyx mori*, *Apis mellifera*, *Periplaneta americana*
Marine	*Euphausia* sp. (Krill), *Paralithodes* sp. (King crab), *Chionoecetes* sp., *Carcinus* sp. (Crab), *Callinectes* sp. (Blue crab), *Loliginidae* sp. (Squid), *Pandalus* sp. (Shrimp), *Nephrops* sp. (Lobster), *Lepas* sp. (Goose Barnacle), *Penaeus* sp. (Prawn), *Crassostrea* sp. (Oyster)
Microorganisms	*Mucor rouxii*, *Mucor racemosus*, *Cunninghamella echinulate*, *Aspergillus niger*, *Rhizopus oryzae*, *Mucor circinelloides*, *Pleurotus ostreatus*, *Absidia coerulea*, *Lentinus edodes*, *Auricularia auricula-judae*

Source: Annu et al., 2017

Even though chitin is biocompatible, biodegradable, it is insoluble in water, solvents, and even acids due to its crystalline structure. It thus becomes imperative to modify chitin to a more amorphous form to improve its applications. One such derivative of chitin is chitosan. In contrast to chitin, the presence of free amine groups along the chitosan chain allows this macromolecule to dissolve in diluted aqueous acidic solvents due to the protonation of these groups, rendering the corresponding chitosan salt in solution (Aranaz et al., 2009).

3.4.2 CHITOSAN

Chitosan is the deacetylated derivative of chitin and is comprised of N-acetyl-2-amino-2-deoxy-d-glucopyranose (acetylated unit) and 2-amino 2-deoxy-glucopyranose (deacetylated unit), where the repeating units are linked by β-(1-4)-glycosidic bonds. The removal of the acetate group imparts a positive charge to the polymer. Chitosan thus binds strongly to negatively charged surfaces and is responsible for many observed biological activities (Chatterjee et al., 2005; Logesh et al., 2012).

3.4.2.1 SOURCES AND DERIVATIVE

Chitin is the primary source of chitosan. The commercial and general production of Chitosan is achieved by the deacetylation of chitin from the exoskeletons of crustaceans, shells of shrimp, crab, krill and lobsters The commercial modification of chitin to chitosan involves the use of chemicals which results in random pattern of deacetylation and non-reproducibility in the products (Smid et al., 1992; Percot et al., 2003). Research is undertaken to expedite the substitutions of chemicals with a greener alternative of enzymes like chitin deacetylases (EC 3.5.1.41, ChDa) (Kaczmarek et al., 2019).

Unlike chitin, chitosan is reactive because of the presence of the protonated amino group, which favors the adsorption of many substrates to it. The protonated state of the amino group is between the pH of 6.2 to 7.0. Hence, chitosan is soluble in acidic solutions (Madihally & Matthew, 1999). In addition to the solubility issue, degree of polymerization (DP) and molecular weight (MW) also plays a significant role in the biological application. Recent research states that chitosan with low MW (below 30 kDa) and DP (below 20 kDa) is better suited for anti-inflammation, immunostimulation, antioxidant antitumor, drug carrier, applications (Liaqat & Eltem, 2018). Thus, the conversion of chitosan to its derivatives named chito oligosaccharides (COS) is desirable (Liaqat & Eltem, 2018). Unlike chitosan, COS are soluble in all types of solvents and have a good membrane permeable property.

3.4.3 CHITO-OLIGOSACCHARIDES (COS)

Chitosan with a degree of polymerization (DP less than 20 and an average MW less than 3.9 kDa are called chitooligosaccharides (COS), chitosan oligomers, or chito-oligomers. Chitosan can be hydrolyzed into oligosaccharide by hydrolyzing agents, which will produce chitosan oligomers with a flexible DP. This can be achieved by chemical, physical, or enzymatic methods.

3.4.3.1 STRUCTURE AND PROPERTIES

Chito oligosaccharides generally consist of GlcNAc or GlcN units linked by a β-1,4-O-glycoside bond, the same as chitosan. However, COS is produced with defined DP and different degree of acetylation (DA). Based on the DP, DA the application of COS varies. Based on the distribution of the monomeric units, COS can be classified as homo-chitooligosaccharides or hetero-chitooligosaccharide is generally two types based on its structure. The homo chitooligosaccharides comprises of GlcN monomer units only. In contrast, hetero chitooligosaccharides has both GlcN and GlcNAc units as its constituents. Unlike chitin and chitosan, some unique COS properties, such as water solubility, cell membrane penetrability, ease in absorption (Liaqat & Eltem, 2018). Based on the second carbon position's functional group, the COS is classified into amino COS, Carboxylated COS, Gallyl COS, and sulfated COS (Lodhi et al., 2014). The DP and MW and the arrangement of the acetyl group of COS are the critical factors which directly affect their physicochemical properties and its biological activities (Harish & Tharanathan, 2007). Hydrolysis of glycosidic bond present in chitosan is the crucial factor for developing COS. The three main types of methods used for this hydrolysis fall into chemical, physical, and enzymatic modifications. The physical method involves hydrolysis using ultrasonic irradiation in which the option could be rapid specific scission of the polymer creating smaller DP units or short polymer chains and oligomers (Popa-Nita et al., 2009; Mourya et al., 2011). However, this method is not commercially viable. The chemical method is used in large-scale industries. It involves hydrolysis by hydrochloric acid with electrolytes. Other acids like nitrous acid, phosphoric acid, hydrofluoric acid, lactic acid, trichloroacetic acid, formic acid, and acetic acid are used. In addition, hydrogen peroxide or persulfate can be used. Using acids DA of 0–90% and DP between 2 and 12 with a mean of DP 5 can be produced in the chemical method. Using hydrogen peroxide, COS of DP 2–9 can be produced (Xia et al., 2013). However, chemical methods are challenging, and the contamination of products by chemicals is high. So nowadays, research and production are focused on the enzymatic process. Recently chitosanase has been reported for the development of COS (Muzzarelli et al., 1994; Vårum et al., 1996, Hamer et al., 2015). Even though some other carbohydrases and proteases can hydrolyze chitosan to produce COS, the specificity is always debatable. The use of the chitosanase can also result in a patterned hydrolysate (Weikert et al., 2017). Some of the considerable health applications are summarized in Table 3.3.

TABLE 3.3 Different Application of Chitooligosaccharides

Application	Description	References
Antibacterial and antifungal activity	In fungal and bacteria, COS will react with negative groups in the cell wall and affect the microbial cell wall's permeability and cause leakage of cell constituents and cause bacteria's death. COS works against various bacteria such as *Escherichia coli*, *Vibrio parahaemolyticus*, *Salmonella* species, *Streptococcus mutants*, *Lactobacillus plantarum*, *Bacillus* species, etc.	Jeon et al. (2001); Gerasimenko et al. (2004); Park et al. (2004); Rahman et al. (2014); Sánchez et al. (2017)
Antitumor activity	COS possess a positive charge and bind to negative charge moieties on the tumor cells and alter tumor cells' permeability. Also suppress tumor angiogenesis by inhibiting the expression of MMP-9, which consequently suppresses the expression of VEGF.	Qin et al. (2002); Han et al. (2015); Luo et al. (2016)
Anti-inflammatory	COS suppresses lipopolysaccharide (LPS) induced inflammatory gene expression associated with reduced NF-κB in B-cells. COS also inhibits the LPS-induced Interleukin 6 (IL-6) and TNF-α production in macrophages by down regulating the MAPK.	Vo et al. (2017); Liang et al. (2016); Azuma et al. (2015); Li et al. (2014)
Antioxidant	The amino groups of COS react with unstable free radicals and form stable products. Antioxidant activities are affected by DD and MW of COS. Moreover, antioxidant activities increase with a decrease in MW.	Laokuldilok et al. (2017); Yang et al. (2016); Vo et al. (2017)
Immunostimulant and wound healing	COS, by interacting with macrophages' membrane receptors, enhance macrophages migration by their chemotactic effects on macrophages and promoting gene transcription and protein secretion of cytokines by COS. Thus, by the above mechanism, it stimulates an immune response. This also helps in wound healing.	Xing et al. (2017); Zheng et al. (2016); Wu et al. (2015); Zhang et al. (2014)
Food preservatives	Inhibitory effects of COS against different microbes encourage their use in food, like preservatives and packaging material.	Fernández-de Castro et al. (2016); Rao et al. (2008)

3.5 ROLE OF CHITOSAN DERIVATIVES ON BLOOD SENTINEL CELLS

3.5.1 ANTI-INFLAMMATORY PROPERTIES

Inflammation is a significant immune response in humans against microbial invasion. Extreme and prolonged inflammation may lead to serious health issues and diseases such as chronic asthma, sepsis, organ failure by septic shock, rheumatoid arthritis, Endometriosis, multiple sclerosis, inflammatory bowel disease, and cancer, etc. Chito oligosaccharides (COS) provide several applications. Its properties, such as DP and MW, make it suitable for various applications and unique from other products (Ngo et al., 2015). Lipopolysaccharide (LPS) is a significant component for initiating inflammation, which is present and abundant in bacteria. TLR4 receptors present in macrophages and neutrophils will be activated by LPS and activates NF-κB and inflammasome-mediated signal transduction pathways. Finally, pro-inflammatory mediators, such as IL-1β, IL-6, IL-8, TNF-α, and platelet-activating factor, are released (Ngkelo et al., 2012). However, COS inhibits the above mechanism. In a study by Sang-Hoon Lee et al., high molecular weight COS (MW 5 to 10 kDa) with 50% (50-HMWCOS) and 90% (90-HMWCOS) deacetylation was prepared. Both chitooligo components at different concentrations were incubated with the macrophage cells, and significant essential cytokines for inflammation and apoptosis, such as IL-6 and TNF-α, respectively, were estimated. Interestingly, on the addition of LPS to cell solution, the release of IL-6 and TNF-α was reduced by COS-treated cells as compared to the untreated cells. Another observation found from the study was higher inhibition with 90-HMWCOS treatment as compared to 50-HMWCOS. In summary, the study concludes the effectiveness of COS with a higher degree of deacetylation (DD) in inhibiting the LPS-mediated inflammation pathway (Lee et al., 2009; Liang et al., 2016).

In another study performed by Ying Qiao et al., the protective efficiency of COS against sepsis was tested. Sepsis is a fatal condition caused by undesirable acute inflammation by macrophages leading to multiple organ failure finally culminating in death. In this study, the COS with 95% DD and DP ranging between 2 and 6 were treated with mice, and then LPS was injected in blood with different concentrations (20 mg/kg and 40 mg/kg) creating a sepsis model. The study revealed the positive impact of COS on the septic model of mice with an improvement of life expectancy by 54% as compared to the non-treated control. A reduction of the pro-inflammatory cytokines such as IL-1β, IL-6, IL-8, TNF-α was also exposed post-LPS challenge. The study also mentioned a reduction in ROS and reduces oxidative stress in the treated animals (Qiao et al., 2011). Shun-Hsien Chang et al. reported a similar observation, 2019 on the NO secretion in LPS-challenged mice model with chitooligosaccharides. Higher MW chitosan inhibited NF-κB activation and iNOS expression by binding to the CR3 and TLR4 receptor, whereas smaller MW chitosan stimulates NF-κB and iNOS expression by binding to CD14, TLR4, and CR3 receptors (Chang et al., 2019).

An extension to the 3D nano-COS is usage in scaffolds where different degrees of acetylation (DA), such as 4% and 15% or 96% and 85% of degree of deacetylation (DD) (Vasconcelos et al., 2019). They observed an inhibition of NRPL3 Inflammasome-dependent inflammation process.

The above studies show COS either with high MW and high DD inhibits the LPS mediated inflammation and oxidative stress mechanism. This favor treating blood macrophages dependent on inflammation diseases like sepsis, inflammatory bowel diseases, cancer, etc., using biocompatible and biodegradable COS.

3.5.2 IMMUNOSTIMULATORY PROPERTIES

The immune system in our body is essential for eradicating foreign pathogens from the body. However, some individuals possess inadequate immune responses. So, immunostimulating medicines and nutraceuticals are of particular interest nowadays. One such component is COS; recent studies have revealed the potential of COS as an immunostimulatory agent. In the study by Wu & Tsai (2007), the efficiency of immunostimulation by LMWCOS (20 kDa) with 95% DD and 1 to 6 DP. Here, the murine macrophages cells were cultured, and immunostimulation was studied by estimating NO synthesis and the iNOS gene expression by NF-κB activation pathway. The result of this study reveals that treating cells with LMWCOS alone will inhibit the secretion of NO and iNOS gene expression because COS possesses high anti-inflammatory properties, as we know. However, if LMWCOS in combination with IFN-γ, is treated with cells, the NF-κB mechanism was activated by migrating into the nucleus and strongly bind with the DNA segment. This stimulates the NO synthesis and the iNOS gene expression efficiently. In addition, the treatment of IFN-γ in combination with LMWCOS, anti-CD14, anti-TLR4, and anti-CR3 macrophages inhibited the NO production. This is proven that IFN-γ plus LMWCOS induced NF-κB activation and NO production by binding with the receptors of CD14, TLR4, and CR3 in macrophages (Wu & Tsai, 2007).

In another two studies, it was proven that COS's immunostimulating properties depend on its molecular weight and concentration gradient or dosage of COS provided. Lower molecular weight COS (LMWCOS below 3 kDa) was best for stimulating the pinocytic activity and production of the TNF-α, IL-6, IFN-γ, NO, and expression of iNOS, TNF-α gene expression. In addition, apoptosis was stimulated by activating TRAF6, JNK, AP-1 mediated pathways. At the same time, higher molecular weight COS inhibits the above mechanism (Mei et al., 2013; Wu et al., 2015; Zheng et al., 2016). In another study by Ronge Xing et al., chitooligosaccharides were prepared by three methods, such as traditional methods, microwave irradiation, and enzymatic hydrolysis. Moreover, the immunostimulation properties of these three types of COS were studied individually. However, COS prepared by the microwave irradiation method comprises disaccharides to hexasaccharides with or without acetyl group. It possesses the best immuno stimulation properties than the other two types (Xing et al., 2017; Chang et al., 2019).

From the above various studies, it is clear that different properties of COS depend on Molecular weight (MW), degree of deacetylation (DD), and degree of polymerization (DP). Thus, future research of physiological properties like MW, DP, and DD can be focused more and make COS feasible for various applications.

3.6 CONCLUSIONS AND PERSPECTIVES

This chapter provides detailed information about different types of blood sentinel cells and their immune response mechanisms. Macrophages and neutrophils are essential sentinel cells that perform effectively as second line of defense against various microbial infections and damage signals. The signal produced by these cells on disease activates other immune responses like inflammation, cellular responses, phagocytosis, etc. At the same time, dendritic cells play a role in activating T lymphocytes and induce cell-mediated inflammation response. However, overexpression or performance will lead to autoimmune and hypersensitive problems, leading to severe diseases like sepsis, cancer, chronic inflammatory bowel diseases, etc. However, several drugs are available in this chapter; various studies reveal Chitosan's efficiency and its derivatives as anti-inflammatory and immunostimulation properties, which can effectively treat immune and inflammation-related diseases. Moreover, insists researchers focus on COS for various healthcare applications.

KEYWORDS

- **blood sentinel cells**
- **chitooligosaccharides**
- **chitosan**
- **degree of deacetylation (DD)**
- **degree of polymerization (DP)**
- **immunostimulatory**
- **inflammation**

REFERENCES

Acosta, N., Jiménez, C., Borau, V., et al., (1993). Extraction and characterization of chitin from crustaceans. *Biomass and Bioenergy, 5*(2), 145–153. doi: 10.1016/0961-9534(93)90096-M.

Ahmed, M. J., Hameed, B. H., & Hummadi, E. H., (2020). Review on recent progress in chitosan/chitin-carbonaceous material composites for the adsorption of water pollutants. *Carbohydrate Polymers*, 116690.

Ahmed, S., & Saiqa, I., (2015). Chitosan and its derivatives: A review in recent innovations. *International Journal of Pharmaceutical Sciences and Research, 6*(1), 14–30. doi: 10.13040/IJPSR.0975-8232.6(1).14-30.

Annu, S. A., Shakeel, A., & Saiqa, I., (2017). Chitin and chitosan: History, composition and properties. *Chitosan*, 1–24. doi: 10.1002/9781119364849.ch1.

Aranaz, I., Mengíbar, M., Harris, R., Paños, et al., (2009). Functional characterization of chitin and chitosan. *Current Chemical Biology, 3*(2), 203–230.

Azuma, K., Tomohiro, O., Seiji, K., et al., (2015). Anti-inflammatory effects of orally administered glucosamine oligomer in an experimental model of inflammatory bowel disease. *Carbohydrate Polymers, 115*, 448–456. doi: 10.1016/j.carbpol.2014.09.012.

Blackwell, J., (1969). Parallel chain systems. *Biopolymers, 7*, 281–298.

Carman, C. V., Peter, T. S., Tracey, E. S., et al., (2007). Transcellular diapedesis is initiated by invasive podosomes. *Immunity, 26*(6), 784–797. Springer. doi: 10.1016/j.immuni.2007.04.015.

Carmona-Rivera, C., & Kaplan, M. J., (2016). *Neutrophil Biology: Encyclopedia of Cell Biology, 3*. Elsevier Ltd. doi: 10.1016/B978-0-12-394447-4.30112-2.

Carvalho, L. O., Aquino, E. N., & Fontes, W., (2015). The neutrophil nucleus and its role in neutrophilic function. *Journal of Cellular Biochemistry, 116*(9), 1831–1836. doi: 10.1002/jcb.25124.

Chang, S. H., Yi, Y. L., Wu, G. J., Huang, C. H., et al., (2019). Effect of chitosan molecular weight on anti-inflammatory activity in the RAW 264.7 macrophage model. *International Journal of Biological Macromolecules, 131*, 167–175. doi: 10.1016/j.ijbiomac.2019.02.066.

Chatterjee, S., Adhya, M., Guha, A. K., et al., (2005). Chitosan from *Mucor rouxii*: Production and physico-chemical characterization. *Process Biochemistry, 40*(1), 395–400. doi: 10.1016/j.procbio.2004.01.025.
Chistiakov, D. A., Bobryshev, Y. V., & Orekhov, A. N., (2015). Neutrophil's weapons in atherosclerosis. *Experimental and Molecular Pathology, 99*(3), 663–671. doi: 10.1016/j.yexmp.2015.11.011.
Clark, G. J., Angel, N., Kato, M., et al., (2000). The role of dendritic cells in the innate immune system. *Microbes and Infection, 2*(3), 257–272. doi: 10.1016/S1286-4579(00)00302-6.
Epelman, S., Kory, J. L., & Gwendalyn, J. R., (2014). origin and functions of tissue macrophages. *Immunity, 41*(1), 21–35. doi: 10.1016/j.immuni.2014.06.013.
Faurschou, M., & Borregaard, N., (2003). Neutrophil granules and secretory vesicles in inflammation. *Microbes and Infection, 5*(14), 1317–1327. doi: 10.1016/j.micinf.2003.09.008.
Fernández-De, C. L., Mengíbar, M., Sánchez, A., et al., (2016). films of chitosan and chitosan-oligosaccharide neutralized and thermally treated: Effects on its antibacterial and other activities. *LWT – Food Science and Technology, 73*, 368–374. doi: 10.1016/j.lwt.2016.06.038.
Fujiwara, N., & Kobayashi, K., (2005). Macrophages in inflammation. *Current Drug Targets-Inflammation and Allergy, 4*(3), 281–286.281–286.
Gerasimenko, D. V., Avdienko, I. D., Bannikova, G. E., et al., (2004). Antibacterial effects of water-soluble low-molecular-weight chitosans on different microorganisms. *Applied Biochemistry and Microbiology, 40*(3), 253–257. doi: 10.1023/B:ABIM.0000025947.84650.b4.
Hamer, S. N., Cord-Landwehr, S., Biarnés, X., et al., (2015). Enzymatic production of defined chitosan oligomers with a specific pattern of acetylation using a combination of chitin oligosaccharide deacetylases. *Scientific Reports, 5,* 8716.
Han, F. S., Cui, B. H., You, X. F., et al., (2015). Anti-proliferation and radio sensitization effects of chitooligosaccharides on human lung cancer line HepG2. *Asian Pacific Journal of Tropical Medicine, 8*(9), 757–761. doi: 10.1016/j.apjtm.2015.07.025.
Harish, P. K. V., & Tharanathan, R. N., (2007). Chitin/chitosan: Modifications and their unlimited application potential: An overview. *Trends in Food Science and Technology, 18*(3), 117–131. doi: 10.1016/j.tifs.2006.10.022.
Hoffmann, K., Sperling, K., Olins, A. L., et al., (2007). The granulocyte nucleus and Lamin B receptor: Avoiding the ovoid. *Chromosoma, 116*(3), 227–235. doi: 10.1007/s00412-007-0094-8.
Jang, M. K., Kong, B. G., Jeong, Y. I., et al., (2004). Physicochemical characterization of α-chitin, β-chitin, and γ-chitin separated from natural resources. *Journal of Polymer Science, Part A: Polymer Chemistry, 42*(14), 3423–3432. doi: 10.1002/pola.20176.
Jeon, Y. J., Park, P. J., & Kim, S. K., (2001). Antimicrobial effect of chitooligosaccharides produced by bioreactor. *Carbohydrate Polymers, 44*(1), 71–76. doi: 10.1016/S0144-8617(00)00200-9.
Jones, M., Kujundzic, M., John, S., et al., (2020). Crab vs. mushroom: A review of crustacean and fungal chitin in wound treatment. *Marine Drugs, 18*(1), 64.
Kaczmarek, M. B., Struszczyk-Swita, K., Li, X., Szczęsna-Antczak, M., & Daroch, M., (2019). Enzymatic modifications of chitin, chitosan, and chitooligosaccharides. *Frontiers in Bioengineering and Biotechnology, 7*, 243. doi: 10.3389/fbioe.2019.00243.

Kasahara, T., & Matsushima, K., (2001). Macrophage signaling, apoptosis, lectins and leukocyte trafficking. *Trends in Immunology, 22*(11), 593, 594. doi: 10.1016/S1471-4906(01)02057-9.

Kumar, M. N. R., (2000). A review of chitin and chitosan applications. *Reactive and Functional Polymers, 46*(1), 1–27.

Kunes, P., Krejsek, J. M. B., Mandak, J., et al., (2009). Neutrophil apoptosis by Fas/FasL: Harmful or advantageous in cardiac surgery? *Thoracic and Cardiovascular Surgeon, 57*(1), 1–6. doi: 10.1055/s-2008-1039060.

Laokuldilok, T., Potivas, T., Kanha, N., et al., (2017). Physicochemical, antioxidant, and antimicrobial properties of chitooligosaccharides produced using three different enzyme treatments. *Food Bioscience, 18*, 28–33. doi: 10.1016/j.fbio.2017.03.004.

Lee, S. H., Senevirathne, M., Chang, B. A., Kim, S. K., & Je, J. Y., (2009). Factors affecting anti-inflammatory effect of chitooligosaccharides in lipopolysaccharides-induced RAW264.7 macrophage cells. *Bioorganic and Medicinal Chemistry Letters, 19*(23), 6655–6658. doi: 10.1016/j.bmcl.2009.10.007.

Li, Y., Liu, H., Xu, Q. S., et al., (2014). Chitosan oligosaccharides block LPS-induced o-glcnacylation of Nf-κb and endothelial inflammatory response. *Carbohydrate Polymers, 99*, 568–578. doi: 10.1016/j.carbpol.2013.08.082.

Liang, T. W., Chen, W. T., Lin, Z. H., et al., (2016). An amphiprotic novel chitosanase from bacillus mycoides and its application in the production of chitooligomers with their antioxidant and anti-inflammatory evaluation. *International Journal of Molecular Sciences, 17*(8), 1–14. doi: 10.3390/ijms17081302.

Liaqat, F., & Eltem, R., (2018). Chitooligosaccharides and their biological activities: A comprehensive review. *Carbohydrate Polymers, 184*, 243–259. doi: 10.1016/j.carbpol.2017.12.067.

Liu, K., & Nussenzweig, M. C., (2010). Origin and development of dendritic cells. *Immunological Reviews, 234*(1), 45–54. doi: 10.1111/j.0105-2896.2009.00879.x.

Lodhi, G., Kim, Y. S., Hwang, J. W., et al., (2014). Chitooligosaccharide and its derivatives: Preparation and biological applications. *BioMed Research International., 2014*. doi: 10.1155/2014/654913.

Logesh, A. R., Thillaimaharani, K. A., Sharmila, K., et al., (2012). Production of chitosan from endolic henic fungi isolated from mangrove environment and its antagonistic activity. *Asian Pacific Journal of Tropical Biomedicine, 2*(2), 140–143. doi: 10.1016/S2221-1691(11)60208-6.

Luo, Y., Deng, L., Deng, Q. J., et al., (2016). Comparative study of the chitooligosaccharides effect on the proliferation inhibition and radiosensitization of three types of human gastric cancer cell line. *Asian Pacific Journal of Tropical Medicine, 9*(6), 601–605. doi: 10.1016/j.apjtm.2016.04.014.

Madihally, S. V., & Howard, W. T. M., (1999). Porous chitosan scaffolds for tissue engineering. *Biomaterials, 20*(12), 1133–1142. doi: 10.1016/S0142-9612(99)00011-3.

Medzhitov, R., (2010). Inflammation 2010: New adventures of an old flame. *Cell, 140*(6), 771–776.

Mei, Y. X., Chen, H. X., Zhang, J., et al., (2013). Protective effect of chitooligosaccharides against cyclophosphamide-induced immunosuppression in mice. *International Journal of Biological Macromolecules, 62*, 330–335. doi: 10.1016/j.ijbiomac.2013.09.038.

Merad, M., Sathe, P., Helft, J., et al., (2013). The dendritic cell lineage: Ontogeny and function of dendritic cells and their subsets in the steady state and the inflamed setting. *Annual Review of Immunology, 31*, 563–604. doi: 10.1146/annurev-immunol-020711-074950.

Mourya, V. K., Inamdar, N. N., & Choudhari, Y. M., (2011). Chitooligosaccharides: Synthesis, characterization and applications. *Polymer Science – Series A, 53*(7), 583–612. doi: 10.1134/S0965545X11070066.

Muzzarelli, R. A. A., Ilari, P., Tarsi, R., et al., (1994). Chitosan from *Absidia coerulea. Carbohydrate Polymers, 25*(1), 45–50. doi: 10.1016/0144-8617(94)90161-9.

Nathan, C. F., (1987). Secretory products of macrophages. *Journal of Clinical Investigation, 79*(2), 319–326. doi: 10.1172/JCI112815.

Nathan, C., & Aihao, D., (2010). Nonresolving inflammation. *Cell, 140*(6), 871–882.

Ngkelo, A., Meja, K., Yeadon, M., et al., (2012). LPS induced inflammatory responses in human peripheral blood mononuclear cells is mediated through NOX4 and G Iα dependent PI-3kinase signaling. *Journal of Inflammation, 9*, 2–8. doi: 10.1186/1476-9255-9-1.

Ngo, D. H., Vo, T. S., Ngo, D. N., et al., (2015). Biological effects of chitosan and its derivatives. *Food Hydrocolloids, 51*, 200–216. doi: 10.1016/j.foodhyd.2015.05.023.

Nwe, N., Furuike, T., & Tamura, H., (2010). Chitin and chitosan from terrestrial organisms. *Chitin, Chitosan, Oligosaccharides and Their Derivatives: Biological Activities and Applications*, 3–9. doi: 10.1201/EBK1439816035.

Onai, N., Obata-Onai, A., Schmid, M. A., et al., (2007). Identification of clonogenic common flt3+m-csfr+ plasmacytoid and conventional dendritic cell progenitors in mouse bone marrow. *Nature Immunology, 8*(11), 1207–1216. doi: 10.1038/ni1518.

Park, P. J., Lee, H. K., & Kim, S. K., (2004). Preparation of hetero-chitooligosaccharides and their antimicrobial activity on vibrio parahaemolyticus. *Journal of Microbiology and Biotechnology, 14*(1), 41–47.

Percot, A., Viton, C., & Domard, A., (2003). Optimization of chitin extraction from shrimp shells. *Biomacromolecules, 4*, 12–18.

Phillipson, M., Heit, B., Colarusso, P., et al., (2006). Intraluminal crawling of neutrophils to emigration sites: A molecularly distinct process from adhesion in the recruitment cascade. *Journal of Experimental Medicine, 203*(12), 2569–2575. doi: 10.1084/jem.20060925.

Popa-Nita, S., Lucas, J. M., Ladavière, C., et al., (2009). Mechanisms involved during the ultrasonically induced depolymerization of chitosan: Characterization and control. *Biomacromolecules, 10*(5), 1203–1211. doi: 10.1021/bm8014472.

Ptaschinski, C., & Lukacs, N. W., (2018). Acute and chronic inflammation induces disease pathogenesis. In: *Molecular Pathology: The Molecular Basis of Human Disease* (pp. 25–43). Elsevier Inc. doi: 10.1016/B978-0-12-802761-5.00002-X.

Qiao, Y., Bai, X. F., & Yu Guang, D., (2011). Chitosan oligosaccharides protect mice from LPS challenge by attenuation of inflammation and oxidative stress. *International Immunopharmacology, 11*(1), 121–127. doi: 10.1016/j.intimp.2010.10.016.

Qin, C., Du, Y., Xiao, L., et al., (2002). Enzymic preparation of water-soluble chitosan and their antitumor activity. *International Journal of Biological Macromolecules, 31*(1–3), 111–117. doi: 10.1016/S0141-8130(02)00064-8.

Rahman, M. H., Hjeljord, L. G., Aam, B. B., et al., (2014). Antifungal effect of chito-oligosaccharides with different degrees of polymerization. *European Journal of Plant Pathology, 141*(1), 147–158. doi: 10.1007/s10658-014-0533-3.

Rao, M. S., Chander, R., & Sharma, A., (2008). Synergistic effect of chitooligosaccharides and lysozyme for meat preservation. *LWT – Food Science and Technology, 41*(10), 1995–2001. doi: 10.1016/j.lwt.2008.01.013.

Roberts, G. A. F., (2008). Thirty years of progress in chitin and chitosan. *Progress in the Chemistry and Application of Chitin and Its Derivatives, 13*, 7–15.

Rotondo, R., Bertolotto, M., Barisione, G., et al., (2011). Exocytosis of azurophil and arginase, *1*-containing granules by activated polymorphonuclear neutrophils is required to inhibit t lymphocyte proliferation. *Journal of Leukocyte Biology, 89*(5), 721–727. doi: 10.1189/jlb.1109737.

Sánchez, M. M., Rivera-Rodríguez, G., Moerchbacher, B., et al., (2017). The effect of preparation processes on the physicochemical characteristics and antibacterial activity of chitooligosaccharides. *Carbohydrate Polymers, 157*, 251–257. doi: 10.1016/j.carbpol.2016.09.055.

Selders, G. S., Fetz, A. E., Radic, M. Z., et al., (2017). An overview of the role of neutrophils in innate immunity, inflammation and host-biomaterial integration. *Regenerative Biomaterials, 4*(1), 55–68. doi: 10.1093/rb/rbw041.

Shimokado, K., Raines, E. W., Madtes, D. K., et al., (1985). A significant part of macrophage-derived growth factor consists of at least two forms of PDGF. *Cell, 43*(1), 277–286. doi: 10.1016/0092-8674(85)90033-9.

Smid, P., Jorning, W. P. A., Van, D. A. M. G., et al., (1992). Applications of chitosan and chitosan derivatives in drug delivery. *Journal of Carbohydrate Chemistry, 11*(7), 849–865. doi: 10.1080/07328309208018274.

Smith, J. A., (1994). Neutrophils, host defense, and inflammation: A double-edged sword. *Journal of Leukocyte Biology, 56*(6), 672–686. doi: 10.1002/jlb.56.6.672.

Synowiecki, J., & Al-Khateeb, N. A., (2003). *Production, Properties, and Some New Applications of Chitin and Its Derivatives*, 145–171.

Takeda, K., Kaisho, T., & Akira, S., (2003). Toll-like receptors. *Annual Review of Immunology, 21*(1), 335–376.

Tharanathan, R. N., & Farooqahmed, S. K., (2003). Chitin – the undisputed biomolecule of great potential. *Critical Reviews in Food Science and Nutrition, 43*(1), 61–87. doi: 10.1080/10408690390826455.

Vårum, K. M., Holme, H. K., Izume, M., et al., (1996). Determination of enzymatic hydrolysis specificity of partially n-acetylated chitosans. *Biochimica et Biophysica Acta – General Subjects, 1291*(1), 5–15. doi: 10.1016/0304-4165(96)00038-4.

Vasconcelos, D. P., De Torre-Minguela, C., Gomez, A. I., et al., (2019). 3D chitosan scaffolds impair nlrp3 inflammasome response in macrophages. *Acta Biomaterialia, 91*, 123–134. doi: 10.1016/j.actbio.2019.04.035.

Vo, T. S., Ngo, D. H., Bach, L. G., et al., (2017). The free radical scavenging and anti-inflammatory activities of gallate-chitooligosaccharides in human lung epithelial A549 cells. *Process Biochemistry, 54*, 188–194. doi: 10.1016/j.procbio.2017.01.001.

Wang, J., & Arase, H., (2014). Regulation of immune responses by TGF. *Annals of the New York Academy of Sciences, 1319*(1), 66–81. doi: 10.1111/nyas.12445.

Waskow, C., Liu, K., Darrasse-Jèze, G., et al., (2008). The receptor tyrosine kinase flt3 is required for dendritic cell development in peripheral lymphoid tissues. *Nature Immunology, 9*(6), 676–683. doi: 10.1038/ni.1615.

Weikert, T., Niehues, A., Cord-Landwehr, S., et al., (2017). Reassessment of chitosanase substrate specificities and classification. *Nature Communications, 8*(1), 1–11.

Wen, J., Nikolaos, G., Chaisuparat, R., et al., (2011). Secretory leukocyte protease inhibitor (SLPI) expression and tumor invasion in oral squamous cell carcinoma. *American Journal of Pathology, 178*(6), 2866–2878. doi: 10.1016/j.ajpath.2011.02.017.

Woodfin, A., Voisin, M. B., & Nourshargh, S., (2010). Recent developments and complexities in neutrophil transmigration. *Current Opinion in Hematology, 17*(1), 9–17. doi: 10.1097/MOH.0b013e3283333930.

Woodfin, A., Voisin, M. B., Imhof, B. A., et al., (2009). Endothelial cell activation leads to neutrophil transmigration as supported by the sequential roles of ICAM-2, JAM-A, and PECAM-1. *Blood, 113*(24), 6246–6257. doi: 10.1182/blood-2008-11-188375.

Wu, G. J., & Tsai, G. J., (2007). Chitooligosaccharides in combination with interferon-γ increase nitric oxide production via nuclear factor-κB activation in murine RAW264.7 Macrophages. *Food and Chemical Toxicology, 45*(2), 250–258. doi: 10.1016/j.fct.2006.07.025.

Wu, N., Wen, Z. S., Xiang, X. W., et al., (2015). Immunostimulative activity of low molecular weight chitosans in RAW264.7 macrophages. *Marine Drugs, 13*(10), 6210–6225. doi: 10.3390/md13106210.

Wynn, T. A., (2003). IL-13 effector functions. *Annual Review of Immunology, 21*, 425–456. doi: 10.1146/annurev.immunol.21.120601.141142.

Xia, Z., Wu, S., & Chen, J., (2013). Preparation of water soluble chitosan by hydrolysis using hydrogen peroxide. *International Journal of Biological Macromolecules, 59*, 242–245. doi: 10.1016/j.ijbiomac.2013.04.034.

Xing, R., Liu, Y., Li, K., et al., (2017). Monomer composition of chitooligosaccharides obtained by different degradation methods and their effects on immunomodulatory activities. *Carbohydrate Polymers, 157*, 1288–1297. doi: 10.1016/j.carbpol.2016.11.001.

Yago, T., Shao, B., Miner, J. J., et al., (2010). E-selectin engages PSGL-1 and CD-44 through a common signaling pathway to induce integrin αlβ2-mediated slow leukocyte rolling. *Blood, 116*(3), 485–494. doi: 10.1182/blood-2009-12-259556.

Yang, F., Qian, D., Hou, D. H. M., et al., (2016). Prevalence of cardiovascular disease risk factor clustering in Chinese adults. *Clinical Trials and Regulatory Science in Cardiology, 15*, 1–6. doi: 10.1016/j.ctrsc.2016.01.007.

Yang, L., Kowalski, J. R., Yacono, P., et al., (2006). Endothelial cell cortactin coordinates intercellular adhesion molecule-1 clustering and actin cytoskeleton remodeling during polymorphonuclear leukocyte adhesion and transmigration. *The Journal of Immunology, 177*(9), 6440–6449. doi: 10.4049/jimmunol.177.9.6440.

Younes, I., & Marguerite, R., (2015). Chitin and chitosan preparation from marine sources. Structure, properties and applications. *Marine Drugs, 13*(3), 1133–1174. doi: 10.3390/md13031133.

Zhang, P., Liu, W., Peng, Y., Han, B., et al., (2014). Toll like receptor 4(TLR4) mediates the stimulating activities of chitosan oligosaccharide on macrophages. *International Immunopharmacology, 23*(1), 254–261. doi: 10.1016/j.intimp.2014.09.007.

Zheng, B., Wen, Z. S., Huang, Y. J., et al., (2016). Molecular weight-dependent immunostimulative activity of low molecular weight chitosan via regulating NF-KB and AP-1 signaling pathways in RAW264.7 macrophages. *Marine Drugs, 14*(9), 1–13. doi: 10.3390/md14090169.

CHAPTER 4

Proteoglycans and Glycosaminoglycans Regulating Functions of the Hematopoietic Stem Cell Niche

SAMIKSHA WASNIK,[1] LAKSHMI MANASA CHATURVEDULA,[2] and CHATURVEDULA TRIPURA[3]

[1]*Division of Regenerative Medicine, Department of Medicine, Loma Linda University, Loma Linda, CA, USA*

[2]*Independent Researcher, South Plainfield, NJ, USA*

[3]*Medical Biotechnology Campus, Annexe-II, CSIR-Centre for Cellular and Molecular Biology, Hyderabad, Telangana, India*

ABSTRACT

Hematopoiesis occurs in specialized soft regions in the cavity of the bone – the marrow. Hematopoietic stem cells (HSCs) are organized within the marrow in two functionally distinct zones; the endosteal niche and the vascular niche, surrounded by supporting cells that secrete soluble and extracellular factors which regulate hematopoiesis. Proteoglycans (PGs) form an essential component of the HSC niche and influence the maintenance, proliferation, differentiation, and mobilization of HSCs. PGs are post-translationally modified proteins with highly negatively charged glycan side chains, known as the glycosaminoglycans (GAGs), which could be either heparan sulfate, chondroitin sulfate, dermatan sulfate, or keratan sulfate. The heparan and chondroitin sulfate proteoglycans constitute a significant component of the HSC niche, and by binding to various cytokines and growth factors and their receptors, they regulate multiple signaling cascades of the hematopoietic cells. In this chapter, the organization of the HSC niche is described in brief,

Hematopoiesis: Biochemical, Cellular, Molecular, and Genomic Perspectives. Ravi Kumar Gutti (Ed.)

followed by elaboration of the various PGs of the niche and their role in regulating the HSC functions. Finally, an overview of the broad scope of applications of GAGs and PGs with a specific emphasis on the therapeutic application for hematopoietic cells is presented.

4.1 INTRODUCTION

The hematopoeitic stem cell (HSC) resides in a specified microenvironment surrounded by supporting cells within specialized zones of the bone cavity. The term HSC niche was first proposed by Schofield for this specialized microenvironment (Schofield, 1978). The supporting cells of the niche play a crucial role in sustaining HSC and regulating hematopoiesis via cellular, molecular, and extracellular matrix (ECM) interactions with the HSCs. Hematopoiesis in adult humans, occurs in the shaft (diaphyseal) and the epiphysis regions of the long appendicular bones and the ribs. With aging, hematopoiesis shifts to the vertebra bones. Within the bone, hematopoiesis is localized to the soft tissue that fills hard bones' cavities – the bone marrow (BM) (Zhao et al., 2016). Much of the marrow's spatiotemporal organization, cellular and molecular composition of the HSC niche is now understood. With advacements in the information regarding the localization of the hematopoietic cells, the details of their regulation and lineage specific differentiation are now becoming increasingly apparent. It is now clear that the fate of HSC residing in a niche is governed by cellular and extracellular components, such as cell surface receptors, chemokines, cytokines, and growth factors, secreted by the niche's cells (Wasnik et al., 2012, 2017).

PGs are post-translationally modified proteins with highly negatively charged glycan side chains, commonly known as glycosaminoglycans (GAGs). GAGs consist of repeating disaccharides capable of forming polymers of varying sizes. Depending on specific disaccharide composition, GAGs can be grouped as heparan, heparan sulfate (HS), chondroitin sulfate (CS), dermatan sulfate (DS), keratan sulfate (KS), and hyaluronic acid (HA) (Dodge et al., 2001). PGs and GAGs are versatile peri-cellular and extracellular components produced by most eukaryotic cells and belong to many different protein families. They act as tissue organizers, influencing specialized tissues' cell growth and maturation. The primary biological functions of PGs include serving as biological filters, modulating growth-factor activities by binding to a variety of cytokines and growth factors, and regulating matrix assembly, such as collagen fibrillogenesis and skin tensile strength, affecting tumor cell growth and invasion, influencing corneal transparency and neurite outgrowth, etc. (Bi et al., 2005). Due to their complex structure, PGs have remained a largely unexplored class of macromolecules. Recent studies elucidating PGs' broad and significant role in regulating a myriad of functions such as proliferation, self-renewal, differentiation, mobilization, and HSCs homing in normal and disease conditions have widened their research and therapeutic applications. This chapter is majorly focused on the PGs of the hematopoietic niche. Keeping within the scope, the concept of the BM HSC niche and its organization is described, followed by the classes/different types of PGs and their synthesis in brief (for a comprehensive review on the various PGs readers are suggested (Iozzo et al., 2015). Next, the PGs present in the hematopoietic niche and their role in regulating the HSC niche functions are elaborated. Finally, an overview of the broad scope of GAGs and PGs *in vitro* use and their therapeutic applications with emphasis in the context of hematopoietic cells is presented.

4.2 BONE MARROW ORGANIZATION AND THE HEMATOPOIETIC NICHE

The BM occupies the cavity of the cancellous bone and is made up of specialized soft and spongy tissue. The hematopoietic islands, stromal cells, adipocytic cells, endothelial cells, and smooth muscle cells make up the cellular constituents of the BM and are embedded in an ECM made up of collagen and other matrix proteins, glycoproteins, and PGs (Travlos, 2006). A balanced interaction between the different cell types and the ECM regulates the complex HSC niche (Wasnik et al., 2017). The functional constituents and the physicochemical factors (variable calcium content and O_2 levels) vary throughout the marrow space (Mendez-Ferrer et al., 2010; Morrison et al., 2014; Szade et al., 2018). Studies of stem cell niches in non-vertebrates, such as *Drosophila melanogaster*, revealed several specific functions of the HSC niche, such as regulating self-renewal, cell survival, supporting asymmetric cell division by providing signals, anchoring HSC in proximity to the physicochemical factors such as nutrient supply and O_2 levels, etc. (Tran et al., 2000). Thus, HSC niches play a crucial role by not only just giving the structural and mechanical support but also provide specific 3D topographical information and physiological cues to the HSC. Further, within the stem cell niche, the molecular cues comprised of growth factors are essential for stem cell functions (Morrison et al., 2008).

The osteo-hematopoietic niche is compartmentalized into two specialized regions: (i) the endosteal niche; and (ii) the vascular niche. The endosteal niche is linked with the endosteum and is comprised of overlapping zones of osteogenesis and hematopoiesis. In contrast, the vascular niche is located in the marrow's stromal part, linked to the blood vessels' outer lining, and shows hematopoietic or multipotent progenitors (MPPs) exclusively (Wasnik et al., 2012). The schematic organization of the BM niche is depicted in Figure 4.1.

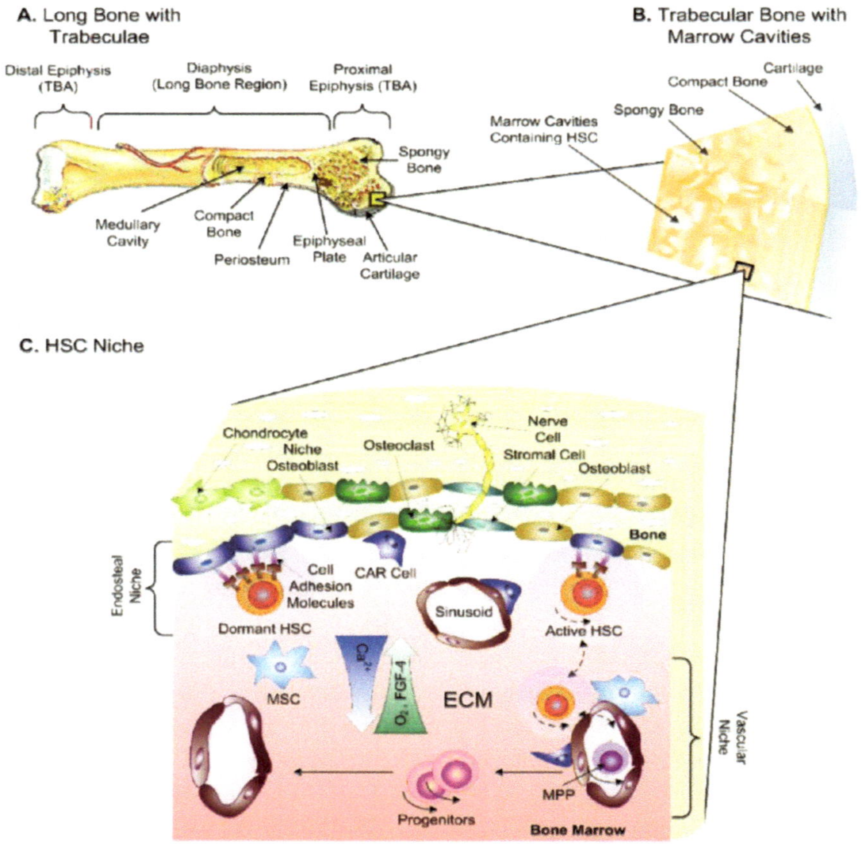

FIGURE 4.1 Cellular and molecular architecture of hematopoietic stem cell niche in the bone marrow.

HSC niche is comprised of two distinct cellular compartments, namely endosteal and vascular niche. ECM in the BM-HSC niche provides structural support and biological cues for HSC quiescence, proliferation, differentiation.

Abbreviations: ECM: Extracellular Matrix, MPP: Multipotent progenitors, MSC: Mesenchymal stem cells, HSC: Hematopoietic stem cells

Source: Reprinted with permission from Wasnik et al. (2012). © Elsevier.

4.2.1 THE ENDOSTEAL NICHE, THE VASCULAR NICHE, AND THEIR INTERACTION

HSC are arranged spatially within the two niches; the endosteal niche comprises the slow dividing/quiescent cells and the vascular niche comprises the more differentiated cells (Calvi et al., 2003). The endosteal/osteoblastic niche maintains the HSC in a quiescent state by providing anchorage through cell-cell interactions and inhibits differentiation of the HSC (Hosokawa et al., 2010). It is shown that HSC isolated from the endosteum and not from any other part of the BM, are more efficient for *in vivo* and *in vitro* hematopoiesis (Arai et al., 2009). The HSC regulation within the endosteal niche is very complex. A diverse set of cellular, acellular, growth factor, and cytokine interactions ensure that the HSC are maintained in quiescence. This multiparameter regulation of HSC ensures that the HSC pool is always maintained and the disturbance of one factor does not lead to a total loss of HSCs.

The vascular niche is enriched with thin-walled blood vessels; the vascular sinusoids, and adjacent to these sinusoids are found the hematopoietic stem and progenitor cells (HSPC). These cells are in an active cell cycle and display a lower repopulation capacity. The vascular niche is more oxygenated and heterogeneous than the less oxygenated quiescent endosteal niche (Rafii et al., 1997; Zhao et al., 2016). The surrounding niche components, including HSC, maintain the integrity of these sinusoidal cells. HSC moves into and out of the BM via the sinusoids, and the sinusoidal cells provide cues for survival, self-renewal, and differentiation. In situ imaging of HSC within the endothelial niches show that HSC are in close contact with the sinusoidal endothelium. Consistent with this, endothelial cells isolated and cultured from the heart and liver supported *in vitro* HSC activity (Li et al., 2004).

Based on these observations, it was proposed that while the endosteal niche conditions favor quiescence and maintain HSC, the vascular niche conditions are more supportive for HSC proliferation (Kopp et al., 2005). Though the two niches differ in their major cellular and physiological compositions, the cell types, such as reticular and stromal cell populations are the standard components in both and thus could share features. For example, CXCL12-abundant reticular cells (CAR cells), are present in both the endosteal and the vascular niches and are shown to maintain the stem cell pool as well as regulate HSC trafficking by CXCL12/CXCR4 axis (Sugiyama et al., 2006). Similarly, the endothelial cells, are scattered throughout the BM and while they are a major cellular component of the vascular niche, they can be found in the endosteal region as well, where they could play important role in osteogenesis and vasculogenesis (Coultas et al., 2005; Nakamura et al.,

2010). Thus, the *in vivo* maintenance, proliferation and differentiation of the HSC population throughout life is regulated by the intrinsic components of the endosteal and vascular niche of the BM.

4.2.2 SOLUBLE AND EXTRA CELLULAR MATRIX COMPONENTS OF THE HSC NICHE

Apart from the cellular components, the soluble secretory cytokines, growth stimulators, or inhibitors that are secreted by the BM cells influence the fate of the HSC and HSPC. The abundant growth factors and cytokines such as SCF, TPO, IL-6, Flt3-L, VEGF, FGF, PDGF, TGF-β; chemokines such as CXCL12 and CXCL2; Granulocyte colony-stimulating factor (G-CSF), Ang-1, OPN, and VCAM together with their receptors activate or repress multiple cellular pathways (reviewed in Wasnik et al., 2012). For instance, the Wnt and Notch signaling act synergistically where notch plays a role in maintaining the undifferentiated state of HSCs and Wnt enhances the self-renewal and maintains the viability of the HSCs (Duncan et al., 2005; Anton et al., 2007); similarly, the fibroblast growth factor (FGF) signaling is shown to preserve the expansion of HSCs *in vivo* (Yeoh et al., 2006). Interestingly, these elements can induce these effects via binding to the heparan sulfate (HS) chains of the PGs directly or via sulfations that induce specific remodeling of the HS chains for effective interactions (Turnbull et al., 1992; Pellegrini et al., 2000; Morimoto-Tomita et al., 2002; Yeoh et al., 2006). The ECM surrounding the cells provides structural integrity, anchorage, and binding of these above-mentioned growth factors and regulates HSC differentiation or maintenance (Sugiyama et al., 2006). There are three main classes of ECM proteins: (i) the structural proteins (collagen and elastin); (ii) specialized proteins (fibronectin and laminin); and (iii) the PGs. Due to their unique location advantage on the cell membrane and in the ECM, PGs, and GAGs can regulate many processes.

4.3 PROTEOGLYCANS AND GLYCOSAMINOGLYCANS

The proteoglycan is a large family of genes, at least 43 of which are known, and their protein products may be classified as intracellular, cell surface, pericellular/basement membrane, and extracellular based on their location and availability (Iozzo et al., 2015). The largest group is the extracellular PGs, categorized into three subgroups: Hyalectins (hyaluronan and lectin binding), Small Leucine-Rich Proteoglycans (SLRPs), and the Spock family. PGs are complex macromolecules with a multidomain protein core attached to different GAG chains. PGs could carry a similar type of GAG side chain or a hybrid of the two different types of GAG chains depending on the time of expression and tissue location. GAGs occur either as free or bound to proteins (as PGs) and regulate a wide range of cellular processes such as signaling, cell proliferation, and cell adhesion by interaction with the extracellular ligands, receptors, and matrix components (Linhardt et al., 2004). Hyaluronan is the most widely available, non-sulfated GAG to which multiple PGs bind to form large aggregates and gives tensile strength to tissues. Based on the sulfated glycan side chains attached, PGs may be identified as heparin/heparan sulfate proteoglycans (HSPG), chondroitin sulfate/dermatan sulfate proteoglycans CSPG/DSPG, and keratan sulfate proteoglycans (KSPG) (Esko et al., 2009). Apart from the glycanated PGs, few SLRP family proteoglycans do not have any GAG side chains attached. However, they are categorized under SLRPs based on their protein homology and domain structure. PGs may be classified based on either their location or their side chain attached, as listed in Table 4.1.

TABLE 4.1 Classification of PGs based on GAG Side Chains and Their Location

GAG	Proteoglycan	Location/Group/Class	References
Heparan sulfate	Syndecan-2 and Syndecan-4	Cell surface – single-pass transmembrane	De Rossi et al. (2014)
	Glypicans 1-6	Cell surface – GPI anchored	Filmus et al. (2014)
	Perlecan/HSPG2	Basement membrane	Iozzo et al. (2001)
	Collagen XVIII	Basement membrane	Dong et al. (2003)
Chondroitin sulfate	CSPG4/NG2	Cell surface – single pass transmembrane	Price et al. (2011)
	Phosphocan	Cell surface – single pass transmembrane	Maurel et al. (1994)
	Versican	Extracellular-hyalectins	Zimmermann et al. (1989)
	Neurocan	Extracellular-hyalectins	Retzler et al. (1996)
Chondroitin/ dermatan sulfate	Decorin	Class I SLRPs – extracellular	Krusius et al. (1986)
	Biglycan	Class I SLRPs – extracellular	Fisher et al. (1989)
	Epiphycan	Class III SLRP – extracellular	Johnson et al. (1997)
	Osteoglycin	Class III SLRP – extracellular	Funderburgh et al. (1997)

TABLE 4.1 *(Continued)*

GAG	Proteoglycan	Location/Group/Class	References
Keratan sulfate	Fibromodulin	Class II A SLRP – extracellular	Antonsson et al. (1989)
	Lumican	Class II A SLRP – extracellular	Chakravarti et al. (1995)
	Keratocan	Class II-B SLRP – extracellular	Liu et al. (2003)
	Osteoadherin	Class II-C SLRP – extracellular	Wendel et al. (1998)
	Chondroadherin	Class IV SLRP – extracellular	Haglund et al. (2011)
Heparan and chondroitin sulfate mixed	Serglycin	Cytosolic-secretory granules	Douaiher et al. (2014)
	Syndecan-1 and Syndecan-3	Cell surface – single-pass transmembrane	Stanford et al. (2009)
	Beta glycan/ TGFβIII receptor	Cell surface – single-pass transmembrane	Lopez-Casillas et al. (1991)
	Agrin	Basement membrane	Nitkin et al. (1987)
	Collagen XV	Basement membrane	Eklund et al. (2001)
	Testican 1, 2, and 3	Spock family – extracellular PGs	Alliel et al. (1993)
	Aggrecan	Extracellular-hyalectins	Wight et al. (2014)
No GAG side chains (proteins not glycanated but categorized under SLRP)	Asporin	Class I SLRP – extracellular	Lorenzo et al. (2001)
	ECM2		
	ECMX		
	PRELP	Class II B SLRP – extracellular	Bengtsson et al. (1995)
	Opticin	Class III SLRP – extracellular	Monfort et al. (2008)
	Nyctalopin, tsukushi	Class IV SLRP – extracellular	Bech-Hansen et al. (2000)
	Podocan, podocan-like	Class V SLRP – extracellular	Ross et al. (2003)
Occasional glycated PGs	Brevican	Extracellular-hyalectins	Yamada et al. (1994)
	CD 44	Transmembrane	Purushothaman et al. (2014)
	Neuropilin-1	Single-pass transmembrane	Fukasawa et al. (2007)

4.3.1 STRUCTURE AND BIOSYNTHESIS

PGs structure is open to extensive modulation during cellular expression. GAG chains are the primary source of structural variation that vary in number, length, and pattern of sulfation resulting in the switching of different chain types with different properties and selective expression of sulfated chain sequences with specific functions (Hardingham et al., 1992). The HSPGs and the CSPGs form the largest group of PGs whose biosynthesis pathway is very complicated, involving several enzymes at each stage and in different tissues enabling their wide variation in structure and functions.

4.3.1.1 HEPARAN SULFATE BIOSYNTHESIS

HS chains are made up of repeating units (50–200) of D-glucuronic (GlcA) or L-iduronic (IdoA) acid linked to 2-amino-2-deoxy-D-glucopyranose (glucosamine, GlcN). The sulfate groups may be attached to these disaccharide units at C-3 and C-6 of Glc-N or C-2 of the uronic acid. Glc-N amine's function may be changed to sulfamate, acetamide or remain unsubstituted (Bedini et al., 2019). Biosynthesis of HS in humans occurs primarily in the Golgi apparatus and involves about 20 different HS biosynthetic enzymes. These enzymes are mainly the type II transmembrane proteins that make highly heterogeneous structures by initiating, extending, and modifying the HS chains (Papy-Garcia et al., 2017). Biosynthesis is initiated by Xylosyltransferase (XylT), where UDP-Xyl serves as a donor, and the acceptor is the PG core protein. There is no specific amino acid sequence for xylosylation except a glycine residue, which constantly lies at the carboxy-terminal targeting serine residues. This is followed by the formation of linkage tetrasaccharide, glucuronic acid-galactose-galactose-xylose catalyzed by the other glycosyltransferases (Gal transferases I, II, and GlcA transferase). The enzymes exostosin (EXT1 and EXT2) further extend the chain by alternating GlcA β1-4 and *N*-acetyl-D-galactosamine (GlcNAc)1-4 residues. Finally, the various HS sulfotransferases (HSSTs) carry out the sequential sulfation's at different residues giving rise to the multiple variations in the final sulfated structure of the PG that imparts unique interaction and function to the HSPG (Papy-Garcia et al., 2017).

4.3.1.2 CHONDROITIN SULFATE BIOSYNTHESIS

Biochemical characterization of purified PGs showed that CSPGs are synthesized along with HSPGs. In CS, the disaccharide is composed of GlcA and GlcNAc, which are joined by alternating β-1→3 and β-1→4 glycosidic bonds (Mikami et al., 2013). CS GAG chain is generated when β–1,4-linked GalNAc is added to the linkage sequence at the non-reducing terminal GlcA unit (Li et al., 2016). By the action of GalNAc- and

GlcA-transferases, polymerization comprises the step-by-step alternating transfer of GlcA and GalNAc residues to the non-reducing end of the nascent polymer chain from the activated precursor. CSPGs are predominantly large biomacromolecules constituting 1 to 100 CS side chains that could be sulfated by highly specific sulfotransferases to different extents (Mikami et al., 2013), thereby imparting the variations in the structure and functions of CSPG.

4.4 PROTEOGLYCANS OF THE HEMATOPOIETIC NICHE AND THEIR BIOLOGICAL FUNCTIONS

The PGs of the HSC niche support the self-renewal and differentiation of human and murine hematopoietic stromal cells. The HSC retention within the niche is important for maintaining stemness and inhibiting excessive differentiation. When treated with a glucocorticoid, BM stromal cells show binding affinity to the blast forming cells and myeloid progenitors compared to non-treated stromal cells. This binding of early HSC progenitors to stroma was observed to depend on HSPG. Biochemical and functional analysis revealed that glucocorticoid-treated stromal cells secrete HSPG with a higher degree of sulfation than non-treated stromal cells (Siczkowski et al., 1992). Differences observed in the degree of HS sulfation and their effects on the HSC binding suggest that the structural variation in HS can determine the cell-specific binding to various HSC/HSPCs. BM-derived HSPG has been shown to mediate hematopoietic progenitor cells' adhesion to various cytokines (Bruno et al., 1995). The cooperative interaction of HSPG with growth and ECM proteins augments progenitor cell localization within the vascular and endosteal niches (Bruno et al., 1995; Gupta et al., 1998). HSPGs play a crucial role in HSC niche maintenance (Sarrazin et al., 2011). Deletion of one of the HSPG biosynthesis enzyme EXT1 in osteolineage cells leads to defective cell retention and adhesion, increased circulation of HSC/HSPC in spleen and peripheral blood, indicating that HSPGs are a critical structural component of the BM microenvironment and regulate the HSPC maintenance (Hoggatt et al., 2014; Saez et al., 2014). The major functions of the important PGs of the hematopoietic niche is given in table 4.2. A graphical representation of the cellular/extracellular location of PGs of the hematopoietic niche is depicted in Figure 4.2.

TABLE 4.2 Major Proteoglycans of the HSC Niche and Their Functions

Proteoglycan	Function in Hematopoietic Stem Cell Niches	References
Syndecan	HSC proliferation.	McDermott et al. (2007)
Glypican	Binding to HSC and bone regulator Wnt, Hedgehog, FGF, and BMPs.	Viviano et al. (2005)
Perlecan	HSC proliferation.	Noonan et al. (1991)
Agrin	Myeloid and erythroblast development.	Mazzon et al. (2012)
Serglycin	Lymphoid cellular adhesion.	Toyama-Sorimachi et al. (1997)
Neuron-glial antigen 2	HSC quiescence.	Smith et al. (1996)
Betaglycan	Co-receptor for TGF-β. MSC stemness. Erythropoiesis.	Dong et al. (2006)
Versican	Lymphoid and myeloid cell adhesion.	Wight et al. (2014)
Biglycan	BM stromal cell functions, ossification.	Kiel et al. (2007)
Decorin	Extracellular matrix assembly. Collagen fibril formation.	Ichii et al. (2012)
CD44	Cell adhesion – HSC transportation, homing, and colonization-leukemia-initiating cells.	Wagner et al. (2008)

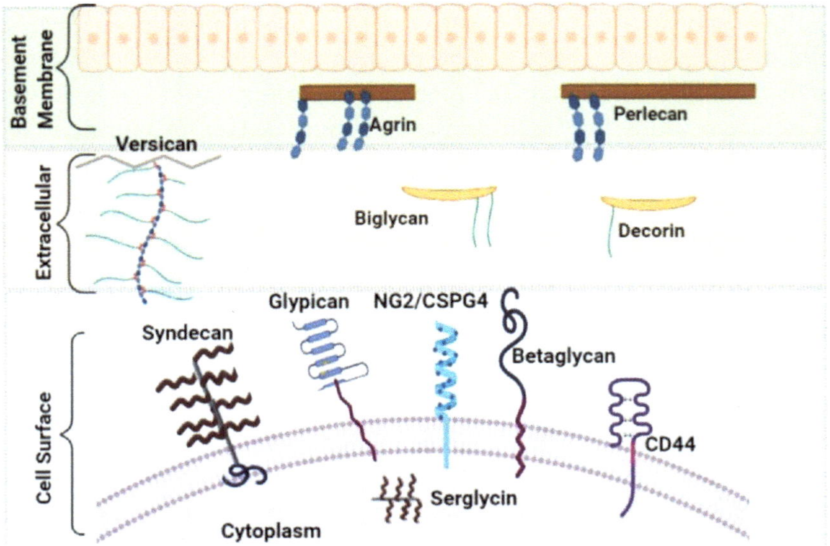

FIGURE 4.2 Cellular and extracellular location of the major PGs of the HSC niche. PGs are located in the cytoplasm, cell surface (transmembrane or GPI anchored), extracellular space, and basement membrane.

Source: Figure created with BioRender.com.

4.4.1 HEPARAN SULFATE PROTEOGLYCANS: SPECIFIC FUNCTIONS

HS mediates adhesive interactions between human HSPCs and the niche's ECM components (Gordon et al., 1988). The PGs carrying either pure HS side chain or a combination of GAGs with predominantly HS are grouped under HSPG. HSPGs are essential components of the HSC niche and are involved in HSC and stromal cells interaction, thus playing a critical role in the HSC niche maintenance (Drzeniek et al., 1999; Sarrazin et al., 2011). HSPGs form the most abundant and significant class of PGs of the HSC niche. Due to their sizeable structural variation, HS can bind to at least 400 different types of proteins known as heparin-binding proteins (HBPs). The HBPs include the chemokines, cytokines, cell adhesion molecules, enzymes, and growth factor receptors (Kreuger et al., 2006; Lindahl & Li, 2009) through which the PGs regulate the multiple cellular processes. Predominant HSPGs present in the ECM of the BM stromal cells are the syndecan (transmembrane PGs), glypican (GPI anchored PGs), perlecan, and agrin (basement membrane PGs).

4.4.1.1 SYNDECAN

Syndecans are transmembrane cell-surface HSPGs that regulate cell-cell signaling pathways and cell-matrix interactions. Syndecans regulate cell proliferation, growth, and adhesion through interactions with neighboring proteins within the plasma membrane or cytoplasmic interface. Syndecan-1 and syndecan-3 are expressed on the surface of $< 10\%$ of murine HSPCs, whereas syndecan-4 is expressed on 50% of lineage negative (Lin^-) progenitor cells and 98% of $c\text{-}Kit^+Sca\text{-}1^+Lin^-$ (KSL) HSPC, KSL short-term HSCs, and 100% long-term HSCs (Termini et al., 2018). Loss of Syndecan-2 (*sdc2*) disrupts the earliest events of definitive hematopoiesis. Syndecan-2 is expressed in 36% of long-term HSCs that are $CD34^-CD48^-CD150^+$ suggesting that syndecan-2 expression marks a highly proliferative population of HSCs with increased multilineage repopulating capacity. Syndecan-2^+ HSCs can be readily isolated to enhance the efficacy of hematopoietic cell transplantation (Termini et al., 2018). As mentioned earlier, syndecan-1 is expressed by a subset of HSCs, affecting HSC function and their susceptibility to malignant transformation. Syndecan/CD138 null mice have defective myeloid cell proliferation and exhibit excessive extramedullary hematopoiesis. One of the central cellular mechanisms governing HSPG interaction with various ECM structural components is regulated by Integrins. Syndecan 1 is shown to enhance the binding of integrin $\alpha 2\beta 1$ to the collagen matrix, thereby activating signaling pathways related to collagen matrix assembly, FAK signaling, MMP gene regulation, and cell cycle regulation (Vuoriluoto et al., 2008). Syndecan/CD138 knockout (KO) mice have reduced numbers of HSC in the BM compared to the wild type (WT) mice, and HSCs from these KO mice are functionally defective in competitive reconstitution assays in WT recipients (McDermott et al., 2007). Syndecan is an integral regulator of B-lymphocytes-matrix interaction. Experimental results obtained with multiple myeloma cell lines revealed that syndecan regulates the high-affinity cell-binding specifically to collagen-I, but not type IV collagen, laminin, and fibronectin (Sanderson et al., 1992). During B cell differentiation, the matrix binding properties of syndecan control the B cell localization within the specific microenvironments (Sanderson et al., 1992). The expression of syndecans/CD138 in multiple myeloma cells is inhibited upon culturing with bone stromal cells, which leads to increased activity of kinases involved in cellular adhesion (Fuhler et al., 2010).

4.4.1.2 GLYPICAN

Glypicans are bound to the outer surface of the plasma membrane by a glycosyl-phosphatidylinositol (GPI) anchor and have six members in their family (GPC 1-6). Glypican's primary function is to bind Wnt, Hedgehog, FGF, and bone morphogenic protein (BMP) (Vuoriluoto et al., 2008). The stimulatory or inhibitory activity of glypican on signaling is dependent on the context. For example, glypicans facilitate and stabilize the interaction of Wnt with their signaling receptors, the frizzled proteins and thus show a stimulatory mechanism. On the other hand, Hedgehog protein signaling during development is inhibited by Glypican 3 (GPC-3) by competing with Patched, the Hedgehog receptor (Filmus et al., 2008). GPC-3 functions in hematopoiesis were demonstrated by studies on *Gpc-3* deficient animals, which exhibited abnormally delayed bone growth, specifically in the tibia and trabecular regions. The delay in average bone growth was associated with delayed recruitment of differentiated osteoclasts, which regulates the normal bone resorption, an essential step of bone remodeling. A defect can explain the reduction in osteoclast production in lineage-specific differentiation of the hematopoietic system affecting the typical progenitors of both monocyte/macrophages and neutrophils (Viviano et al., 2005). These indicate the direct influence of GPC-3 on hematopoiesis-mediated skeletal development.

In terms of functions of GPC-3 in HSC/HSPC homing, loss of GPC-3 was associated with an increase in numbers of circulating hematopoietic progenitors. It has been reported that GPC-3 regulates the cell retention in the niche by inhibiting the peptidase activity of CD26, the enzyme that cleaves and inactivates CXCL12 to deplete chemokine activity (Khurana et al., 2013). Previously, much evidence supported the role of aorta-gonad-mesonephros (AGM) as regulatory in supporting primitive hematopoiesis. Glypican-1 (*Gpc-1*) was a highly expressed gene in AGM region-derived stromal cells. Functional properties of non-supportive stromal cells were modified with overexpression of GPC-1, making them HSC/HSPC growth supportive, which indicates GPC-1 plays essential role in creating stem cell supporting niche (Nagao et al., 2008). These studies show that the GPC family plays a significant role in the HSC niche and the hematopoietic system by impacting osteogenesis.

4.4.1.3 PERLECAN/HSPG2

During the proliferation and differentiation of HSPCS, the adhesion of these cells to the BM niche has to be a tightly controlled process. The interactions between the various adhesion and non-adhesion molecules of the stromal cells and their ECM, modulate the adhesive strength, thereby regulating the egress of mature cells. Perlecan (HSPG2) is one of the vascular ECM and basement membrane's significant components and is abundantly expressed by the BM mesenchymal tissues and cultured endothelial cells (Noonan et al., 1991; Nugent et al., 2000). Expression of HSPG2 is often associated with cell invasion potential, adhesion, proliferation, angiogenesis, and differentiation, such as observed in prostate cancer (Knox et al., 2006; Grindel et al., 2018). Regarding the hematopoietic niche, perlecan/HSPG2 is actively synthesized by the BM stromal cells, the major cellular components of the HSC niche, and critical regulators of vascular homeostasis (Nugent et al., 2000). Perlecan directly or indirectly influences the HSC differentiation. For example, perlecan binds and stores the morphogenic factors and heparin-binding growth factors and keeps these factors ready for release when required by hematopoietic progenitor cells (Knox et al., 2006; Decarlo et al., 2012). To support the growth factor binding property, a series of *in vitro* experiments by the semi-solid colony-forming assays (CFUs) showed that plastic culture dishes coated with perlecan bind to exogenously added GM-CSF; once bound, the GM-CSFs are presented to the HSCs (Forsten et al., 1997; Nugent et al., 2000). Endothelial cells derived perlecan exhibit the potent inhibitory activity of smooth muscle cell proliferation. Enzymatic treatment with heparinase can diminish the perlecan antiproliferation functions, indicating perlecan's crucial roles in maintaining the vascular architecture (Nugent et al., 2000).

A prominent study suggested the anti-adhesive and repulsive property of perlecan core protein towards hematopoietic cells, such as leukemic cell lines (Kgla, K562, HL60, and U937), BM-MNC, and peripheral blood-derived lymphocytes, while displaying adhesion affinity to fibroblast and endothelial cells (Klein et al., 1995; Lorente-Gea et al., 2017). Integrins mediate the perlecan adhesive interaction with cells, whereas the antiadhesive action of perlecan is concentration-dependent. *In vitro* evidence suggests that when perlecan is mixed with the BM-derived matrix protein fibronectin, the capability to adhere to a monocytic cell line diminishes with a gradual increase in perlecan concentration in the perlecan: fibronectin mixture. Similar experimental results were obtained when perlecan was combined with collagen type VI, another HSC cell adhesion matrix protein. Thus, perlecan may modulate a specific microenvironment depending on the prevailing conditions and components involved (Klein et al., 1995). Furthermore, perlecan interacts with other ECM molecules, for example, Collagen-X formed by hypertrophic chondrocytes at the chondro-osseous junction of the spongeous bone, thereby stabilizing the structural architecture of the bone microenvironment. Mutations in either perlecan or collagen-X indirectly attenuate hematopoiesis via disruption of the endochondral ossification processes, resulting in marrow aplasia and lymphopenia (Rodgers et al., 2008; Sweeney et al., 2011). In the context of Perlecan's role on HSC proliferation, the substantial evidence comes from studies with Trol (Terribly Reduced Optic Lobes). Trol is the *Drosophila melanogaster* homolog of the vertebrate protein perlecan. Loss of Trol in the lymph gland leads to topographical changes in the ECM, which negatively regulates hematopoiesis by inhibiting cell proliferation (Grigorian et al., 2013).

4.4.1.4 AGRIN

Agrin, is a HSPG that is critical in controlling the survival and proliferation of HSCs. Agrin is one of the essential components of the endosteal niche as it is expressed by both MSC and osteoblastic cells (Calvi, 2011). Mazzon and group showed that agrin's genetic ablation caused lethal neuromuscular defects that can be rescued by overexpression of agrin in skeletal muscle cells receptors. Agrin deficient MSCs do not support HSPC proliferation, and agrin KO mice exhibit defective hematopoiesis and immature apoptosis of HSPCs (Mazzon et al., 2011). $Lin^{-}Sca1^{+}c\text{-}Kit^{+}$ (LSK) cells express the receptor for agrin (α-dystroglycan (α-DG), and agrin is shown to regulate monocyte and macrophage development via this receptor (Mazzon et al., 2011, 2012). The α-DG receptor is also expressed on human $CD34^+$ cells (Steidl et al., 2004), suggesting a similar role for agrin in human hematopoiesis. It was recently demonstrated that agrin is a novel regulator of Ephrin (Eph) receptor signaling and regulates cell-cell adhesion and red cell development within the erythroid niche. Eph receptor activation is induced by agrin in erythroblasts, which leads to the activation of master cell adhesion regulator integrin α5β1v (Anselmo et al., 2016).

4.4.2 CHONDROITIN SULFATE PROTEOGLYCAN: SPECIFIC FUNCTIONS

The PGs carrying either pure CS side chain or a combination of GAGs with predominantly CS are grouped under CSPGs. The CSPGs of the niche include serglycin (intracellular); CSPG2/NG2, betaglycan (transmembrane); versican (extracellular); biglycan, and decorin (SLRP Class I); and CD44 (occasionally glycated PG).

4.4.2.1 SERGLYCIN

Serglycin, also known as hematopoietic PG core protein or secretory granule PG core protein (Kolset et al., 1990; Kolset et al., 2008), is the only known intracellular PG. The GAG heparin is generally associated with serglycin in intracellular granules of mast cells. The CS-modified serglycin acts as a ligand for CD44 and contributes to lymphoid cells' cellular adhesion function and activation. Extensive molecular assays, including immunoprecipitation using antibodies against the core protein of CSPG serglycin, revealed that cells such as hematopoietic, lymphoid, myeloid, megakaryocytes, platelets, and specific tumor cells express different types of serglycin that have CD44 binding capacity (Toyama-Sorimachi et al., 1997). Serglycin may serve as a primary ligand for CD44 in various events in the lymphohematopoietic system. The primary function of serglycin is to form storage granules, store critical regulators of inflammation, and play an essential role in inflammatory reactions (Kolset et al., 2008). With the advancement of molecular technologies for marker analysis, serglycin has been detected as a marker for hematological malignancy, specifically for acute myeloid leukemia (AML), anaplastic large cell lymphoma (ALCL), natural killer cell (NK cell) lymphoma, and acute lymphoblastic leukemia (ALL) (Niemann et al., 2007).

Genetically modified mouse strains that lack serglycin expression demonstrated that it is essential for HSC storage molecules (Schick, 2010). The serglycin deficient mice studies showed that platelet aggregation is also defective, which indicates that loss of serglycin is responsible for the defects. Platelets interact with many cell types, such as HSC, macrophages, monocytes, leukocytes, and tumor cells during various cellular coagulation and inflammation processes. The platelet interaction with various cell types is regulated by platelet-derived protein PF-4 binding to CSPG (Kolset et al., 1996; Schick, 2010). In serglycin KO mice, the expression of PF-4 is drastically reduced in the platelets (Woulfe et al., 2008). An educated guess would be that the loss of serglycin could affect these and other factors in the α-granules of platelets.

4.4.2.2 NG2/CSPG4

Neuron-glial antigen 2 (NG2), also known as chondroitin sulfate proteoglycan 4 (CSPG4), is a regulator of angiogenesis. NG2/CSPG4 is a surface type I transmembrane core PG. The normal human hematopoietic cells do not express NG2; however, it is expressed by AML blasts. The expression of NG2 on the leukemic cell surface raises the possibility that the NG2 molecule is generally expressed at some stage of normal human embryonic hematopoietic development (Smith et al., 1996). The HSC niche's major structural and functional constituents are BM mesenchymal cells, which express NG2 (Kozanoglu et al., 2009). In vascular niches, NG2 is expressed by microvascular pericytes. Pericyte-derived NG2 forms a complex with integrin $\alpha 3\beta 1$ and galectin expressed by endothelial cells to initiate endothelial cell migration (Fukushi et al., 2004). By utilizing the high throughput whole-mount confocal microscopy, the three-dimensional architecture of the BM HSC niche revealed the multicellular interaction of stromal cells, vascular cells, and HSCs. The studies showed that quiescent LT-HSCs are located near the small arterioles in the BM endosteal niche. The arterioles pericytes exclusively express NG2/CSPG4. Genetic disruption of NG2 alters the slow-cycling property of LT-HSC, indicating the crucial role of pericyte-derived NG2 in maintaining the HSC quiescence. Extending to the arterioles derived NG2, it has been reported that depletion of $NG2^+$ arteriolar niche cells selectively depletes Von Willebrand factor (vWF) lymphoid-biased HSCs. Additionally, another subset, the $nestin^+NG2^+$ pericytes, promotes HSC expansion (Pinho et al., 2018).

4.4.2.3 BETAGLYCAN

Betaglycan is a transmembrane CSPG/HSPG, also known as Transforming growth factor-beta receptor III (TGFβR3/TBRIII). One of the most important functions of betaglycan is to bind various members of the TGF-β family and the β-FGF family. TGF-β is a pleiotropic factor that plays a crucial role in hematopoiesis (Dong et al., 2006) and skeletal development (Balooch et al., 2005). Betaglycan is the most abundant co-receptor for TGF-β and has two separate TGF-β binding sites (Boyd et al., 1990; Massague et al., 1990). Previous reports suggest that betaglycan has both positive and negative regulation activity for TGF-β. Betaglycan can be found on the plasma membrane or in the extracellular domains, from where it is secreted as a soluble form of PG. TBRIII, in the soluble proteoglycan form, has been shown to inhibit the TGF-β functions (Sun et al., 1997).

Betaglycan is expressed by the normal human megakaryocytes (Schick, 2010). In terms of the association with hematopoiesis, it has been reported that betaglycan deficient mice have lethal consequences due to failure in normal erythropoiesis (Stenvers et al., 2003). Further evidence of betaglycan's indirect roles comes from the study that shows that it has regulatory effects on MSC cellular properties. The deletion of betaglycan enhances TGF-β signaling in MSC, increases stemness, and inhibits differentiation into osteoblasts. Notably, betaglycan deficient mice show complexity in bone development due to defective osteoblastic differentiation (Cook et al., 2019). Furthermore, endoglin (CD105), which is considered to be a surface marker for receptor-targeted gene transfer into human long-term repopulating HSCs (Kays et al., 2015) together with betaglycan, acts as a TGF-βR accessory molecule; however, its functions in the receptor-ligand interactions are still poorly understood (Parker et al., 2003).

4.4.2.4 VERSICAN

Versican, an extracellular PG, is expressed by cultured fibroblasts, smooth muscle cells, cartilage, and blood vessels. Versican exists in the long-term culture of mouse BM cells (Senda et al., 2016). As a structural component of ECM, versican plays regulatory roles in cellular adhesion, migration, and differentiation. The cell-binding functions of versican are regulated mainly by interaction with integrin and non-integrin molecules. Versican interacts with lymphoid and myeloid cell types and contributes to cell adhesion.

Versican binds to hyaluronan and other ECM molecules, such as fibronectin and several chemokines, thereby influencing leucocyte function. During inflammation process, the leukocytes migrating from the inner blood vessels into the surrounding damaged tissues, encounter a matrix highly enriched in versican, which can interact with many receptors on the surface of immune cells, including CD44, P-selectin glycoprotein-1, and Toll-like receptors (Toyama-Sorimachi et al., 1997). The increased expression and secretion of versican by stromal cells leads to enhanced leukocyte adhesion to diminish the inflammation (Wight et al., 2014).

A recent study corroborated the versican-ECM interaction with vasculogenesis and primitive hematopoietic events. The authors showed a failure of vascular structure formation and initiation of hematopoiesis in versican deficient mice. Further analysis of hematopoiesis in wild type reveals co-regulation of versican and hyaluronan in FLK^+ cell type. FLK^+ cells are common progenitors of vascular and hematopoietic cells, migrating from the yolk sac to the mesoderm during development. However, in versican deficient mice, the expression and synthesis of hyaluronan in both yolk sac and embryo are drastically decreased, indicating the importance of versican in ECM, such as hyaluronan stabilization (Nandadasa et al., 2021).

4.4.2.5 BIGLYCAN

Biglycan is an SLRP class I PG found in ECM tissues, such as bone. The functional importance of biglycan in the HSC niche in BM has remained mostly unknown. Biglycan is a significant product of skeletal cells such as osteoblasts and chondroblasts. Biglycan deletion results in a reduced bone volume in terms of trabecular; however, it does not affect the number of HSCs detrimentally in the BM (Kiel et al., 2007). Another critical study showed that biglycan deficient mice show impaired BM stromal cell functions. BM stromal cells are considered one of the most critical structural and functional components of HSC niche in normal conditions. Biglycan deficient osteoblast cells (OBs) are low in alkaline phosphatase, a biomarker of normal osteoblastic cells, and lower CFU-Ob forming capacity. Also, biglycan deficient OBs exhibit abnormal and premature apoptosis and have defects in interacting with another important pleiotropic factor TGF-β, which leads to the decreased synthesis of collagen-I, a critical component of the endosteal niche (Bi et al., 2005). A novel biological function of the soluble PG biglycan, mediated by its interaction with Toll-like receptor-2, is also reported (Moreth et al., 2014). Biglycan regulates erythropoietin synthesis; abnormal biglycan expression leads to secondary polycythemia disease where a selective increase in circulating mature erythrocytes is seen (Frey et al., 2017).

4.4.2.6 DECORIN

Decorin is also an SLRP class I PG, which is multifunctional, and structurally resembles biglycan (Iozzo et al., 2010). Decorin is directly involved in modulating the signaling pathway of Wnt3a, thus, shaping the niche supportive of hematopoiesis. Decorin is synthesized by endothelial cells, osteoblasts, MSC (Ichii et al., 2012), and $CD169^{+}$ macrophages in the BM (Chow et al., 2011). *In vitro* cell culture experiments supported decorin as an essential component of stem and progenitor cell niches. Studies with animal models reveal that decorin deficient mice form defective collagen fibril and eventually develop lymphoma. In the decorin deficient mice, the hematopoietic niche in BM is unaffected in the cellular architecture of the HSC, lymphoid, myeloid, erythroid immature, and mature cells. However, a specific subset of HSC defined as $Lin^{-}Sca1^{+}c\text{-}Kit^{hi}CD34^{-}Flk\text{-}2^{-}CD150^{+}$ HSCs number was increased compared to the WT. These cells have lower colony-forming unit capacity when cultured in the standard CFU-s culture (Ichii et al., 2012). Decorin is also involved in the ECM assembly through interactions with ECM proteins, including certain collagen types (Kamma-Lorger et al., 2016), fibronectin (Schmidt et al., 1991), thrombospondins 1 (Winnemoller et al., 1992), and tenascin-X (Elefteriou et al., 2001). Apart from the regulatory role of collagen fibril formation, the decorin's primary function is to bind TGF-β. It is suggested that decorin may play a role in bone formation by binding to TGF-β and enhancing its bio-activities.

4.4.1.3 CD44

CD44, in its CSPG form, is an essential regulator of the hematopoietic niche and is responsible for HSC adhesion to mesenchymal stromal cells (Wagner et al., 2008). CD44 is a receptor for HA and is known as a part-time proteoglycan due to its existence as a non-glycated structure. Binding HSCs to the niche's cellular components is essential for maintaining stemness and prohibiting excessive cell differentiation. Side population cells (SP cells) correspond to a specific subpopulation of drug-resistant, quiescent HSPC cell populations.

SP phenotype acquisition/reactivation by circulating negative lineage cells depends on interactions with MSCs through VLA-4/α4β1-integrin and CD44 (Malfuson et al., 2014). Several studies also report that cellular adhesion of hematopoietic CFC to BM stromal cells is mediated through α4β1 and α5β1 integrins and the CD44 receptor. Integrins and CD44 receptors regulate the HSPC cell proliferation and differentiation via interactions between progenitors and stromal cells or ECM components. It is shown that, CD44 and α4β1 integrin contributes to the adhesion of HSC to the osteogenic niche and on using anti-CD44 and anti-α4β1, HSCs are dislodged from the BM niche (Christ et al., 2001; Zohren et al., 2008). CD44/CS proteoglycan and α4β1 integrins also mediate the CFC adhesion to stromal fibronectins and inhibit normal CFC proliferation (Lundell et al., 1997). HSC adhesion to the BM niche is mainly regulated by Hyaluronan, which also regulates CD44 biological activities. CD44 contributes to the generation of the Hyaluronan rich BM niches by regulating the matrix assembly. In addition, CD44 mediates the adhesion of HSC to the HA-coated endothelial cells to facilitate stem cell homing and colonization (Szczepanek et al., 2008; Kuhn et al., 2010; Winkler et al., 2012). Apart from HSC regulation, CD44 also regulates the transportation, homing, and colonization of leukemia-initiating cells in the HSC niche, as evident by the experiments with anti-CD44 antibodies. These anti-CD44 antibodies alter the niche's CD44 expression, resulting in increased cell differentiation (Zoller et al., 2007).

4.5 *IN VITRO* AND *IN VIVO* APPLICATIONS OF GLYCOSAMINOGLYCANS AND PROTEOGLYCANS

Glycan therapeutics refers to the application of PG core proteins, GAGs (natural or synthetic), or neo-proteoglycans to address issues concerning human health and disease. Heparin, the long-chain heterogeneous canonical GAG, remains the most widely known and applied glycotherapeutic (Paderi John et al., 2018). Due to the high-affinity interaction of heparin with the HBPs (chemokines, growth factors, and receptors), heparin is considered an anti-inflammatory, antiangiogenic, anti-metastatic, antibacterial, and antiviral agent (Hao et al., 2019). Heparin being a commercially available alternative to HS and owing to their structural similarity is often used in experimental work in place of HS (Mulloy et al., 2006). With its modifications that reduce its anticoagulation activity and improve its properties, such as half-life extension and oral availability, heparin has led to the development of several novel glycans apart from heparin derivatives (Magnani et al., 2009). For example, Roneparstat (SST0001) and Sevuparin are chemically treated heparin derivatives used in clinical trials for treating multiple myelomas and vaso occlusive crisis in sickle cell anemia, respectively (Galli et al., 2018). Heparin-inspired synthetic glycans such as Pixatimod (PG545) and Muparfostat (PI-88) are in clinical trials for treating solid tumors/hepatocellular carcinomas (Hirsh et al., 2001; Brennan et al., 2016; Rivara et al., 2016). Heparin is used in extracorporeal treatments such as hemodialysis and membrane oxygenation. Preclinical testing of extracorporeal devices having heparin has shown a reduction in a load of viruses such as Zika virus, CMV, and adenoviruses and most recently SARSCoV-2 as well in *in-vitro* studies (Seffer et al., 2020), suggesting the vast array of applications where the GAG heparin can be used. Other GAGs such as CS (SB-061) or carboxymethylated dextran sulfate (Caclicol20; Calcipliq20) are being tested for osteoarthritis and corneal or dermal wound repairs (reviewed in (Paderi John et al., 2018). GAGs such as HA, CS, and DS are also being explored in hydrogels, coatings, or scaffolds for improving the delivery and bioavailability of several growth factors and chemokines such as FGF-2, VEGF, PDGF, HB-EGF, BMP-2, NGF, opening up new possibilities for the therapeutic application of these GAGs (Hachim et al., 2019).

Neoproteoglycans or PG mimetic graft copolymers is an evolving field involving PGs, glycopolymers, and nanoparticle assemblies where GAG presenting materials are used in drug delivery, biosensors, and tissue engineering applications (Weyers et al., 2013). Neoproteoglycans with varying-sized GAG chains of HA, CS, and heparin were successfully tested for their ability to bind and deliver FGF-2 to MSCs, opening up new possibilities for the incorporation of GAGs into implants for efficient delivery of the growth factor(s) to the sites of healing (Place et al., 2014). The most widely appreciated therapeutic application of PGs is seen in wound healing where glypican 1, syndecan 4, decorin, and lumican together enhance angiogenesis in diabetic/ischemic and corneal wound healing (Das et al., 2016; Hill et al., 2018). The other PGs such as aggrecan, lubricin, perlecan, and fibromodulin have been used to treat osteoarthritis, cartilage, bone regeneration, etc. (reviewed in Walimbe et al., 2020).

4.5.1 THERAPEUTIC APPLICATIONS OF GAGs AND PGs IN HEMATOPOIETIC CELLS

The egress and mobilization of HSCs from BM is an essential process for HSC transplantation where the stromal-cell derived factor (SDF-1α)/

CXCR-4 mediated migration is crucial for homing and engraftment (Peled et al., 1999). Several studies report that exogenously added/*in vivo* administered HSPG can induce rapid cell mobilization in the peripheral blood. In contrast, overexpression of heparinase decreases the mobilization and increases cell accumulation in BM (Spiegel et al., 2008). G-CSF (the most commonly used mobilization agent) induces HSCs to be released from the niche and egress into the peripheral blood. It was observed that HSPC mobilization and an increase in $CD34^+$ cells were seen when heparin was used as an anticoagulant in leukapheresis (Frenette et al., 2000; Sweeney et al., 2002; Di Giacomo et al., 2012). Synthetic HS mimetics have also been tested for their ability to induce mobilization of HSPCs to mature leukocytes (Albanese et al., 2009). In another study, hyaluronan fragments were found to enhance the mobility effects of SDF-1α on $CD34^+$ HSPCs in an *in vitro* chemotactic system (Sbaa-Ketata et al., 2002). GPC-3 has been reported to enhance the HSC/HSPC mobilization from BM to the peripheral blood by activating the tissue-factor pathway inhibitor (TFPI). Treatment with TFPI enhanced $Gpc3^+$ HSPC mobilization specifically and not of $Gpc3^-$ HSPCs, which leads to the decrease in the primitive HSC pool in BM (Khurana et al., 2013). Synthetic HSPGs, known as mimetics, also induce the mobilization of HSC, HSPCs, B, and T lymphocytes for up to 3-fold in association with GM-CSF/AMD3100. The HSPG mimetics became an efficient therapeutic strategy in inducing HSC mobilization during hematopoietic reconstitution in animal models and maintained long-term hematopoiesis (Di Giacomo et al., 2012). In the case of non-malignant hematopoietic conditions, G-CSF administration and heparin led to the mobilization of distinct HSPC that had more rapid hematopoietic reconstitution and increased secondary transplant ability, suggesting that inhibiting HS production could be a way to minimize the toxicities of radiation or chemotherapy (Saez et al., 2014). Further supporting clinical evidence showed that patients receiving heparin for other conditions achieved a higher HSC mobilization than matched controls or those receiving plerixafor (Merter et al., 2017). To treat non-malignant hematologic diseases requiring allogeneic haploidentical or autologous stem cell transplantation, HS inhibition could help overcome the stem cell dose barriers. BM transplant conditions where the release of stem cells is needed, HSPGs limit this process. Using HS mimetics such as heparin induces a loss of function of HSPGs by competitive binding and leads to the release of the much-needed cells into circulation. Whereas, HSPCs accumulated in the BM, upon overexpression of the HS-cleaving enzyme heparanase (Spiegel et al., 2008).

Blocking of adhesion receptors by naturally sulfated glycans such as fucoidan induced mobilization of the HSPC, and the mechanism was proposed via the blockage of the endothelial selectin (Frenette et al., 2000). The mechanism was further deciphered and shown that upon treatment with synthetic sulfated glycans, the SDF-1 that is anchored to HSPG and thus sequestered in the BM is displaced and moves into the plasma, thereby creating a concentration gradient that aids in the migration of both WBCs and HSPCs into the peripheral circulation (Sweeney et al., 2002). Homing of HSCs to the BM has also been attributed to the expression of a glycoform of the HA receptor, CD44 (HCELL), that contains sialyl Lewis x (sLex+), which binds to endothelial E-selectin (Dimitroff et al., 2001). Furthermore, it was shown that human cord blood when treated with exogenous α1-3 fucosyltransferase VI for surface fucosylation led to increased cell surface sLex. This enabled better binding to endothelial E-selectin and the transplanted cells can home significantly better to the BM (Xia et al., 2004). These studies are very encouraging and open up newer possibilities of using synthetic and natural glycans to engineer cells *in vitro*/*in vivo* to achieve hematopoietic cells' mobilization, homing, and engraftment.

4.6 CONCLUSIONS

The HSC niche of the BM is a complex network of cellular components and their secreted factors, together with their signaling cascades balance and fine-tune the hematopoietic system. PGs and GAGs are an essential component of the HSC niche, regulating the maintenance, differentiation, and migration of HSCs. Due to their complex structure and with the advent of protein and DNA technologies that dominated the research, glycans have largely been ignored for many years. However, the last few years have seen a change in this trend, and GAGs and PGs are slowly resurfacing, showing promise in various therapeutic areas. As newer studies are beginning to unravel novel functions of the PGS in the hematopoietic niche, glycan therapy is slowly gaining importance. The future will see a rise in a number of novel therapeutics for better hematopoietic disease management and will aid in improved homing and engraftment in transplantation therapies. GAGs and PGs, either alone or combined with cell therapy, promise a new generation of regenerative treatment strategies.

KEYWORDS

- **chondroitin sulfate**
- **glycosaminoglycans**
- **hematopoiesis**
- **hematopoietic stem cells**
- **heparan sulfate**
- **proteoglycans**

REFERENCES

Albanese, P., Caruelle, D., Frescaline, G., et al., (2009). Glycosaminoglycan mimetics–induced mobilization of hematopoietic progenitors and stem cells into mouse peripheral blood: Structure/function insights. *Experimental Hematology, 37*(9), 1072–1083.

Alliel, P. M., Perin, J. P., Jolles, P., & Bonnet F. J., (1993). Testican, a multidomain testicular proteoglycan resembling modulators of cell social behavior. *European Journal of Biochemistry, 214*(1), 347–350.

Anselmo, A., Lauranzano, E., Soldani, C., et al., (2016). Identification of a novel agrin-dependent pathway in cell signaling and adhesion within the erythroid niche. *Cell Death and Differentiation, 23*(8), 1322–1330.

Anton, R., Kestler, H. A., & Kühl, M., (2007). β-Catenin signaling contributes to stemness and regulates early differentiation in murine embryonic stem cells. *FEBS Letters, 581*(27), 5247–5254.

Antonsson, P., Heinegård, D., & Oldberg, Å., (1989). The keratan sulfate-enriched region of bovine cartilage proteoglycan consists of a consecutively repeated hexapeptide motif. *Journal of Biological Chemistry, 264*(27), 16170–16173.

Arai, F., Yoshihara, H., Hosokawa, K., et al., (2009). Niche regulation of hematopoietic stem cells in the endosteum: The role of thrombopoietin/MPL signaling in the maintenance of quiescent hematopoietic stem cells. *Annals of the New York Academy of Sciences, 1176*(1), 36–46.

Balooch, G., Balooch, M., Nalla, R. K., et al., (2005). TGF-β regulates the mechanical properties and composition of bone matrix. *Proc. Natl. Acad. Sci., 102*(52), 18813–18818.

Bech-Hansen, N. T., Naylor, M. J., Maybaum, T. A., et al., (2000). Mutations in NYX, encoding the leucine-rich proteoglycan nyctalopin, cause X-linked complete congenital stationary night blindness. *Nature Genetics, 26*, 319–323.

Bedini, E., Corsaro, M. M., Fernández-Mayoralas, A., & Iadonisi, A., (2019). Chondroitin, dermatan, heparan, and keratan sulfate: Structure and functions. In: *Extracellular Sugar-Based Biopolymers Matrices* (pp. 187–233). Springer, Cham.

Bengtsson, E., Neame, P. J., Heinegård, D., & Sommarin, Y., (1995). The primary structure of a basic leucine-rich repeat protein, PRELP, found in connective tissues. *The Journal of Biological Chemistry, 270*(43), 25639–25644.

Bi, Y., Stuelten, C. H., Kilts, T., et al., (2005). Extracellular matrix proteoglycans control the fate of bone marrow stromal cells. *Journal of Biological Chemistry, 280*(34), 30481–30489.

Boyd, F. T., Cheifetz, S., Andres, J., Laiho, M., & Massagué, J., (1990). Transforming growth factor-beta receptors and binding proteoglycans. *Journal of Cell Science, 13S*, 131–138. https://doi.org/10.1242/jcs.1990.Supplement_13.12.

Brennan, T. V., Lin, L., Brandstadter, J. D., et al., (2016). Heparan sulfate mimetic PG545-mediated anti-lymphoma effects require TLR9-dependent NK cell activation. *The Journal of Clinical Investigation, 126*(1), 207–219.

Bruno, E., Luikart, S. D., Long, M. W., & Hoffman, R., (1995). Marrow-derived heparan sulfate proteoglycan mediates the adhesion of hematopoietic progenitor cells to cytokines. *Experimental Hematology, 23*(11), 1212–1217.

Calvi, L. M., (2011). Agrin complicates the niche. *Blood, 118*(10), 2641–2642.

Calvi, L. M., Adams, G. B., Weibrecht, K. W., et al., (2003). Osteoblastic cells regulate the haematopoietic stem cell niche. *Nature, 425*(6960), 841–846.

Chakravarti, S., Stallings, R. L., Sundar, R. N., Cornuet, P. K., & Hassell, J. R., (1995). Primary structure of human lumican (keratan sulfate proteoglycan) and localization of the gene (LUM) to chromosome 12q21. 3-q22. *Genomics, 27*(3), 481–488.

Chow, A., Lucas, D., Hidalgo, A., et al., (2011). Bone marrow CD169+ macrophages promote the retention of hematopoietic stem and progenitor cells in the mesenchymal stem cell niche. *Journal of Experimental Medicine, 208*(2), 261–271.

Christ, O., Kronenwett, R., Haas, R., & Zöller, M., (2001). Combining G-CSF with a blockade of adhesion strongly improves the reconstitutive capacity of mobilized hematopoietic progenitor cells. *Experimental Hematology, 29*(3), 380–390.

Cook, L. M., Frieling, J. S., Nerlakanti, N., et al., (2019). Beta glycan drives the mesenchymal stromal cell osteogenic program and prostate cancer-induced osteogenesis. *Oncogene, 38*(44), 6959–6969.

Coultas, L., Chawengsaksophak, K., & Rossant, J., (2005). Endothelial cells and VEGF in vascular development. *Nature, 438*(7070), 937–945.

Das, S., Monteforte, A. J., Singh, G., et al., (2016). Syndecan-4 enhances therapeutic angiogenesis after hind limb ischemia in mice with type 2 diabetes. *Advanced Healthcare Materials, 5*(9), 1008–1013.

De Carlo, A. A., Belousova, M., Ellis, A. L., et al., (2012). Perlecan domain 1 recombinant proteoglycan augments BMP-2 activity and osteogenesis. *BMC Biotechnology, 12*(1), 1–16.

De Rossi, G., Evans, A. R., Kay, E., et al., (2014). Shed syndecan-2 inhibits angiogenesis. *Journal of Cell Science, 127*(21), 4788–4799.

Di Giacomo, F., Lewandowski, D., Cabannes, E., et al., (2012). Heparan sulfate mimetics can efficiently mobilize long-term hematopoietic stem cells. *Haematologica, 97*(4), 491–499.

Dimitroff, C. J., Lee, J. Y., Rafii, S., Fuhlbrigge, R. C., & Sackstein, R., (2001). CD44 is a major E-selectin ligand on human hematopoietic progenitor cells. *The Journal of Cell Biology, 153*(6), 1277–1286.

Dimitroff, C. J., Lee, J. Y., Schor, K. S., Sandmaier, B. M., & Sackstein, R., (2001). Differential L-selectin binding activities of human hematopoietic cell L-selectin ligands, HCELL and PSGL-1. *Journal of Biological Chemistry, 276*(50), 47623–47631.

Dodge, G. R., & Heimer, R., (2001). Proteoglycans analyzed by composite gel electrophoresis and immunoblotting. *Methods in Molecular Biology, 171*, 149–158.

Dong, M., & Blobe, G. C., (2006). Role of transforming growth factor-β in hematologic malignancies. *Blood, 107*(12), 4589–4596.

Dong, S., Cole, G. J., & Halfter, W., (2003). Expression of collagen XVIII and localization of its glycosaminoglycan attachment sites. *Journal of Biological Chemistry, 278*(3), 1700–1707.

Douaiher, J., Succar, J., Lancerotto, L., et al., (2014). Development of mast cells and importance of their tryptase and chymase serine proteases in inflammation and wound healing. *Advances in Immunology, 122*, 211–252.

Drzeniek, Z., Stöcker, G., Siebertz, B., et al., (1999). Heparan sulfate proteoglycan expression is induced during early erythroid differentiation of multipotent hematopoietic stem cells. *Blood, 93*(9), 2884–2897.

Duncan, A. W., Rattis, F. M., DiMascio, L. N., et al., (2005). Integration of notch and Wnt signaling in hematopoietic stem cell maintenance. *Nature Immunology, 6*(3), 314–322.

Eklund, L., Piuhola, J., Komulainen, J., et al., (2001). Lack of type XV collagen causes a skeletal myopathy and cardiovascular defects in mice. *Proceedings of the National Academy of Sciences USA, 98*(3), 1194–1199.

Elefteriou, F., Exposito, J. Y., Garrone, R., & Lethias, C., (2001). Binding of tenascin-X to decorin. *FEBS Letters, 495*(1, 2), 44–47.

Esko, J. D., Kimata, K., & Lindahl, U., (2009). Proteoglycans and sulfated glycosaminoglycans. In: Varki, A., & Cummings, R. D., (eds.), *Essentials of Glycobiology*. Cold Spring Harbor Laboratory Press, New York.

Filmus, J., & Capurro, M., (2014). The role of glypicans in hedgehog signaling. *Matrix Biology, 35*, 248–252.

Filmus, J., Capurro, M., & Rast, J. G., (2008). Glypicans. *Genome Biology, 9*, 224.

Fisher, L. W., Termine, J. D., & Young, M. F., (1989). Deduced protein sequence of bone small proteoglycan I (biglycan) shows homology with proteoglycan II (decorin) and several nonconnective tissue proteins in a variety of species. *Journal of Biological Chemistry, 264*(8), 4571–4576.

Forsten, K. E., Courant, N. A., & Nugent, M. A., (1997). Endothelial proteoglycans inhibit bFGF binding and mitogenesis. *Journal of Cellular Physiology, 172*(2), 209–220.

Frenette, P. S., & Weiss, L., (2000). Sulfated glycans induce rapid hematopoietic progenitor cell mobilization: Evidence for selectin-dependent and independent mechanisms. *Blood, 96*(7), 2460–2468.

Frey, H., Moreth, K., Hsieh, L. T. H., et al., (2017). A novel biological function of soluble biglycan: Induction of erythropoietin production and polycythemia. *Glycoconjugate Journal, 34*(3), 393–404.

Fuhler, G. M., Baanstra, M., Chesik, D., et al., (2010). Bone marrow stromal cell interaction reduces syndecan-1 expression and induces kinomic changes in myeloma cells. *Experimental Cell Research, 316*(11), 1816–1828.

Fukasawa, M., Matsushita, A., & Korc, M., (2007). Neuropilin-1 interacts with integrin β1 and modulates pancreatic cancer cell growth, survival and invasion. *Cancer Biology and Therapy, 6*(8), 1184–1191.

Fukushi, J. I., Makagiansar, I. T., & Stallcup, W. B., (2004). NG2 proteoglycan promotes endothelial cell motility and angiogenesis via engagement of galectin-3 and α3β1 integrin. *Molecular Biology of the Cell, 15*(8), 3580–3590.

Funderburgh, J. L., Corpuz, L. M., Roth, M. R., Funderburgh, M. L., Tasheva, E. S., & Conrad, G. W., (1997). Mimecan, the 25-kDa corneal keratan sulfate proteoglycan, is a product of the gene producing osteoglycin. *Journal of Biological Chemistry, 272*(44), 28089–28095.

Galli, M., Chatterjee, M., Grasso, M., et al., (2018). Phase I study of the heparanase inhibitor roneparstat: An innovative approach for ultiple myeloma therapy. *Haematologica, 103*(10), e469–472.

Gordon, M. Y., Riley, G. P., & Clarke, D., (1988). Heparan sulfate is necessary for adhesive interactions between human early hemopoietic progenitor cells and the extracellular matrix of the marrow microenvironment. *Leukemia, 2*(12), 804–809.

Grigorian, M., Liu, T., Banerjee, U., & Hartenstein, V., (2013). The proteoglycan Trol controls the architecture of the extracellular matrix and balances the proliferation and differentiation of blood progenitors in the *Drosophila* lymph gland. *Developmental Biology, 384*(2), 301–312.

Grindel, B. J., Martinez, J. R., Tellman, T. V., et al., (2018). Matrilysin/MMP-7 cleavage of perlecan/HSPG2 complexed with semaphoring 3A supports FAK-mediated stromal invasion by prostate cancer cells. *Scientific Reports, 8*(1), 1–14.

Gupta, P., Oegema, Jr. T. R., Brazil, J. J., Dudek, A. Z., Slungaard, A., & Verfaillie, C. M., (1998). Structurally specific heparan sulfates support primitive human hematopoiesis by the formation of a multimolecular stem cell niche. *Blood, 92*(12), 4641–4651.

Hachim, D., Whittaker, T. E., Kim, H., & Stevens, M. M., (2019). Glycosaminoglycan-based biomaterials for growth factor and cytokine delivery: Making the right choices. *Journal of Controlled Release, 313*, 131–147.

Haglund, L., Tillgren, V., Addis, L., Wenglén, C., Recklies, A., & Heinegård, D., (2011). Identification and characterization of the integrin α2β1 binding motif in chondroadherin mediating cell attachment. *Journal of Biological Chemistry, 286*(5), 3925–3934.

Hao, C., Xu, H., Yu, L., & Zhang, L., (2019). Heparin: An essential drug for modern medicine. *Progress in Molecular Biology and Translational Science: Glycans and Glycosaminoglycans as Clinical Biomarkers and Therapeutics-Part B, 163*, 1–19.

Hardingham, T. E., & Fosang, A. J., (1992). Proteoglycans: Many forms and many functions. *The FASEB Journal, 6*(3), 861–870.

Hill, L. J., Moakes, R. J., Vareechon, C., et al., (2018). Sustained release of decorin to the surface of the eye enables scarless corneal regeneration. *Regenerative Medicine, 3*(1), 1–12.

Hirsh, J., Warkentin, T. E., Shaughnessy, S. G., et al., (2001). Heparin and low-molecular-weight heparin mechanisms of action, pharmacokinetics, dosing, monitoring, efficacy, and safety. *Chest, 119*(1), 64S–94S.

Hoggatt, J., Tate, T. A., & Pelus, L. M., (2014). Hematopoietic stem and progenitor cell mobilization in mice. In: *Hematopoietic Stem Cell Protocols* (pp. 43–64). Humana Press, New York, NY.

Hosokawa, K., Arai, F., Yoshihara, H., et al., (2010). Knockdown of N-cadherin suppresses the long-term engraftment of hematopoietic stem cells. *Blood, 116*(4), 554–563.

Ichii, M., Frank, M. B., Iozzo, R. V., & Kincade, P. W., (2012). The canonical Wnt pathway shapes niches supportive of hematopoietic stem/progenitor cells. *Blood, 119*(7), 1683–1692.

Iozzo, R. V., & San, A. J. D., (2001). Heparan sulfate proteoglycans: Heavy hitters in the angiogenesis arena. *The Journal of Clinical Investigation, 108*(3), 349–355.

Iozzo, R. V., & Schaefer, L., (2010). Proteoglycans in health and disease: Novel regulatory signaling mechanisms evoked by the small leucine-rich proteoglycans. *The FEBS Journal, 277*(19), 3864–3875.

Iozzo, R. V., & Schaefer, L., (2015). Proteoglycan form and function: A comprehensive nomenclature of proteoglycans. *Matrix Biology, 42*, 11–55.

Johnson, H. J., Rosenberg, L., Choi, H. U., Garza, S., Hook, M., & Neame, P. J., (1997). Characterization of epiphycan, a small proteoglycan with a leucine-rich repeat core protein. *Journal of Biological Chemistry, 272*(30), 18709–18717.

Kamma-Lorger, C. S., Pinali, C., Martínez, J. C., et al., (2016). Role of decorin core protein in collagen organization in congenital stromal corneal dystrophy (CSCD). *PLoS One, 11*(2), e0147948.

Kays, S. K., Kaufmann, K. B., Abel, T., et al., (2015). CD105 is a surface marker for receptor-targeted gene transfer into human long-term repopulating hematopoietic stem cells. *Stem Cells and Development, 24*(6), 714–723.

Khurana, S., Margamuljana, L., Joseph, C., Schouteden, S., Buckley, S. M., & Verfaillie, C. M., (2013). Glypican-3–mediated inhibition of CD26 by TFPI: A novel mechanism in hematopoietic stem cell homing and maintenance. *Blood, 121*(14), 2587–2595.

Kiel, M. J., Radice, G. L., & Morrison, S. J., (2007). Lack of evidence that hematopoietic stem cells depend on N-cadherin-mediated adhesion to osteoblasts for their maintenance. *Cell Stem Cell, 1*(2), 204–217.

Klein, G., Conzelmann, S., Beck, S., Timpl, R., & Müller, C. A., (1995). Perlecan in human bone marrow: A growth-factor-presenting, but anti-adhesive, extracellular matrix component for hematopoietic cells. *Matrix Biology, 14*(6), 457–465.

Knox, S. M., & Whitelock, J. M., (2006). Perlecan: How does one molecule do so many things? *Cellular and Molecular Life Sciences, 63*(21), 2435–2445.

Kolset, S. O., & Gallagher, J. T., (1990). Proteoglycans in haemopoietic cells. *Biochimica et Biophysica Acta (BBA) – Reviews on Cancer, 1032*(2, 3), 191–211.

Kolset, S. O., & Tveit, H., (2008). Serglycin–structure and biology. *Cellular and Molecular Life Sciences, 65*(7), 1073–1085.

Kolset, S. O., Mann, D. M., Uhlin-Hansen, L., Winberg, J. O., & Ruoslahti, E., (1996). Serglycin-binding proteins in activated macrophages and platelets. *Journal of Leukocyte Biology, 59*(4), 545–554.

Kopp, H. G., Avecilla, S. T., Hooper, A. T., & Rafii, S., (2005). The bone marrow vascular niche: Home of HSC differentiation and mobilization. *Physiology, 20*(5), 349–356.

Kozanoglu, I., Boga, C., Ozdogu, H., et al., (2009). Human bone marrow mesenchymal cells express NG2: Possible increase in discriminative ability of flow cytometry during mesenchymal stromal cell identification. *Cytotherapy, 11*(5), 527–533.

Kreuger, J., Spillmann, D., Li, J. P., & Lindahl, U., (2006). Interactions between heparan sulfate and proteins: The concept of specificity. *The Journal of Cell Biology, 174*(3), 323.

Krusius, T., & Ruoslahti, E., (1986). Primary structure of an extracellular matrix proteoglycan core protein deduced from cloned cDNA. *Proceedings of the National Academy of Sciences USA 83*(20), 7683–7687.

Kuhn, N. Z., & Tuan, R. S., (2010). Regulation of stemness and stem cell niche of mesenchymal stem cells: Implications in tumorigenesis and metastasis. *Journal of Cellular Physiology, 222*(2), 268–277.

Li, J. P., & Kusche-Gullberg, M., (2016). Heparan sulfate: Biosynthesis, structure, and function. *International Review of Cell and Molecular Biology, 325*, 215–273.

Li, W., Johnson, S. A., Shelley, W. C., & Yoder, M. C., (2004). Hematopoietic stem cell repopulating ability can be maintained *in vitro* by some primary endothelial cells. *Experimental Hematology, 32*(12), 1226–1237.

Lindahl, U., & Li, J. P., (2009). Interactions between heparan sulfate and proteins-design and functional implications. *International Review of Cell and Molecular Biology, 276*, 105–159.

Linhardt, R. J., & Toida, T., (2004). Role of glycosaminoglycans in cellular communication. *Accounts of Chemical Research, 37*(7), 431–438.

Liu, C. Y., Birk, D. E., Hassell, J. R., Kane, B., & Kao, W. W. Y., (2003). Keratocan-deficient mice display alterations in corneal structure. *Journal of Biological Chemistry, 278*(24), 21672–21677.

López-Casillas, F., Cheifetz, S., Doody, J., Andres, J. L., Lane, W. S., & Massague, J., (1991). Structure and expression of the membrane proteoglycan betaglycan, a component of the TGF-β receptor system. *Cell, 67*(4), 785–795.

Lorente-Gea, L., García, B., Martín, C., Quirós, L. M., & Fernández-Vega, I., (2017). Heparan sulfate proteoglycans and heparanases in Alzheimer's disease: Current outlook and potential therapeutic targets. *Neural Regeneration Research, 12*(6), 914.

Lorenzo, P., Aspberg, A., Önnerfjord, P., Bayliss, M. T., Neame, P. J., & Heinegård, D., (2001). Identification and characterization of asporin: A novel member of the leucine-rich repeat protein family closely related to decorin and biglycan. *Journal of Biological Chemistry, 276*(15), 12201–12211.

Lundell, B. I., McCarthy, J. B., Kovach, N. L., & Verfaillie, C. M., (1997). Activation of β1 integrins on CML progenitors reveals cooperation between β1 integrins and CD44 in the regulation of adhesion and proliferation. *Leukemia, 11*(6), 822–829.

Magnani, J. L., & Ernst, B., (2009). Glycomimetic drugs - a new source of therapeutic opportunities. *Discovery Medicine, 8*(43), 247–252.

Malfuson, J. V., Boutin, L., Clay, D., et al., (2014). SP/drug efflux functionality of hematopoietic progenitors is controlled by mesenchymal niche through VLA-4/CD44 axis. *Leukemia, 28*(4), 853–864.

Massague, J., Cheifetz, S., Boyd, F. T., & Andres, J. L., (1990). TGF-β receptors and TGF-β binding proteoglycans: Recent progress in identifying their functional properties. *Annals of the New York Academy of Sciences, 593*(1), 59–72.

Maurel, P., Rauch, U., Flad, M., Margolis, R. K., & Margolis, R. U., (1994). Phosphacan, a chondroitin sulfate proteoglycan of the brain that interacts with neurons and neural cell-adhesion molecules, is an extracellular variant of a receptor-type protein tyrosine phosphatase. *Proceedings of the National Academy of Sciences USA 91*(7), 2512–2516.

Mazzon, C., Anselmo, A., Cibella, J., et al., (2011). The critical role of agrin in the hematopoietic stem cell niche. *Blood, 118*(10), 2733–2742.

Mazzon, C., Anselmo, A., Soldani, C., et al., (2012). Agrin is required for survival and function of monocytic cells. *Blood, 119*(23), 5502–5511.

McDermott, S. P., Ranheim, E. A., Leatherberry, V. S., Khwaja, S. S., Klos, K. S., & Alexander, C. M., (2007). Juvenile syndecan-1 null mice are protected from carcinogen-induced tumor development. *Oncogene, 26*(10), 1407–1416.

Méndez-Ferrer, S., Michurina, T. V., Ferraro, F., et al., (2010). Mesenchymal and haematopoietic stem cells form a unique bone marrow niche. *Nature, 466*(7308), 829–834.

Merter, M., Sahin, U., Yildirim, Y., Ilhan, O., & Beksac, M., (2017). Heparin enhances CD34+ content of stem cells collected following G-CSF+/-chemotherapy. *Cytotherapy, 19*(5), e12.

Mikami, T., & Kitagawa, H., (2013). Biosynthesis and function of chondroitin sulfate. *Biochimica et Biophysica Acta (BBA)-General Subjects, 1830*(10), 4719–4733.

Monfort, J., Tardif, G., Roughley, P., et al., (2008). Identification of opticin, a member of the small leucine-rich repeat proteoglycan family, in human articular tissues: A novel target for MMP-13 in osteoarthritis. *Osteoarthritis and Cartilage, 16*(7), 749–755.

Moreth, K., Frey, H., Hubo, M., et al., (2014). Biglycan-triggered TLR-2-and TLR-4-signaling exacerbates the pathophysiology of ischemic acute kidney injury. *Matrix Biology, 35*, 143–151.

Morimoto-Tomita, M., Uchimura, K., Werb, Z., Hemmerich, S., & Rosen, S. D., (2002). Cloning and characterization of two extracellular heparin-degrading endosulfatases in mice and humans. *Journal of Biological Chemistry, 277*(51), 49175–49185.

Morrison, S. J., & Scadden, D. T., (2014). The bone marrow niche for haematopoietic stem cells. *Nature, 505*(7483), 327–334.

Morrison, S. J., & Spradling, A. C., (2008). Stem cells and niches: Mechanisms that promote stem cell maintenance throughout life. *Cell, 132*(4), 598–611.

Mulloy, B., & Rider, C. C., (2006). Cytokines and proteoglycans: An introductory overview. *Biochemical Society Transactions, 34*(3), 409–413.

Nagao, K., Ohta, T., Hinohara, A., et al., (2008). Expression profile analysis of aorta-gonad-mesonephros region-derived stromal cells reveals genes that regulate hematopoiesis. *Biochemical and Biophysical Research Communications, 377*(1), 205–209.

Nakamura, Y., Arai, F., Iwasaki, H., et al., (2010). Isolation and characterization of endosteal niche cell populations that regulate hematopoietic stem cells. *Blood, 116*(9), 1422–1432.

Nandadasa, S., O'Donnell, A., Murao, A., et al., (2021). The versican-hyaluronan complex provides an essential extracellular matrix niche for Flk1+ hematoendothelial progenitors. *Matrix Biology, 97*, 40–57.

Niemann, C. U., Kjeldsen, L., Ralfkiaer, E., Jensen, M. K., & Borregaard, N., (2007). Serglycin proteoglycan in hematologic malignancies: A marker of acute myeloid leukemia. *Leukemia, 21*(12), 2406–2410.

Nitkin, R. M., Smith, M. A., Magill, C., et al., (1987). Identification of agrin, a synaptic organizing protein from torpedo electric organ. *The Journal of Cell Biology, 105*(6), 2471–2478.

Noonan, D. M., Fulle, A., Valente, P., et al., (1991). The complete sequence of perlecan, a basement membrane heparan sulfate proteoglycan, reveals extensive similarity with laminin A chain, low density lipoprotein-receptor, and the neural cell adhesion molecule. *Journal of Biological Chemistry, 266*(34), 22939–22947.

Nugent, M. A., Nugent, H. M., Iozzo, R. V., Sanchack, K., & Edelman, E. R., (2000). Perlecan is required to inhibit thrombosis after deep vascular injury and contributes to endothelial cell-mediated inhibition of intimal hyperplasia. *Proceedings of the National Academy of Sciences USA, 97*(12), 6722–6727.

Paderi, J., Prestwich, G. D., Panitch, A., Boone, T., & Stuart, K., (2018). Glycan therapeutics: Resurrecting an almost pharma-forgotten drug class. *Advanced Therapeutics, 1*(8), 1800082.

Papy-Garcia, D., & Albanese, P., (2017). Heparan sulfate proteoglycans as key regulators of the mesenchymal niche of hematopoietic stem cells. *Glycoconjugate Journal, 34*(3), 377–391.

Parker, W. L., Goldring, M. B., & Philip, A., (2003). Endoglin is expressed on human chondrocytes and forms a heteromeric complex with betaglycan in a ligand and type II TGFβ receptor independent manner. *Journal of Bone and Mineral Research, 18*(2), 289–302.

Peled, A., Petit, I., Kollet, O., et al., (1999). Dependence of human stem cell engraftment and repopulation of NOD/SCID mice on CXCR4. *Science, 283*(5403), 845–848.

Pellegrini, L., Burke, D. F., Von, D. F., Mulloy, B., & Blundell, T. L., (2000). Crystal structure of fibroblast growth factor receptor ectodomain bound to ligand and heparin. *Nature, 407*(6807), 1029–1034.

Pinho, S., Marchand, T., Yang, E., Wei, Q., Nerlov, C., & Frenette, P. S., (2018). Lineage-biased hematopoietic stem cells are regulated by distinct niches. *Developmental Cell, 44*(5), 634–641.

Place, L. W., Kelly, S. M., & Kipper, M. J., (2014). Synthesis and characterization of proteoglycan-mimetic graft copolymers with tunable glycosaminoglycan density. *Biomacromolecules, 15*(10), 3772–3780.

Price, M. A., Colvin, W. L. E., Yang, J., et al., (2011). CSPG4, a potential therapeutic target facilitates malignant progression of melanoma. *Pigment Cell and Melanoma Research, 24*(6), 1148–1157.

Purushothaman, A., & Toole, B. P., (2014). Serglycin proteoglycan is required for multiple myeloma cell adhesion, *in vivo* growth, and vascularization. *Journal of Biological Chemistry, 289*(9), 5499–5509.

Rafii, S., Mohle, R., Shapiro, F., Frey, B. M., & Moore, M. A., (1997). Regulation of hematopoiesis by microvascular endothelium. *Leukemia and Lymphoma, 27*(5, 6), 375–386.

Retzler, C., Wiedemann, H., Kulbe, G., & Rauch, U., (1996). Structural and electron microscopic analysis of neurocan and recombinant neurocan fragments. *Journal of Biological Chemistry, 271*(29), 17107–17113.

Rivara, S., Milazzo, F. M., & Giannini, G., (2016). Heparanase: A rainbow pharmacological target associated to multiple pathologies, including rare diseases. *Future Medicinal Chemistry, 8*(6), 647–680.

Rodgers, K. D., San, A. J. D., & Jacenko, O., (2008). Heparan sulfate proteoglycans: A GAGgle of skeletal-hematopoietic regulators. *Developmental Dynamics, 237*(10), 2622–2642.

Ross, M. D., Bruggeman, L. A., Hanss, B., et al., (2003). Podocan, a novel small leucine-rich repeat protein expressed in the sclerotic glomerular lesion of experimental HIV-associated nephropathy. *Journal of Biological Chemistry, 278*(35), 33248–33255.

Saez, B., Ferraro, F., Yusuf, R. Z., Cook, C. M., Yu, V. W., Pardo-Saganta, A., Sykes, S. M., et al., (2014). Inhibiting stromal cell heparan sulfate synthesis improves stem cell mobilization and enables engraftment without cytotoxic conditioning. *Blood, 124*(19), 2937–2947.

Sanderson, R. D., Sneed, T. B., Young, L. A., Sullivan, G. L., & Lander, A. D., (1992). Adhesion of B lymphoid (MPC-11) cells to type I collagen is mediated by integral membrane proteoglycan, syndecan. *The Journal of Immunology, 148*(12), 3902–3911.

Sarrazin, S., Lamanna, W. C., & Esko, J. D., (2011). Heparan sulfate proteoglycans. *Cold Spring Harbor Perspectives in Biology, 3*(7), a004952.

Sbaa-Ketata, E., Courel, M. N., Delpech, B., & Vannier, J. P., (2002). Hyaluronan-derived oligosaccharides enhance SDF-1-dependent chemotactic effect on peripheral blood hematopoietic CD34+ cells. *Stem Cells, 20*(6), 585–587.

Schick, B. P., (2010). Serglycin proteoglycan deletion in mouse platelets: Physiological effects and their implications for platelet contributions to thrombosis, inflammation, atherosclerosis, and metastasis. *Progress in Molecular Biology and Translational Science, 93*, 235–287.

Schmidt, G., Hausser, H., & Kresse, H., (1991). Interaction of the small proteoglycan decorin with fibronectin. Involvement of the sequence NKISK of the core protein. *Biochemical Journal, 280*(2), 411–414.

Schofield, R., (1978). The relationship between the spleen colony-forming cell and the haemopoietic stem cell. *Blood Cells, 4*(1, 2), 7–25.

Seffer, M. T., Cottam, D., Forni, L. G., & Kielstein, J. T., (2021). Heparin 2.0: A new approach to the infection crisis. *Blood Purification, 50*(1), 28–34.

Senda, M., Fukuyama, R., & Nagasaka, T., (2016). Kinetics of versican-expressing macrophages in bone marrow after cord blood stem cell transplantation for treatment of acute myelogenous leukaemia. *Journal of Clinical Pathology, 69*(10), 906–911.

Siczkowski, M., Clarke, D., & Gordon, M. Y., (1992). Binding of primitive hematopoietic progenitor cells to marrow stromal cells involves heparan sulfate. *Blood, 80*(4), 912–919.

Smith, F., (1996). The human homologue of rat NG2, a chondroitin sulfate proteoglycan, is not expressed on the cell surface of normal hematopoietic cells but is expressed by acute myeloid leukemia blasts from poor-prognosis patients with abnormalities of chromosome band 11q23. *Blood, 87*, 1123–1133.

Spiegel, A., Zcharia, E., Vagima, Y., et al., (2008). Heparanase regulates retention and proliferation of primitive Sca-1+/c-Kit+/Lin− cells via modulation of the bone marrow microenvironment. *Blood, 111*(10), 4934–4943.

Stanford, K. I., Bishop, J. R., Foley, E. M., et al., (2009). Syndecan-1 is the primary heparan sulfate proteoglycan mediating hepatic clearance of triglyceride-rich lipoproteins in mice. *The Journal of Clinical Investigation, 119*(11), 3236–3245.

Steidl, U., Bork, S., Schaub, S., et al., (2004). Primary human CD34+ hematopoietic stem and progenitor cells express functionally active receptors of neuromediators. *Blood, 104*(1), 81–88.

Stenvers, K. L., Tursky, M. L., Harder, K. W., et al., (2003). Heart and liver defects and reduced transforming growth factor β2 sensitivity in transforming growth factor β type III receptor-deficient embryos. *Molecular and Cellular Biology, 23*(12), 4371–4385.

Sugiyama, T., Kohara, H., Noda, M., & Nagasawa, T., (2006). Maintenance of the hematopoietic stem cell pool by CXCL12-CXCR4 chemokine signaling in bone marrow stromal cell niches. *Immunity, 25*(6), 977–988.

Sun, L., & Chen, C., (1997). Expression of transforming growth factor β type III receptor suppresses tumorigenicity of human breast cancer MDA-MB-231 cells. *Journal of Biological Chemistry, 272*(40), 25367–25372.

Sweeney, E. A., Lortat-Jacob, H., Priestley, G. V., Nakamoto, B., & Papayannopoulou, T., (2002). Sulfated polysaccharides increase plasma levels of SDF-1 in monkeys and mice: Involvement in mobilization of stem/progenitor cells. *Blood, 99*(1), 44–51.

Sweeney, E., Roberts, D., & Jacenko, O., (2011). Altered matrix at the chondro–osseous junction leads to defects in lymphopoiesis. *Annals of the New York Academy of Sciences, 1237*(1), 79–87.

Szade, K., Gulati, G. S., Chan, C. K., et al., (2018). Where hematopoietic stem cells live: The bone marrow niche. *Antioxidants and Redox Signaling, 29*(2), 191–204.

Szczepanek, K., Kieda, C., & Cichy, J., (2008). Differential binding of hyaluronan on the surface of tissue-specific endothelial cell lines. *Acta Biochimica Polonica, 55*(1), 35–42.

Termini, C., Li, M., Kim, J., Zhao, L., & Chute, J. P., (2018). Syndecan-2 surface expression identifies hematopoietic stem cells with increased repopulating capacity. *Blood, 132*(S1), 1273. https://doi.org/10.1182/blood-2018-99-110701.

Toyama-Sorimachi, N., Kitamura, F., Habuchi, H., Tobita, Y., Kimata, K., & Miyasaka, M., (1997). Widespread expression of chondroitin sulfate-type serglycins with CD44 binding ability in hematopoietic cells. *Journal of Biological Chemistry, 272*(42), 26714–26719.

Tran, J., Brenner, T. J., & DiNardo, S., (2000). Somatic control over the germline stem cell lineage during *Drosophila* spermatogenesis. *Nature, 407*(6805), 754–757.

Travlos, G. S., (2006). Normal structure, function, and histology of the bone marrow. *Toxicologic Pathology, 34*(5), 548–565.

Turnbull, J. E., Fernig, D. G., Ke, Y., Wilkinson, M. C., & Gallagher, J. T., (1992). Identification of the basic fibroblast growth factor binding sequence in fibroblast heparan sulfate. *Journal of Biological Chemistry, 267*(15), 10337–10341.

Viviano, B. L., Silverstein, L., Pflederer, C., Paine-Saunders, S., Mills, K., & Saunders, S., (2005). Altered hematopoiesis in glypican-3-deficient mice results in decreased osteoclast differentiation and a delay in endochondral ossification. *Developmental Biology, 282*(1), 152–162.

Vuoriluoto, K., Jokinen, J., Kallio, K., Salmivirta, M., Heino, J., & Ivaska, J., (2008). Syndecan-1 supports integrin α2β1-mediated adhesion to collagen. *Experimental Cell Research, 314*(18), 3369–3381.

Wagner, W., Wein, F., Roderburg, C., et al., (2008). Adhesion of human hematopoietic progenitor cells to mesenchymal stromal cells involves CD44. *Cells Tissues Organs, 188*(1, 2), 160–169.

Walimbe, T., & Panitch, A., (2020). Proteoglycans in biomedicine: Resurgence of an underexploited class of ECM molecules. *Frontiers in Pharmacology, 10*, 1661.

Wasnik, S., Chen, W., Ahmed, A. S., Zhang, X. B., Tang, X., & Baylink, D. J., (2017). HSC niche: Regulation of mobilization and homing. In: *Biology and Engineering of Stem Cell Niches* (pp. 63–73). Academic Press.

Wasnik, S., Tiwari, A., Kirkland, M. A., & Pande, G., (2012). Osteohematopoietic stem cell niches in bone marrow. *International Review of Cell and Molecular Biology, 298*, 95–133.

Wendel, M., Sommarin, Y., & Heinegård, D., (1998). Bone matrix proteins: Isolation and characterization of a novel cell-binding keratan sulfate proteoglycan (osteoadherin) from bovine bone. *The Journal of Cell Biology, 141*(3), 839–847.

Weyers, A., & Linhardt, R. J., (2013). Neoproteoglycans in tissue engineering. *The FEBS Journal, 280*(10), 2511–2522.

Wight, T. N., Kang, I., & Merrilees, M. J., (2014). Versican and the control of inflammation. *Matrix Biology, 35*, 152–161.

Winkler, C. W., Foster, S. C., Matsumoto, S. G., et al., (2012). Hyaluronan anchored to activated CD44 on central nervous system vascular endothelial cells promotes lymphocyte extravasation in experimental autoimmune encephalomyelitis. *Journal of Biological Chemistry, 287*(40), 33237–33251.

Winnemöller, M., Schön, P., Vischer, P., & Kresse, H., (1992). Interactions between thrombospondin and the small proteoglycan decorin: Interference with cell attachment. *European Journal of Cell Biology, 59*(1), 47–55.

Woulfe, D. S., Lilliendahl, J. K., August, S., et al., (2008). Serglycin proteoglycan deletion induces defects in platelet aggregation and thrombus formation in mice. *Blood, 111*(7), 3458–3467.

Xia, L., McDaniel, J. M., Yago, T., Doeden, A., & McEver, R. P., (2004). Surface fucosylation of human cord blood cells augments binding to P-selectin and E-selectin and enhances engraftment in bone marrow. *Blood, 104*(10), 3091–3096.

Yamada, H., Watanabe, K., Shimonaka, M., & Yamaguchi, Y., (1994). Molecular cloning of brevican, a novel brain proteoglycan of the aggrecan/versican family. *Journal of Biological Chemistry, 269*(13), 10119–10126.

Yeoh, J. S., Van, O. R., Weersing, E., et al., (2006). Fibroblast growth factor-1 and-2 preserve long-term repopulating ability of hematopoietic stem cells in serum-free cultures. *Stem Cells, 24*(6), 1564–1572.

Zhao, M., & Li, L., (2016). Dissecting the bone marrow HSC niches. *Cell Research, 26*(9), 975, 976.

Zimmermann, D. R., & Ruoslahti, E., (1989). Multiple domains of the large fibroblast proteoglycan, versican. *The EMBO Journal, 8*(10), 2975–2981.

Zohren, F., Toutzaris, D., Klärner, V., Hartung, H. P., Kieseier, B., & Haas, R., (2008). The monoclonal anti–VLA-4 antibody natalizumab mobilizes CD34+ hematopoietic progenitor cells in humans. *Blood, 111*(7), 3893–3895.

Zöller, M., Rajasagi, M., Vitacolonna, M., & Luft, T., (2007). Thymus repopulation after allogeneic reconstitution in hematological malignancies. *Experimental Hematology, 35*(12), 891–905.

CHAPTER 5

Epigenetics of Hematopoiesis: Role of HATs and HDACs

ARUNASREE M. KALLE and DEBASMITA NAIK

Department of Animal Biology, School of Life Sciences, University of Hyderabad, Hyderabad, Telangana, India

ABSTRACT

The generation of different types of blood cells from a single hematopoietic stem cell is a well-coordinated process involving epigenetics in cell fate decisions and lineage commitments. DNA methylation, histone protein modifications, and chromatin remodeling are the three main epigenetic mechanisms by which histone protein modifications play a prominent role in gene regulation during development and differentiation. Among the several histone protein modifications, acetylation and deacetylation of lysine residues in the N-terminal region of histone proteins are crucial for cell fate commitment. The role of the enzymes involved in this modification during hematopoietic lineage commitment has been explored.

5.1 INTRODUCTION

Hematopoiesis is a process by which all blood cells are generated from a single stem cell, the hematopoietic stem cells (HSC). The development and differentiation of HSCs into a specific blood cell type require coordinated efforts of several transcriptional repressors and activators to inhibit a particular lineage of cell development and activation of another cell lineage development. Epigenetics plays a crucial role in the development and differentiation

Hematopoiesis: Biochemical, Cellular, Molecular, and Genomic Perspectives. Ravi Kumar Gutti (Ed.)

of HSCs into multiple cell types. Of the several epigenetic mechanisms, the posttranslational modifications (PTMs) at the N-terminus of the histone protein are involved in chromatin remodeling. Although several PTMs of histone proteins are known to epigenetically regulate the gene expression, acetylation and deacetylation by histone acetyl transferases (HATs) and histone deacetylases (HDACs), respectively, are the most studied. This review focuses on the role of HATs and HDACs in hematopoiesis.

5.2 HISTONE ACETYL TRANSFERASES (HATs) AND HEMATOPOIESIS

Both histone and non-histone proteins undergo acetylation on their lysine residues. The addition or removal of acetyl groups (C_2H_3O) from lysine residues of non-histone proteins results either in functional activation or repression of the protein. The addition of acetyl group to lysine by HATs, results in conversion of positively charged amine group into a neutral amide resulting in "open" conformation of DNA and thus transcriptional activation. Due to the presence of positively charged nitrogen in lysine amino acid, it can bind to negatively charged phosphate groups of DNA and thus help in compact coiling of DNA ("tight" conformation) around the histone nucleosome leading to transcriptional inactivity.

HATs catalyze the transfer of acetyl group from acetyl-coA enzyme onto the lysine residue of proteins. Based on the subcellular localization, HATs are of mainly two types: Type A are nuclear and thus regulate the gene expression by acetylating the histone and other nuclear proteins, and Type B are cytoplasmic and hence regulate the protein function and localization. Type A HATs are further classified into four families based on their sequence homology (i) The Gcn5-related N-acetyltransferases (GNAT) family; (ii) the MOZ, YBF2/SAS3, SAS2, and TIP60 (MYST) family; (iii) the p300/CBP family; and (iv) Other HATs – nuclear receptor coactivators. The role of HATs in development and differentiation has been well established. Several recent studies identified that histone H3 Lys9 (H3K9) and Lys14 (H3K14) and H4 Lys8 (H4K8) and Lys12 (H4K12) acetylation mediated by HATs play a crucial role in differentiation (Zhou et al., 2014).

5.2.1 THE GNAT FAMILY

The GNAT family consists of members with conserved bromodomain that helps them to recognize and bind to acetylated histone proteins. There are eight members in GNAT family – Gcn5, PCAF, Hat1, Elp3, ATF-2. Recent study identified the role of Gcn5 in lymphoid cell development and differentiation (Sun et al., 2015) and due to its involvement in tumorigenesis, Gcn5 inhibitors are shown to be promising drug candidates for treating hematological malignancies (Chimenti et al., 2009; Secci et al., 2014). Although the exact role of P300/CBP-associated factor (PCAF or K(lysine) acetyltransferase 2B (KAT2B)) in normal hematopoiesis remains to be identified, its involvement in malignant hematopoiesis is well-established (Santillan et al., 2006). Histone acetyltransferase 1 (HAT1), in general, plays an important role in the formation of chromatin structure, however, studies have demonstrated that HAT1 is enriched in 2 types of stem cells – the hematopoietic and neural stem cells suggesting its role in their development and differentiation (Park et al., 2002; Lim et al., 2006). Elongator Acetyltransferase Complex Subunit 3 (Elp3) is a part of RNA polymerase II elongator complex and is involved in transcription. Activating transcription factor 2 (ATF2) plays an important role in erythroid differentiation by regulating the expression of gamma globin gene (Liu et al., 2013).

5.2.2 THE MYST FAMILY

The MYST family is named after its founding members in yeast, MOZ, YBF2/SAS3, SAS2, and TIP60, and in humans comprises of 5 proteins, Tip60, MOZ, MORF, HBO1 and MOF. The MYST family members consist of a highly conserved MYST domain with acetyl-coA binding motif. Some members may also have a chromodomain, plant homeodomain-linked (PHD) motif or a zinc finger motif (Yang, 2004). These proteins are associated with large multi-subunit protein complexes that are evolutionarily conserved (Yang, 2004; Smith et al., 2005). These proteins act as coactivators for several transcription factors such as RUNX1, RUNX2, P53, etc.

Tat-interactive protein-60 (Tip60) or Lysine acetyl transferase 5 (KAT5) is a part of nucleosomal acetyltransferase of H4 (NuA4) complex and is known to be involved in DNA repair and transcriptional regulation by acetylating Histone H3, H4, and H2A along with its variants (Doyon & Côté, 2004; Squatrito, Gorrini, & Amati, 2006). Recent studies in mice demonstrated that TiP60 is required for HSC maintenance and deletion of TiP60 is associated with decreased acetylation of H2A.Z leading to the reduced expression of MYC-dependent genes necessary for HSC maintenance (Doyon & Côté, 2004).

The monocytic leukemia zinc finger protein (MOZ), also known as Lysine acetyltransferase 6A (KAT6A) and MOZ-related factor (MORF) or lysine acetyl transferase 6B (KAT6B) are a part of a complex that acetylate lysine 9 residue of histone H3. Chromosomal translocations involving MOZ/MORF are associated with acute myeloid leukemia (AML) as it acts as a transcriptional coactivator for AML1 protein. MOZ is required for maintaining HSCs (Katsumoto et al., 2006). Homozygous loss of MOZ in mice was shown to be embryonically lethal and mice with loss of function mutations failed to develop HSCs during embryogenesis (Katsumoto et al., 2006; Thomas et al., 2006; Perez-Campo et al., 2009).

The lysine acetyl transferase 7 (KAT7 or HBO1) regulates DNA replication by specifically acetylating histone H4 at lysine 5, 8 and 12 (H4K5, H4K8, and H4K12) (Iizuka & Stillman, 1999; Saksouk et al., 2009; Miotto & Struhl, 2010; Lalonde et al., 2013). Acetylation of histone H3 at lysine 14 (H3K14) by KAT7 facilitates RNA Pol II processivity (MacPherson et al., 2020). Recent studies demonstrated that KAT7 complexes with Bromodomain-containing protein 1 (Brd1) and regulates the transcription of erythroid-specific genes thus promoting erythroid differentiation (Mishima et al., 2011).

The lysine acetyl transferase 8 (KAT8/MOF/MYST1) mediates Histone H4 lysine 16 (H4K16) acetylation. Studies demonstrated that KAT8 associates with MLL1 and is thus important in development and maintenance of HSCs (Dou et al., 2005). In mice, homozygous deletion of KAT8 affected adult hematopoiesis but not early and midgestational hematopoiesis (Valerio et al., 2017).

5.2.3 THE p300/CBP FAMILY

As the name suggests, the family consists of two closely related proteins: p300 and CBP (CREB-binding protein) that act as coactivators for several transcription factors. Several studies indicated the role of p300/CBP in self renewal of HSCs and lack of these proteins is shown to be embryonically lethal in mice (Heintzman et al., 2009; Visel et al., 2009). The p300/CBP protein has a tumor suppressor role and thus low expression of these proteins is implicated in leukemogenesis (Goodman & Smolik, 2000; Sun et al., 2015; Dutta et al., 2016).

5.2.4 OTHER HATs – NUCLEAR RECEPTOR COACTIVATORS

Apart from the GNAT, MYST, and p300/CBP HAT family of proteins, there are several other proteins such as ACTR/SRC-1, $TAF_{II}250$, TFIIIC, Rtt109, and CLOCK with acetyl transferase activity and do not share any sequence homology with the HAT domain of GNAT, MYST and p300/CBP member proteins (Yuan & Marmorstein, 2013). These proteins generally act as transcriptional coactivators and nuclear receptor coactivators (Ogryzko, 2001). The nuclear receptor coactivators consist of mainly three proteins – ACTR, SRC-1 and TIFII. These are known to function in complex with other HATs, such as p300/CBP and regulate several cellular processes, including hematopoiesis (Torok & Grant, 2004).

5.3 HISTONE DEACETYLASES (HDACs) AND HEMATOPOIESIS

Histone deacetylases (HDACs) are generally a class of enzymes that are well known for their function of lysine deacetylation of both histones as well as non-histone proteins and thus repressive regulation of gene expression (Greco et al., 2011; Seto & Yoshida, 2014; Sahakian et al., 2017). HDACs are involved in the regulation of many biological functions of which opposing the HATs functions is one of the most important. They make histones and DNA interaction stronger in contrast to acetylation by making histones more positively charged, thus inhibiting gene transcription (Seto & Yoshida, 2014; Tan et al., 2011).

HDACs considerably play a critical role in the development of multi-lineage processes that include HSC progenitor lineage, the lineage of erythropoietic cells, the lineage of granulocyte and monocyte, and lineage of lymphoid cells. HDACs mechanistically alter the expression of genes that regulate hematopoiesis *via* different repressor complexes along with transcriptional factors (TFs) and other cofactors (Duan et al., 2005; Chen et al., 2006; Saleque et al., 2007; Fujiwara et al., 2010; Kong et al., 2011; van Oorschot et al., 2019).

The family of HDACs mainly consists of 18 highly conserved genes in mammals (Shah et al., 2013), and based on the presence of homology sequence to yeast and phylogenetic analysis, they are classified into four major classes named Class I (includes HDAC1, HDAC2, HDAC3, and HDAC8), Class IIA (includes HDAC4, HDAC5, HDAC7, and HDAC9), Class IIB (includes HDAC6 and HDAC10), Class III (includes sirtuin family; SIRT1-7) and Class IV (HDAC11) (Varricchio et al., 2014). HDACs of Class I, called Rpd3-like proteins, are known to have ubiquitous expression with predominant localization in the nucleus (Watamoto et al., 2003; Wada et al., 2009) and consisting of nearly 400 amino acids (Micelli &

Rastelli, 2015). However, HDACs of class II (also called Hda1-like proteins) have a selective expression that can actively shuttle between the nucleus and cytoplasm (Wada et al., 2009; Watamoto et al., 2003; Seto & Yoshida, 2014). Class IIa members generally consist of amino acids of 600 to 1,200 in number having regulatory domain at N-terminal that mediates the tissue-specific interaction with transcription factors as well as with corepressors. Additionally, their nuclear export is determined with the presence of two or three conserved residues of serine at the N-terminal domain which can be phosphorylated by the action of kinases like PKD (protein kinase D) (Parra & Verdin, 2010; Delehanty et al., 2012). In class IIb HDACs, HDAC6, and HDAC10, there is another catalytic domain along with the ubiquitin-binding zinc finger domain at the C-terminal end (Parra & Verdin, 2010; Micelli & Rastelli, 2015). The deacetylases of the sirtuin family under class III are functionally different from HDACs in that their deacetylation mechanism is NAD^+ dependent instead of Zn^{2+} (Micelli & Rastelli, 2015). HDAC11 is the lone member of class IV HDAC and is known as the smallest member of the HDAC family and is predominantly located in the nucleus with 80% of amino acids comprising the catalytic domain (Yanginlar & Logie, 2018).

5.3.1 CLASS I HDACs

Class I HDACs, HDAC1, and HDAC2 are crucial regulators in the formation of hematopoietic stem cells as well as in the maintenance of their homeostasis. Experimentally demonstrated that loss of both HDAC1 and 2 results in deregulation of genes like Dmkn, Nurcks1, and Tpt1 that are associated with the survival of stem cells, including HSCs (Heideman et al., 2014). HDAC1 knockout (KO) promotes myeloid differentiation of progenitor cells (Iwasaki et al., 2006; Yamamura et al., 2006; Wada et al., 2009). Promoter reporession of PARP1 gene by HDAC1 along with E2F4-RBL2-BRM transcription factors (TFs), results in repression of pluripotent-related TFs like NANOG, SOX2, and POU5F1 which adversely affect monocyte differentiation (Wiśnik et al., 2018). Moreover, HDAC1 is known to have a critical role in terminal differentiation in macrophage by transcriptional repression of c-Jun in association with heterodimers of MEF2A and MEF2D (Aude-Garcia et al., 2010). HDAC2 forms a core-pressor complex in association with TFs such as ETO2, SIN3A and hinder the transcription of MCSFR (granulocyte-macrophage colony-stimulating factor) and GMCSFR (granulocyte colony-stimulating factor) (Hu et al.,

2011). Both HDAC1 and HDAC2 are important components of co-repressor complexes that are formed in association with proteins such as CoREST, SIN3A, and NuRD (Ahringer, 2000; Jian et al., 2017), and in undifferentiated macrophage-erythrocyte progenitor cells (MEP) inhibit the expression of the β-globin gene in association with SIN3A and NuRD (Chen & Bieker, 2004; Wada et al., 2009; Gregory et al., 2010; Yang et al., 2012). On the other hand, they transcriptionally suppress c-myc gene *via* the CoREST repressor complex and thereby initiate the differentiation of erythroid cells (Yamamoto et al., 2018). However, HDAC1 in association with erythroid-specific TFs such as Ikaros, GATA1, and FOG1 silence γ-globin during switching to β-globin in adults (Bottardi et al., 2009) and activate γ-globin in association with NF-E4 during fetal development (Zhao et al., 2004). HDAC2 is also known for playing a role in the differentiation of B cells at the Pre-germinal center by inhibiting receptor proteins like S1pr1 and Gpr183, which are responsible for B cell trafficking (Huang et al., 2014). HDAC1/2 maintain the integrity of CD4 lineage by inhibiting CD8 lineage genes (Dovey et al., 2013; Boucheron et al., 2014).

HDAC3, another member of Class I HDACs, functionally identified to negatively regulate the expansion of HSC (Elizalde et al., 2012) and loss of which impair S-phase through inhibiting replication of DNA, in turn, hinder MPP cells progression (Summers et al., 2013). HDAC3 also inhibits genes targeted by GATA2 through directly interacting with it in hematopoietic progenitor cells (HPCs) (Ozawa et al., 2001). Like HDAC1, it is also known to interact with RUNX1 for activation of granulopoiesis (Leong et al., 2016). In association with nuclear receptor corepressor (NCoR), it was identified to repress the gene of γ-globin during terminal erythroid differentiation (Mankidy et al., 2006). Also, HDAC3 plays an important role in the maturation of B cells (Stengel et al., 2017) and crucially impedes B cell terminal differentiation to plasma cells by hindering the transcription of Prdm1 gene (Tanaka et al., 2016). Additionally, found to have a role in multiple stages of dT cell evelopment which includes the advancement of CD4 lineage (Philips et al., 2019), thymocytes positive selection (Philips et al., 2016), transition into an early double-positive stage from double-negative stage 4 (Stengel et al., 2015), and post-thymic maturation in association with NKAP (Hsu et al., 2015). HDAC8, a member of class I, shows higher expression in long term HSCs (LT-HSCs), LMPP (lymphoid-primed multipotential progenitor), and MPP cells and plays a significant functional role in maintaining HSC integrity through p53 deacetylation (Hua et al., 2017).

5.3.2 CLASS II HDACs

A class IIa member, HDAC7 identified as an important decision-making regulator in the life and death of T cells in the thymic selection state. It is responsible for the regulation of genes that determines T cells survival mediating signal transduction between TCR and downstream molecular targets (Kasler et al., 2011). Also, have a crucial regulatory function in the commitment of B-cell progenitors into its lineage from pro B to pre B cell through interacting with a transcription factor called MEF2C (Matthews et al., 2006; Azagra et al., 2016). HDAC5, also another member of class IIa, was revealed to have a functional role in maintaining homeostasis of Treg cells (Xiao et al., 2016). Also identified to play shuttling activity between cytoplasm and nucleus transporting erythroid-specific proteins like GATA1 and EKLF regulating the expression of β-globin gene and inhibition of γ-globin gene, also suppress maturation of erythroid (Watamoto et al., 2003; Delehanty et al., 2012; Varricchio et al., 2014). Moreover, found as a negative regulator of HSC through deacetylation of p65 (Huang et al., 2018).

HDAC9, a member of Class IIb HDACs, identified as having a role in the activation of the γ-globin gene in the maturation of erythroid cells (Muralidhar et al., 2011).

5.3.3 CLASS III HDACs

Members of Class III, SIRT1-SIRT7 functionally discovered to have a role of HSC protection from aging. SIRT1 promotes HSC localization in the nucleus through activating FOXO3 and inhibiting the signaling of mTOR (Leko et al., 2012; Roth et al., 2013; Rimmelé et al., 2014). SIRT3 was identified to reverse the defects in aged HSC by deacetylating mitochondrial proteins decreasing ROS (Brown et al., 2013). Through hindering Wnt target gene transcription as a result of interaction with TCF and LEF1, SIRT6 was revealed to maintain homeostasis of HSC (Wang et al., 2016). SIRT7 was also found to regulate aged HSCs homeostasis as well as prevent its differentiation into myeloid cells by inhibiting the NRF1 gene and PFSmt (Wrighton, 2015).

5.3.4 CLASS IV HDACs

HDAC11, a member of class IV HDAC, was found to negatively regulate phenotype as well as the function of T effector cells (Villagra et al., 2009).

Also play a role in the differentiation of neutrophils from pro-myelocytes (Sahakian et al., 2017).

5.4 CONCLUSION

The development and differentiation of cells is a complex process regulated by several genetic and epigenetic factors in a coordinated temporal and spatial manner. Epigenetic regulation of HSC differentiation into specific lineage is a tightly regulated process of gene expression with simultaneous activation of one lineage-specific genes and repression of other lineage-specific genes. The histone protein acetylation by HATs and deacetylation by HDACs along with coordinated involvement of various transcription factors is the major underlying mechanism in HSC lineage commitment and specification. Although, studies have shown the involvement of all the members of HATs, HDACs, and transcription factors in hematopoiesis (Figure 5.1), there is yet a huge vacuum in the understanding of the HSC lineage commitment from pluripotent stem cell to a terminally differentiated blood cell.

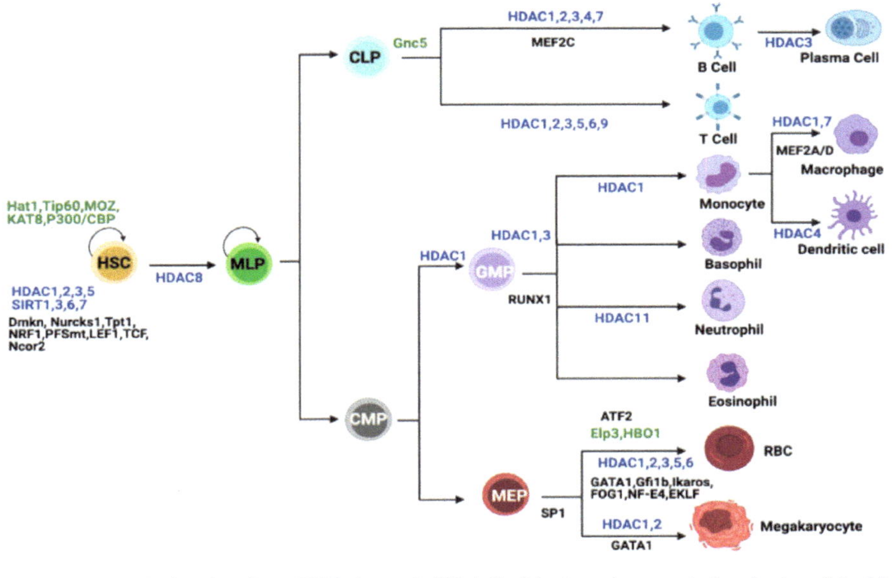

FIGURE 5.1 Role of various HATs (green), HDACs (blue), and transcription factors (black) in lineage commitment during hematopoiesis.

Source: Adapted from Wang, Wang & Liu, 2020

KEYWORDS

- **epigenetics**
- **hematopoiesis**
- **histone acetyltransferases**
- **histone deacetylases**

REFERENCES

Ahringer, J., (2000). NuRD and SIN3 histone deacetylase complexes in development. *Trends Genet., 16*(8), 351–356.

Aude-Garcia, C., Collin-Faure, V., Bausinger, H., Hanau, D., Rabilloud, T., & Lemercier, C., (2010). Dual roles for MEF2A and MEF2D during human macrophage terminal differentiation and c-Jun expression. *Biochem. J., 430*(2), 237–244.

Azagra, A., Román-González, L., Collazo, O., et al., (2016). *In vivo* conditional deletion of HDAC7 reveals its requirement to establish proper B lymphocyte identity and development. *J. Exp. Med., 213*(12), 2591–2601.

Bottardi, S., Ross, J., Bourgoin, V., et al., (2009). Ikaros and GATA-1 combinatorial effect is required for silencing of human gamma-globin genes. *Mol. Cell Biol., 29*(6), 1526–1537.

Boucheron, N., Tschismarov, R., Goeschl, L., et al., (2014). CD4(+) T cell lineage integrity is controlled by the histone deacetylases HDAC1 and HDAC2. *Nat. Immunol., 15*(5), 439–448.

Brown, K., Xie, S., Qiu, X., et al., (2013). SIRT3 reverses aging-associated degeneration. *Cell Rep., 3*(2), 319–327.

Chen, G. Y., Osada, H., Santamaria-Babi, L. F., & Kannagi, R., (2006). Interaction of GATA-3/T-bet transcription factors regulates expression of sialyl Lewis X homing receptors on Th1/Th2 lymphocytes. *Proc. Natl. Acad. Sci. USA 103*(45), 16894–16899.

Chen, X., & Bieker, J. J., (2004). Stage-specific repression by the EKLF transcriptional activator. *Mol. Cell Biol., 24*(23), 10416–10424.

Chimenti, F., Bizzarri, B., Maccioni, E., et al., (2009). A novel histone acetyltransferase inhibitor modulating Gcn5 network: Cyclopentylidene-[4-(4'-chlorophenyl)thiazol-2-yl) hydrazone. *J. Med. Chem., 52*(2), 530–536.

Delehanty, L. L., Bullock, G. C., & Goldfarb, A. N., (2012). Protein kinase D-HDAC5 signaling regulates erythropoiesis and contributes to erythropoietin cross-talk with GATA1. *Blood, 120*(20), 4219–4228.

Dou, Y., Milne, T. A., Tackett, A. J., et al., (2005). Physical association and coordinate function of the H3 K4 methyltransferase MLL1 and the H4 K16 acetyltransferase MOF. *Cell, 121*(6), 873–885.

Dovey, O. M., Foster, C. T., Conte, N., et al., (2013). Histone deacetylase 1 and 2 are essential for normal T-cell development and genomic stability in mice. *Blood, 121*(8), 1335–1344.

Doyon, Y., & Côté, J., (2004). The highly conserved and multifunctional NuA4 HAT complex. *Curr. Opin. Genet. Dev., 14*(2), 147–154.

Duan, Z., Zarebski, A., Montoya-Durango, D., Grimes, H. L., & Horwitz, M., (2005). Gfi1 coordinates epigenetic repression of p21Cip/WAF1 by recruitment of histone lysine methyltransferase G9a and histone deacetylase 1. *Mol. Cell Biol., 25*(23), 10338–10351.

Dutta, R., Tiu, B., & Sakamoto, K. M., (2016). CBP/p300 acetyltransferase activity in hematologic malignancies. *Mol. Genet. Metab., 119*(1, 2), 37–43.

Elizalde, C., Fernández-Rueda, J., Salcedo, J. M., et al., (2012). Histone deacetylase 3 modulates the expansion of human hematopoietic stem cells. *Stem Cells Dev., 21*(14), 2581–2591.

Fujiwara, T., Lee, H. Y., Sanalkumar, R., & Bresnick, E. H., (2010). Building multifunctionality into a complex containing master regulators of hematopoiesis. *Proc. Natl. Acad. Sci. USA, 107*(47), 20429–20434.

Goodman, R. H., & Smolik, S., (2000). CBP/p300 in cell growth, transformation, and development. *Genes Dev., 14*(13), 1553–1577.

Greco, T. M., Yu, F., Guise, A. J., & Cristea, I. M., (2011). Nuclear import of histone deacetylase 5 by requisite nuclear localization signal phosphorylation. *Mol. Cell Proteomics, 10*(2), M110.004317.

Gregory, G. D., Miccio, A., Bersenev, A., et al., (2010). FOG1 requires NuRD to promote hematopoiesis and maintain lineage fidelity within the megakaryocytic-erythroid compartment. *Blood, 115*(11), 2156–2166.

Heideman, M. R., Lancini, C., Proost, N., Yanover, E., Jacobs, H., & Dannenberg, J. H., (2014). Sin3a-associated Hdac1 and Hdac2 are essential for hematopoietic stem cell homeostasis and contribute differentially to hematopoiesis. *Haematologica, 99*(8), 1292–1303.

Heintzman, N. D., Hon, G. C., Hawkins, R. D., et al., (2009). Histone modifications at human enhancers reflect global cell-type-specific gene expression. *Nature, 459*(7243), 108–112.

Hsu, F. C., Belmonte, P. J., Constans, M. M., et al., (2015). Histone deacetylase 3 is required for T cell maturation. *J. Immunol., 195*(4), 1578–1590.

Hu, Z., Gu, X., Baraoidan, K., et al., (2011). RUNX1 regulates corepressor interactions of PU.1. *Blood, 117*(24), 6498–6508.

Hua, W. K., Qi, J., Cai, Q., et al., (2017). HDAC8 regulates long-term hematopoietic stem-cell maintenance under stress by modulating p53 activity. *Blood, 130*(24), 2619–2630.

Huang, C., Gonzalez, D. G., Cote, C. M., et al., (2014). The BCL6 RD2 domain governs commitment of activated B cells to form germinal centers. *Cell Rep., 8*(5), 1497–1508.

Huang, X., Guo, B., Liu, S., Wan, J., & Broxmeyer, H. E., (2018). Neutralizing negative epigenetic regulation by HDAC5 enhances human haematopoietic stem cell homing and engraftment. *Nat. Commun., 9*(1), 2741.

Iizuka, M., & Stillman, B., (1999). Histone acetyltransferase HBO1 interacts with the ORC1 subunit of the human initiator protein. *J. Biol. Chem., 274*(33), 23027–23034.

Iwasaki, H., Mizuno, S., Arinobu, Y., et al., (2006). The order of expression of transcription factors directs hierarchical specification of hematopoietic lineages. *Genes Dev., 20*(21), 3010–3021.

Jian, W., Yan, B., Huang, S., & Qiu, Y., (2017). Histone deacetylase 1 activates PU.1 gene transcription through regulating TAF9 deacetylation and transcription factor IID assembly. *FASEB J., 31*(9), 4104–4116.

Kasler, H. G., Young, B. D., Mottet, D., et al., (2011). Histone deacetylase 7 regulates cell survival and TCR signaling in CD4/CD8 double-positive thymocytes. *J. Immunol., 186*(8), 4782–4793.

Katsumoto, T., Aikawa, Y., Iwama, A., et al., (2006). MOZ is essential for the maintenance of hematopoietic stem cells. *Genes Dev., 20*(10), 1321–1330.

Kong, S., Kim, S. J., Sandal, B., et al., (2011). The type III histone deacetylase Sirt1 protein suppresses p300-mediated histone H3 lysine 56 acetylation at Bclaf1 promoter to inhibit T cell activation. *J. Biol. Chem., 286*(19), 16967–16975.

Lalonde, M. E., Avvakumov, N., Glass, K. C., et al., (2013). Exchange of associated factors directs a switch in HBO1 acetyltransferase histone tail specificity. *Genes Dev., 27*(18), 2009–2024.

Leko, V., Varnum-Finney, B., Li, H., et al., (2012). SIRT1 is dispensable for the function of hematopoietic stem cells in adult mice. *Blood, 119*(8), 1856–1860.

Leong, W. Y., Guo, H., Ma, O., Huang, H., Cantor, A. B., & Friedman, A. D., (2016). Runx1 hosphorylation by SRC increases trans-activation via augmented stability, reduced histone deacetylase (HDAC) binding, and increased DNA affinity, and activated Runx1 favors granulopoiesis. *J. Biol. Chem., 291*(2), 826–836.

Lim, D. A., Suárez-Fariñas, M., Naef, F., et al., (2006). *In vivo* transcriptional profile analysis reveals RNA splicing and chromatin remodeling as prominent processes for adult neurogenesis. *Mol. Cell Neurosci., 31*(1), 131–148.

Liu, L., Karmakar, S., Dhar, R., et al., (2013). Regulation of Gγ-globin gene by ATF2 and its associated proteins through the cAMP-response element. *PLoS One, 8*(11), e78253.

MacPherson, L., Anokye, J., Yeung, M. M., et al., (2020). HBO1 is required for the maintenance of leukaemia stem cells. *Nature, 577*(7789), 266–270.

Mankidy, R., Faller, D. V., Mabaera, R., et al., (2006). Short-chain fatty acids induce gamma-globin gene expression by displacement of a HDAC3-NCoR repressor complex. *Blood, 108*(9), 3179–3186.

Matthews, S. A., Liu, P., Spitaler, M., et al., (2006). Essential role for protein kinase D family kinases in the regulation of class II histone deacetylases in B lymphocytes. *Mol. Cell Biol., 26*(4), 1569–1577.

Micelli, C., & Rastelli, G., (2015). Histone deacetylases: Structural determinants of inhibitor selectivity. *Drug Discov. Today, 20*(6), 718–735.

Miotto, B., & Struhl, K., (2010). HBO1 histone acetylase activity is essential for DNA replication licensing and inhibited by geminin. *Mol. Cell, 37*(1), 57–66.

Mishima, Y., Miyagi, S., Saraya, A., et al., (2011). The Hbo1-Brd1/Brpf2 complex is responsible for global acetylation of H3K14 and required for fetal liver erythropoiesis. *Blood, 118*(9), 2443–2453.

Muralidhar, S. A., Ramakrishnan, V., Kalra, I. S., Li, W., & Pace, B. S., (2011). Histone deacetylase 9 activates gamma-globin gene expression in primary erythroid cells. *J. Biol. Chem., 286*(3), 2343–2353.

Ogryzko, V. V., (2001). Mammalian histone acetyltransferases and their complexes. *Cell Mol. Life Sci., 58*(5, 6), 683–692.

Ozawa, Y., Towatari, M., Tsuzuki, S., et al., (2001). Histone deacetylase 3 associates with and represses the transcription factor GATA-2. *Blood, 98*(7), 2116–2123.

Park, I. K., He, Y., Lin, F., et al., (2002). Differential gene expression profiling of adult murine hematopoietic stem cells. *Blood, 99*(2), 488–498.

Parra, M., & Verdin, E., (2010). Regulatory signal transduction pathways for class IIa histone deacetylases. *Curr. Opin. Pharmacol., 10*(4), 454–460.

Perez-Campo, F. M., Borrow, J., Kouskoff, V., & Lacaud, G., (2009). The histone acetyl transferase activity of monocytic leukemia zinc finger is critical for the proliferation of hematopoietic precursors. *Blood, 113*(20), 4866–4874.

Philips, R. L., Chen, M. W., McWilliams, D. C., Belmonte, P. J., Constans, M. M., & Shapiro, V. S., (2016). HDAC3 is required for the downregulation of RORγt during thymocyte positive selection. *J. Immunol., 197*(2), 541–554.

Philips, R. L., Lee, J. H., Gaonkar, K., et al., (2019). HDAC3 restrains CD8-lineage genes to maintain a bi-potential state in CD4(+)CD8(+) thymocytes for CD4-lineage commitment. *Elife, 8*.

Rimmelé, P., Bigarella, C. L., Liang, R., et al., (2014). Aging-like phenotype and defective lineage specification in SIRT1-deleted hematopoietic stem and progenitor cells. *Stem Cell Reports., 3*(1), 44–59.

Roth, M., Wang, Z., & Chen, W. Y., (2013). Sirtuins in hematological aging and malignancy. *Crit. Rev. Oncog., 18*(6), 531–547.

Sahakian, E., Chen, J., Powers, J. J., et al., (2017). Essential role for histone deacetylase 11 (HDAC11) in neutrophil biology. *J. Leukoc Biol., 102*(2), 475–486.

Saksouk, N., Avvakumov, N., Champagne, K. S., et al., (2009). HBO1 HAT complexes target chromatin throughout gene coding regions via multiple PHD finger interactions with histone H3 tail. *Mol. Cell, 33*(2), 257–265.

Saleque, S., Kim, J., Rooke, H. M., & Orkin, S. H., (2007). Epigenetic regulation of hematopoietic differentiation by Gfi-1 and Gfi-1b is mediated by the cofactors CoREST and LSD1. *Mol. Cell, 27*(4), 562–572.

Santillan, D. A., Theisler, C. M., Ryan, A. S., et al., (2006). Bromodomain and histone acetyltransferase domain specificities control mixed lineage leukemia phenotype. *Cancer Res., 66*(20), 10032–10039.

Secci, D., Carradori, S., Bizzarri, B., et al., (2014). Synthesis of a novel series of thiazole-based histone acetyltransferase inhibitors. *Bioorg. Med. Chem., 22*(5), 1680–1689.

Seto, E., & Yoshida, M., (2014). Erasers of histone acetylation: The histone deacetylase enzymes. *Cold Spring Harb. Perspect. Biol., 6*(4), a018713.

Shah, R. R., Koniski, A., Shinde, M., et al., (2013). Regulation of primitive hematopoiesis by class I histone deacetylases. *Dev. Dyn., 242*(2), 108–121.

Smith, E. R., Cayrou, C., Huang, R., Lane, W. S., Côté, J., & Lucchesi, J. C., (2005). A human protein complex homologous to the *Drosophila* MSL complex is responsible for the majority of histone H4 acetylation at lysine 16. *Mol. Cell Biol., 25*(21), 9175–9188.

Squatrito, M., Gorrini, C., & Amati, B., (2006). Tip60 in DNA damage response and growth control: Many tricks in one HAT. *Trends Cell Biol., 16*(9), 433–442.

Stengel, K. R., Barnett, K. R., Wang, J., et al., (2017). Deacetylase activity of histone deacetylase 3 is required for productive VDJ recombination and B-cell development. *Proc. Natl. Acad. Sci. USA, 114*(32), 8608–8613.

Stengel, K. R., Zhao, Y., Klus, N. J., et al., (2015). Histone deacetylase 3 is required for efficient T cell development. *Mol. Cell Biol., 35*(22), 3854–3865.

Summers, A. R., Fischer, M. A., Stengel, K. R., et al., (2013). HDAC3 is essential for DNA replication in hematopoietic progenitor cells. *J. Clin. Invest., 123*(7), 3112–3123.

Sun, X. J., Man, N., Tan, Y., Nimer, S. D., & Wang, L., (2015). The role of histone Acetyltransferases in normal and malignant hematopoiesis. *Front Oncol., 5*, 108.

Tan, M., Luo, H., Lee, S., et al., (2011). Identification of 67 histone marks and histone lysine crotonylation as a new type of histone modification. *Cell, 146*(6), 1016–1028.

Tanaka, H., Muto, A., Shima, H., et al., (2016). Epigenetic regulation of the blimp-1 gene (PRDM1) in B cells involves bach2 and histone deacetylase 3. *J. Biol. Chem., 291*(12), 6316–6330.

Thomas, T., Corcoran, L. M., Gugasyan, R., et al., (2006). Monocytic leukemia zinc finger protein is essential for the development of long-term reconstituting hematopoietic stem cells. *Genes Dev., 20*(9), 1175–1186.

Torok, M. S., & Grant, P. A., (2004). Histone acetyltransferase proteins contribute to transcriptional processes at multiple levels. *Adv. Protein Chem., 67*, 181–199.

Valerio, D. G., Xu, H., Eisold, M. E., Woolthuis, C. M., Pandita, T. K., & Armstrong, S. A., (2017). Histone acetyltransferase activity of MOF is required for adult but not early fetal hematopoiesis in mice. *Blood, 129*(1), 48–59.

Van, O. R., Hansen, M., Koornneef, J. M., et al., (2019). Molecular mechanisms of bleeding disorder associated GFI1B(Q287*) mutation and its affected pathways in megakaryocytes and platelets. *Haematologica., 104*(7), 1460–1472.

Varricchio, L., Dell'Aversana, C., Nebbioso, A., et al., (2014). Identification of NuRSERY, a new functional HDAC complex composed by HDAC5, GATA1, EKLF and pERK present in human erythroid cells. *Int. J. Biochem. Cell Biol., 50*, 112–122.

Villagra, A., Cheng, F., Wang, H. W., et al., (2009). The histone deacetylase HDAC11 regulates the expression of interleukin 10 and immune tolerance. *Nat. Immunol., 10*(1), 92–100.

Visel, A., Blow, M. J., Li, Z., et al., (2009). ChIP-seq accurately predicts tissue-specific activity of enhancers. *Nature, 457*(7231), 854–858.

Wada, T., Kikuchi, J., Nishimura, N., Shimizu, R., Kitamura, T., & Furukawa, Y., (2009). Expression levels of histone deacetylases determine the cell fate of hematopoietic progenitors. *J. Biol. Chem., 284*(44), 30673–30683.

Wang, H., Diao, D., Shi, Z., et al., (2016). SIRT6 controls hematopoietic stem cell homeostasis through epigenetic regulation of Wnt signaling. *Cell Stem Cell., 18*(4), 495–507.

Watamoto, K., Towatari, M., Ozawa, Y., et al., (2003). Altered interaction of HDAC5 with GATA-1 during MEL cell differentiation. *Oncogene, 22*(57), 9176–9184.

Wiśnik, E., Płoszaj, T., & Robaszkiewicz, A., (2018). Author correction: Downregulation of PARP1 transcription by promoter-associated E2F4-RBL2-HDAC1-BRM complex contributes to repression of pluripotency stem cell factors in human monocytes. *Sci. Rep., 8*(1), 5764.

Wrighton, K. H., (2015). Stem cells: SIRT7, the UPR and HSC ageing. *Nat. Rev. Mol. Cell Biol., 16*(5), 266–267.

Xiao, H., Jiao, J., Wang, L., et al., (2016). HDAC5 controls the functions of Foxp3(+) T-regulatory and CD8(+) T cells. *Int. J. Cancer, 138*(10), 2477–2486.

Yamamoto, R., Kawahara, M., Ito, S., et al., (2018). Selective dissociation between LSD1 and GFI1B by a LSD1 inhibitor NCD38 induces the activation of ERG super-enhancer in erythroleukemia cells. *Oncotarget, 9*(30), 21007–21021.

Yamamura, K., Ohishi, K., Katayama, N., et al., (2006). Pleiotropic role of histone deacetylases in the regulation of human adult erythropoiesis. *Br. J. Haematol., 135*(2), 242–253.

Yang, T., Jian, W., Luo, Y., et al., (2012). Acetylation of histone deacetylase 1 regulates NuRD corepressor complex activity. *J. Biol. Chem., 287*(48), 40279–40291.

Yang, X. J., (2004). The diverse superfamily of lysine acetyltransferases and their roles in leukemia and other diseases. *Nucleic Acids Res., 32*(3), 959–976.

Yanginlar, C., & Logie, C., (2018). HDAC11 is a regulator of diverse immune functions. *Biochim. Biophys Acta Gene Regul. Mech., 1861*(1), 54–59.

Yuan, H., & Marmorstein, R., (2013). Histone acetyltransferases: Rising ancient counterparts to protein kinases. *Biopolymers, 99*(2), 98–111.

Zhao, Q., Cumming, H., Cerruti, L., Cunningham, J. M., & Jane, S. M., (2004). Site-specific acetylation of the fetal globin activator NF-E4 prevents its ubiquitination and regulates its interaction with the histone deacetylase, HDAC1. *J. Biol. Chem., 279*(40), 41477–41486.

Zhou, Y., Peng, J., & Jiang, S., (2014). Role of histone acetyltransferases and histone deacetylases in adipocyte differentiation and adipogenesis. *Eur. J. Cell Biol., 93*(4), 170–177.

CHAPTER 6

Systems Biology Approaches Toward Understanding Human Long Noncoding RNA in Hematopoietic Cells

SWATI DAHARIYA,[1] SANJEEV RAGHUWANSHI,[2] and RAVI KUMAR GUTTI[1]

[1]*Department of Biochemistry, School of Life Sciences, University of Hyderabad, Hyderabad, Telangana, India*

[2]*Department of Pharmacology and Regenerative Medicine, College of Medicine, University of Illinois, Chicago, USA*

ABSTRACT

Human transcriptome consists of various RNA biotypes, including long noncoding RNAs (lncRNAs). lncRNAs are the largest class of noncoding genes in the human genome. Over the past decade, only a small subset of lncRNA class has risen to prominence with newly defined functions, like post-transcriptional regulation of gene expression, chromatin modifications, protein interactions, imprinting, etc., however, the functional significance of a majority of lncRNAs in the regulation of biological processes still remains an enigma. Notably, several lncRNAs have been identified with clinical potential in the diagnosis, prognosis, and treatment of various diseases. Increasing evidence of the implications of lncRNAs in various diseases has further enhanced the need to gain more specific mechanistic understanding of the lncRNA functions. System biology has emerged as an integrative field of study that approach mathematical modeling, computational procedures, hypothesis generation, high-throughput experimental technologies, analytical testing within biological systems to understand how these complex

Hematopoiesis: Biochemical, Cellular, Molecular, and Genomic Perspectives. Ravi Kumar Gutti (Ed.)

interactions applied to the physiological functions, phenotype and behavior of particular biological systems of an organism. Systems biology approaches provide immense opportunities for understanding the functional role of lncRNAs in blood cell development. This chapter presents a comprehensive review of the various system biology approaches and tools with the potential in identifying and annotation of lncRNAs in hemopoietic cells. This chapter also discusses integrative computational approaches used for the functional characterization of lncRNAs.

6.1 NONCODING RNAs

Noncoding RNA (ncRNA) will not usually directly code for protein and were defined as the junk DNA in the genome with little to no functional significance. Recently, several research groups have started focusing on the functional evaluation of ncRNAs in genome regulation. In current research, ncRNAs have become one of the major focuses in the field offunctional genomics. The class of ncRNAs is a functionally diverse set of transcripts, most of them were initially discovered through earlier research investigation studying expressed sequence tags (ETS). Many of them are long noncoding RNA (lncRNA), which contribute 80% to the noncoding transcriptome. The lncRNAs is a class of ncRNAs with distinct molecular functions. They generally act by interacting with other biomolecules (DNA, RNA, proteins or small molecules) in various mechanisms and regulate transcriptional, post-transcriptional, translational or post-translational phases to affect gene activity. These functional characteristics of lncRNAs are broadly classified as guides, decoys, signals, and scaffolds, depending upon the types of molecular interactions in the cells (Wang & Chang, 2011). These interactions of lncRNAs with other biomolecules in the cells results in a wide spectrum of functional outcomes such as chromatin remodeling, RNA interference, or scaffolding of biomolecules. Evidently, lncRNAs play a major role in cell proliferation, apoptosis, differentiation, and several essential biological processes. The lncRNA functional diversity has also been implicated in a number of disease conditions such as cancers, developmental disorders, neurological, metabolic, and immunological disorders (Jalali et al., 2015).

The total number of lncRNAs in human genome, other animals, or plants is still not accurately defined, because the lncRNAs are being identified in different research studies, and the information of novel lncRNAs is getting updated in the reference databases. However, in different databases, the number of identified lncRNAs varies widely. This difference could be because of the differences in the definition of lncRNAs, e.g., the database GENCODE v35 (Frankish et al., 2019) lists 17,957 lncRNA genes, whereas LncBook (Ma et al., 2019) reports as high as 268,848 LncRNA transcripts. On the other hand, comparing with GENCODE, FANTOM CAT generated a comprehensive atlas of 27,919 human lncRNA genes, of which 19,175 were reported to be potentially functional lncRNAs in the human genome (Hon et al., 2017).

The lncRNAs have been reported in different animals and plants, however, the information about lncRNAs is most comprehensive for humans. Moreover, different research groups are trying to dissect the distinct functional roles and molecular interactions involving lncRNAs. In the last two decades, despite large numbers of reported lncRNAs, the functional characterization has been done only for a small subset of lncRNAs. It would be noteworthy to mention that in the future research, more lncRNAs are likely to be discovered and a majority of them are yet to be functionally annotated. The majority of lncRNAs annotated in the past are the reports from the analysis of transcriptome sequencing datasets and the functional information on these transcripts is largely limited to the genomic loci and expression patterns in studies performed in different models and conditions. Thus, there is a large gap in understanding the functional role of lncRNAs in different cellular systems and conditions. This gap in understanding the functional role of lncRNAs is mostly owing to the unavailability of tools to characterize the biomolecular interactions of these transcripts on genome-scale and limited resources to catalog this information. Systems biology approaches toward understanding human long noncoding RNA in hematopoietic cells, and other systems are integrating the information from different genome-scale datasets, expression patterns, and functional and structure annotations. These Computational approaches may provide immense support to uncover the functional role of lncRNAs.

6.2 lncRNA DATA RESOURCES

Several data resources are available online, which contain information about lncRNA sequences, RNA-interactome, and functional annotations. Some of these data resources contain experimentally validated data, others are curated databases, which provide information from literature, and few others report information from in silico predictions. Table 6.1 list out some of the well-known databases and resources containing lncRNA interactome and annotations.

TABLE 6.1 List of Databases and Resources Containing lncRNA Interactome and Annotations

Database Name	Latest Release [Last Modified]	Description	Web Link	References
GENCODE	Human:35 [08.2020]; Mouse: M25 [04.2020]	Encyclopedia of all the experimentally validated annotations of the genes in the human and mouse genome.	http://www.gencodegenes.org/	Frankish et al. (2019)
FANTOM-CAT	FANTOM CAT1.0.0 [08.2017]	FANTOM5 CAGE data performed on multiple transcript collections to generate a comprehensive catalog of 23887 high-confidence 5′ complete human lncRNA genes and their expression profiles across 1,829 samples from the major human primary cell types and tissues.	https://fantom.gsc.riken.jp/cat/	Hon et al. (2017)
FANTOM-6	[07.2020]	Functional annotation of long noncoding RNAs in the human genome.	https://fantom.gsc.riken.jp/6/	Ramilowski et al. (2020)
RNAcentral	Release 16 [18.09.2020]	RNA central is a free, public resource that offers integrated access to a comprehensive and up-to-date set of ncRNA sequences integrated from multiple expert databases. LncRNA relevant databases included are LNCipedia, LncBook, NONCODE, lncRNAdb and LncBase. RNAcentral enables integrated text search, sequence similarity search, bulk downloads, and programmatic data access.	https://rnacentral.org/	RNAcentral Consortium (2020)
LNCipedia	Version 5.2 [02.08.2018]	A database for annotated human lncRNA transcript sequences and structures.	http://www.lncipedia.org/	Volders et al. (2019)
lncBook	Version 1.2 [11.19.2019]	A curated knowledgebase of human long noncoding RNAs.	https://bigd.big.ac.cn/lncbook/index	Ma et al. (2019)
LncRNADisease	Version 2.0	An updated database of long noncoding RNA-associated diseases, the LncRNADisease v2.0 now hosts disease associated LncRNAs and CircRNAs.	http://www.rnanut.net/lncrnadisease/index.php/	Bao et al. (2019)
RNAInter	Version 2.0	RNAInter (RNA interactome database), a repository for RNA-associated interactions.	http://www.rna-society.org/raid/	Lin et al. (2020)

6.3 SYSTEMS BIOLOGY APPROACHES FOR lncRNA ANNOTATION AND EXPRESSION ANALYSIS

In earlier studies, the large-scale annotations of lncRNAs have largely emerged from the annotation of ESTs and from the transcriptome profiles generated using tiling microarrays. These studies have listed the large number of new transcripts which were unknown in previous transcriptome data, many of these newly identified transcripts do not have any obvious potential to code for functional protein (Bertone et al., 2004; ENCODE Project Consortium, 2007; Kapranov et al., 2007). Further, the existence of many of these transcripts were validated using deep sequencing from the ENCODE consortium (Bernstein et al., 2012). Most studies utilize the conventional microarray-based approach to profile the known transcripts, contrary deep sequencing approaches, e.g., RNA-Seq is the platform which facilitates large scale genome wide transcriptome profiling thereby uncover many rare transcripts and alternatively spliced isoforms (Hoeijmakers et al., 2013; Mutz et al., 2013; Hrdlickova et al., 2017). Further in depth and focused approaches would be required to uncover the large number of novel transcripts. A generalized computational analysis pipeline for the annotation of the lncRNA is a process of transcriptome study which completes in three major steps: (i) alignment of reads against a reference genome; (ii) transcripts and isoforms assembly; and (iii) transcripts and isoforms annotation for coding potential. Some other approaches, such as proteogenomics approaches and RNA secondary structure-based approaches, are also being used to annotate lncRNAs (Table 6.2).

TABLE 6.2 Approaches Used to Investigate the Coding Potential of a lncRNA

Methods	Advantages	Disadvantages
Bioinformatic tools (e.g., CPC, CPAT, and CNCI)	Rapid and easy-to-use web interface to assess the protein-coding potential of a transcript.	Result outputs are considered based on probability rather than definitive evidence.
In vitro transcription/ translation assays.	Low hands-on time.	Limitations of overexpression system.
Genetic modification.	Endogenous expression testing.	Time taking process.
Ribosome profiling methods.	Allow genome wide mapping of Ribosome binding sites.	Possibilities of technical concerns.
Mass spectrometry-based proteomics.	Definitive	Require tedious procedures and expensive instrumentation.

6.3.1 ALIGNMENT/MAPPING OF READS AND TRANSCRIPTS ASSEMBLY

The process of taking short nucleotide sequences/sequencing reads and assigning them to specific locations in the reference genome is called read alignment or read mapping. This is the process that determines the location of short reads in the reference genome and the first step in transcriptome analysis pipeline. In a conventional transcriptome analysis process the high-quality reads are aligned to the respective reference genomes using spliced read aligners or mapping tools such as TopHat (Kim et al., 2013) or STAR (Dobin et al., 2013). The individual transcripts from RNA-seq reads that have been aligned to the genome are further reconstructed or assembled using reference guided assemblers like Cufflinks (Trapnell et al., 2010) or Scripture (Guttman et al., 2010). The Cufflinks and Scripture methods are based on two different algorithms, the Cufflinks is a probabilistic approach, on the other hand, Scripture follows a segmentation approach. However, both are methods for transcriptome reconstruction that relies solely on RNA-Seq reads and an assembled genome to build a transcriptome ab initio. The major difference in these two methods is that scripture is a very stringent assembler and doesn't report lowly expressed transcripts, but cufflinks does. Also, there are reports on de novo approaches which were attempted for transcriptome reconstruction in many organisms (Grabherr et al., 2011; Li et al., 2009; Zerbino & Birney, 2008). There are several methods available for spliced alignment and transcript reconstruction, theextensive evaluation and discussion of several RNA-Seq aligners (Engstrom et al., 2013) and transcriptome assemblers (Steijger et al., 2013) is available in recent literature.

6.3.2 ANNOTATION OF LONG NONCODING RNAs (lncRNAs)

The study of lncRNAs involves the identification and documentation of non-protein coding transcripts (>200 nucleotides long) that lack an open reading frame (ORF) of significant length. In the lncRNA annotation study, the main step is to predict ORFs, the tools like GetORF from the EMBOSS suite (Rice et al., 2000) are used to find and extract open reading frames. An ORF may be defined as a region of a specified minimum size between two STOP codons, or between a START and a STOP codon. In public databases such as SwissPort and International Protein Index, more than 95% of the mammalian proteins have an ORF of greater than 100 amino acids. According to the SwissProt and International Protein Index, more than 95% of mammalian proteins have an ORF of greater than 100 amino acids (Frith et al., 2006). In lncRNA annotation studies, the cutoff of 100 amino acids is adapted. However, in some studies the ORF length cutoff ranging from 30 to 100 amino acids is also used. To assess a transcript's coding potential, several features are extracted from the transcript's sequence. A true protein-coding transcript is more likely to have a long and high-quality ORF compared with a noncoding transcript. However, in many recent studies, the translated short peptides from short sORFs and their function as biologically active peptide have also drawn attention (Andrews & Rothnagel, 2014). Other than the ORF length, several other distinguishing features such as ORF conservation, nucleotide composition, residue bias and codon usage are also examined in the lncRNA annotation process. Distinguishing protein-coding from non-protein-coding sequence is the first and most crucial step in genome annotation; programs like PhyloCSF (a comparative genomics method to distinguish protein-coding and noncoding regions) (Lin et al., 2011) and CRITICA (Coding Region Identification Tool Invoking Comparative Analysis; Badger & Olsen, 1999) are the cross-species comparison-based programs to distinguish protein-coding and noncoding regions. On the other hand, different complementary approaches like CSTMiner (Mignone et al., 2003) and RNAcode (Washietl et al., 2011) are also popular, these methods analyze nucleotide substitution and gap patterns to assess the coding potential of a sequence. Apart from these methods, many other popular approaches, such as coding potential calculator (CPC), a support vector machine-based classifier to assess the protein-coding potential of a transcript (Kong et al., 2007), PORTRAIT (Prediction of transcriptomic ncRNA by ab initio methods; Arrial et al., 2009) which uses multiple distinct sequence features and coding potential assessment tool (CPAT) which uses an alignment-free logistic regression model (Wang et al., 2013). Alternately, to detect whether a novel nucleotide sequence is a protein-coding gene or not, there are many sequence similarity search programs like BLASTX (Gish & States, 1993) and HMMER3 (Eddy, 2011). Moreover, apart from systems biology approaches, many recent studies have used ribosome profiling data to assess the protein-coding capacity of lncRNAs (Bazzini et al., 2014).

6.3.3 PROTEOGENOMICS APPROACHES TO EVALUATE CODING POTENTIAL OF TRANSCRIPTS

Many functional studies on lncRNAs have confirmed that some lncRNAs indeed had small sORF (length <300 nt) that could code for a short peptide with key biological functions (Andrews & Rothnagel, 2014; Huang et al., 2017). These lncRNAs could function, with both the RNA and peptide, and therefore suggested to be classified as bifunctional RNAs. Proteogenomics has recently emerged as a powerful technique to enhance our ability to annotate protein-coding genes and sequences that code both the lncRNA and small peptides. In the last five years, efficient methodologies and software pipelines have been developed in proteogenomics (Nesvizhskii, 2014). In these studies, transcripts are scanned for the presence of putative peptides using tandem mass spectrometry studies. For example, the recent ENCODE study by Banfai et al. (2012) analyzed RNA-Seq data along with the proteomics data and identified that ~92% of annotated lncRNAs are not translated.

Recently, different research groups have introduced some novel strategies like ribosome profiling (the deep sequencing of ribosome-protected RNA fragments). In these strategies, the ribosomal occupancy on transcript is utilized to indicate the protein-coding potential of a transcript (Ingolia et al., 2011; Jalali et al., 2015). However, these studies have also demonstrated that ribosomal occupancy alone does not sufficiently indicate the protein-coding potential of a transcript (Brannan et al., 1990; Ingolia et al., 2011). A recently defined metric called ribosome release score (RRS) (Guttman et al., 2013) could be one of the potential proteogenomics approaches to correctly distinguish between coding and noncoding RNA.

6.3.4 STRUCTURAL APPROACHES TO SCREEN FOR lncRNAs

There are several reports on the use of tools that classify transcripts on the basis of their conserved secondary structure. These tools can classify transcripts as lncRNAs based on the presence of conserved predicted RNA secondary structures. There is a list of software tools and web portals used for RNA structure prediction: QRNA (Rivas & Eddy, 2001) – QRNA uses comparative genome sequence analysis to detect conserved RNA secondary structures; RNAFold (Hofacker, 2003) – this is a minimum free energy (MFE) RNA structure prediction algorithm which calculate MFE secondary structures and partition function of RNAs; EvoFold (Pedersen et al., 2006)

– EvoFold is a comparative method for identifying functional RNA structures in multiple-sequence alignments; RNAz (Gruber et al., 2010) – RNAz is a widely used software package for predicting structurally conserved and thermodynamically stable RNA secondary structures in comparative genomics data, it can be used for de novo detection of structured noncoding RNAs. These software packages and online tools have been used widely in several research studies (Jalali et al., 2015) and discussed for their advantages and drawbacks (Gorodkin & Hofacker, 2011). lncRNA prediction based on RNA secondary structure and conservation alone could be misleading, because these conserved secondary structures could also be shared between coding and ncRNAs (Dinger et al., 2008). Because of these limitations, the focus of recent research studies is sifted toward the analysis of RNA structure and binding properties for functional annotation of lncRNAs over their detection (Glazko et al., 2012).

6.4 COMPUTATIONAL APPROACHES TO UNDERSTAND lncRNA FUNCTION

In previous studies, a large number of lncRNAs have been discovered through high-throughput profiling and system biology approaches, but the functional annotation and molecular mechanisms of gene regulation are not fully investigated. Unlike protein-coding gene, the unique and complex properties of lncRNAs, such as functional diversity, tissue-specific peaks in expression, and low conservation among species makes much more puzzling to unravel the cellular functions of lncRNAs.The functional characterization of lncRNAs is still in its infancy and highly complicated layer of lncRNAs regulation, it is because of the dynamic interactions of lncRNAs with various molecules like DNA, RNA, or proteins (Figure 6.1). Evidence support that the functional appearances of lncRNAs is depend on their location and interaction capability with different proteins, chromatin, and other RNAs molecules (Mercer et al., 2009). LncRNAs triggers the genetic manipulation, which specifically targets regulatory regions of genes to control the transcription and translation processes. Genetic manipulation strategies like RNA interference and CRISPR-Cas system-based experimental techniques are a powerful alternative approach to characterize the functional roles of lncRNAs (Wong et al., 2018). Because of its limitations, such as high off-target effects, triggering an immune response resulting in toxicity and the ethical debates have stalled the progress of the conventional methods used to study functional characterization of lncRNAs.

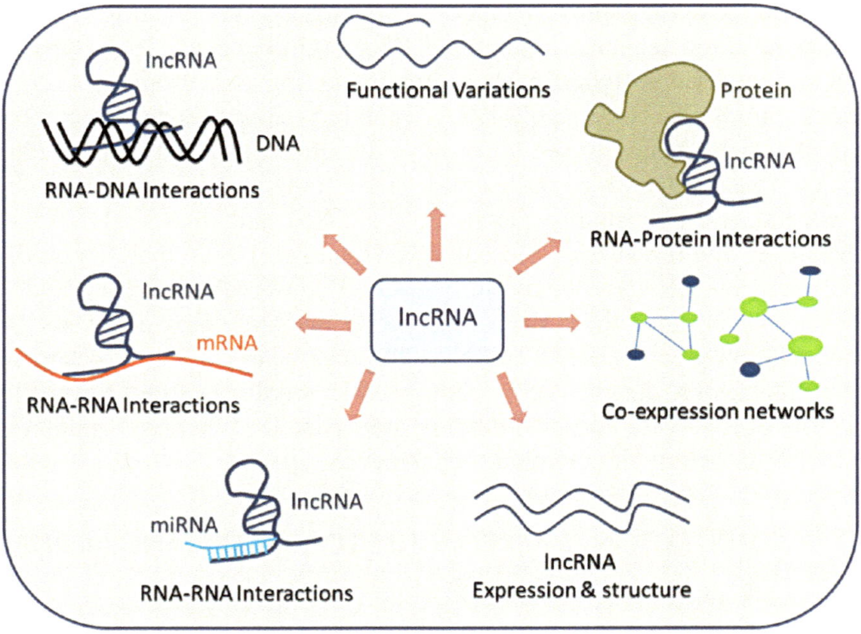

FIGURE 6.1 The schematic representation showing the summary of current approaches used to understand functional characteristics of lncRNAs.

Systems biology uses a wide range of quantitative experimental and computational methodologies to provide systems-level insights to understand the dynamic interaction of lncRNAs with several molecules at various scales, within cells, tissues, organs, and organisms. "Guilt by association" is one of the most popular approaches which can foretell the putative functional roles of lncRNAs by carrying out correlation statics calculation between specific lncRNAs and protein-coding genes expression profiling (Klomp & Furge, 2012; Lee et al., 2004; Wolfe et al., 2005). Generally, to show the correlation between sets of gene co-expressed with target lncRNAs, a gene module map can be constructed. Besides, gene set enrichment analysis correlation has been used to evaluate the genes positively or negatively co-regulated against specific interest lncRNA (Stuart et al., 2003; Subramanian et al., 2005). Later to explore the curated tools like gene ontology (GO) to determine biological processes, Molecular Signatures Database (MSigDB c2 collection) for several signaling pathways and Kyoto Encyclopedia of Genes and Genomes (KEGG), etc., for highly correlated protein coding gene sets are linked to the target lncRNAs function. In addition, a few bioinformatics based web tools and databases have been developed on enrichment strategies to forecast functionality of lncRNAs in *in-silico*. This also offers lncRNAs expression profiles and other information related to lncRNAs like its localization, structures and conservation status. For the functional analysis of lncRNAs, several web-based tools, such as LncRNAdb, LNCipedia, lncRNome, Co-lncRNA, StarBase v2.0, lncRNADisease, lnc2Cancer, and lncRNA-Screen, etc., have emerged, which will eventually make it possible to identify, study and understand the relationship between lncRNAs with other molecules as well as their pathophysiology (Fritah et al., 2014; Gong et al., 2017; Jalali et al., 2012; Washietl et al., 2012; Yotsukura et al., 2017). These databases are web-based platforms that interconnect extensive information related to lncRNAs, which increases our understanding about lncRNAs significance in several biological processes of blood cells (Figure 6.1).

6.5 LncRNAs FUNCTION IN BLOOD CELLS

LncRNAs are involved in several essential biological processes such as cell proliferation, apoptosis, differentiation of blood cells, and cancer cells metabolic reprogramming (Dahariya et al., 2019; Liu et al., 2019). LncRNAs can have several functional features, such as scaffolding, decoying, guiding, and signaling. These functional features depend on their interactions with other biomolecules (DNA, RNA, proteins, and small molecules) in cells (Li et al., 2019; Wang & Chang, 2011; Dahariya et al., 2019). LncRNAs can also act as transcription regulators by regulating RNA splicing, mRNA processing, and transportation (Goff & Rinn, 2015). There are several lncRNAs that were reported as dynamic scaffolds, and they are involved in epigenetic modification on chromatin structure and affect gene expression (Khalil et al., 2009).

The latest developments and advancements in high throughput RNA-sequencing technologies provide the possibility of discovering several types of ncRNA transcripts, in particular lncRNAs. However, the functional features of lncRNAs are yet to be understood. Hence, deciphering the functional role of lncRNAs has become an interesting topic in genome research. In the last two decades, our understanding on the lncRNAs functional aspects at various levels, such as genomics, transcriptomics, proteomics, and metabolomics within the cells, tissues, organs as well as organisms has been greatly expanded due to the rapid growth in high throughput sequencing experiments as well as resulted in massive amounts of data (Guh, 2020; Li et al., 2016). The expression of lncRNAs is the basic information to discover the principle functions of lncRNAs from a complete genome of the cell by using an ensemble approaches of systems biology. Hence, highly advanced throughput technologies combined with bioinformatics analysis in order to determine the core of molecular characterization of lncRNAs in hematopoietic system, which otherwise remains elusive. System biology is a vast growing field with numerous approaches to studying lncRNAs.

6.6 DATABASES AND TOOLS FOR lncRNA STUDY IN HEMATOPOIETIC CELLS

Over the past many years, extensive research and a tremendous amount of expression and epigenetic profiling data is generated from different hematopoietic cells. This data can be used for molecular characterization of different hematopoietic cells and functional elucidation of lncRNAs in different blood cells. System-wide gene expression data can be observed by different experimental techniques such as Serial Analysis of Gene Expression (SAGE), Northern and Western blot, DNA microarray, High throughput RNA-Seq, and rapid amplification of cDNA ends (RACE). The whole transcriptome sequencing provides a comprehensive view of the cellular transcriptional profile at given biological conditions and greatly enhances the power to discover thousands of lncRNAs.

There is a large number of transcriptomic datasets which allows identification of lncRNAs in the whole eukaryotic genomes. The study of RNA-Seq raw data has empowered the documentation of genes which are expressed along with open reading frames (ORFs), also letting the proof of identity for a number of lncRNAs expressed in hematopoietic cells. LncRNAs expression profiles significantly vary in various hematopoietic cells; the expression profiles of lncRNAs can be compared between several committed hematopoietic lineages to conclude their relationships between them. For example, Zhu et al. (2019) has examined the differences in circulating lncRNAs signature expression in eosinophilic and neutrophilic asthma samples from the transcriptome of peripheral whole blood. Immunochip and transcriptome-based analysis across four populations of cord blood-derived T-helper cells and seven populations of peripheral blood leukocytes were compared between each other to infer the association between identified lncRNAs, mapping to specific location shared between autoimmune and immune-related disorders (AID) are significantly enhanced in immune cell types compared to lncRNAs from the whole genome (Ahmad et al., 2020; Hrdlickova et al., 2014).

Performing such type of study on a larger scale, requires specialized database considerations because experimental-based expression profiling of RNA-Seq data is very costly and difficult to run in precision. Existing publicly available hematopoietic databases as BloodExpress, HAEMCODE, CODEX and ImmGen, etc. Table 6.3 reviews the repositories of datasets, restricted

TABLE 6.3 List of Databases and Repositories for lncRNA Research in Blood Cells

Databases	URL	Cell Types	Main Feature	Organism	References
Blueprint	http://dcc.blueprint-epigenome.eu/#/home	Bone marrow, cord blood, lymph node, thymus cells, etc.	RNA-seq, Bisulfite Seq, DNAme-seq, ATAC Seq, DNAseI-seq, ChIP-seq data, epigenomic data.	Human	Fernández et al. (2016)
BloodChIP	http://www.med.unsw.edu.au/CRCWeb.nsf/page/BloodChIP	Human CD34+ and leukemic cells.	Gene expression data, histone and transcription factors ChIP-seq data, DNase-seq data and digital genomic footprinting data.	Human	Chacon et al. (2014)
CODEX	http://codex.stemcells.cam.ac.uk/	Hematopoietic and embryonic stem cells.	NGS data, ChIP-Seq of transcription factors and histones, RNA-Seq, DNase-Seq and ATAC-Seq	Human and mouse	Sánchez-Castillo et al. (2015)
HEMBASE	http://hembase.niddk.nih.gov/	Tyes of blood cells and erythrocytes.	Erythroblast expressed sequenced tags (ESTs), gene annotation.	Human	Goh et al. (2004)
ENCODE	https://www.encodeproject.org/	Primary and hematopoietic cells.	RNA Seq, chip seq and several types of ontology.	Human	Hong et al. (2016)
ENA	https://www.ebi.ac.uk/ena/browser/home	All types of hematopoietic cells.	Nucleotide and raw sequencing data, sequence assembly data and functional annotation.	Human and other vertebrates	–
NCBI GEO	https://www.ncbi.nlm.nih.gov/geo/	All types of hematopoietic cells.	Gene expression data, high-throughput functional genomics data and hybrid array and sequence-based data.	Human and other vertebrates.	Clough & Barrett (2016)
ImmGen	http://www.immgen.org/	Primary hematopoietic cells, myeloid and lymphoid cells.	Gene expression, regulatory module and genetic mutation.	Human and mouse	Heng et al. (2008)

to lncRNAs expression-based studies in blood cells. But they differ based on specific cell types, platform sources, and omics applications. European Nucleotide Archive (https://www.ebi.ac.uk/ena/), UCSC Genome Browser (https://genome.ucsc.edu/), Gene Expression Omnibus (https://www.ncbi.nlm.nih.gov//geo/), European Bioinformatics Institute (http://www.ebi.ac.uk/), National Center for Biotechnology Information (https://www.ncbi.nlm.nih.gov/) are some of the publicly available repository databases which provide transcriptome data.

KEYWORDS

- **blood cells**
- **coding potential**
- **hematopoiesis**
- ***in silico* algorithm**
- **lncRNA**
- **lncRNA annotation**
- **noncoding RNA**

REFERENCES

Ahmad, I., Valverde, A., Ahmad, F., & Naqvi, A. R., (2020). Long noncoding RNA in myeloid and lymphoid cell differentiation, polarization and function. *Cells, 9*(2), 269.

Andrews, S. J., & Rothnagel, J. A., (2014). Emerging evidence for functional peptides encoded by short open reading frames. *Nature Reviews Genetics, 15*(3), 193–204.

Arrial, R. T., Togawa, R. C., & Brigido, M., (2009). Screening noncoding RNAs in transcriptomes from neglected species using PORTRAIT: Case study of the pathogenic fungus *Paracoccidioides brasiliensis*. *BMC Bioinformatics, 10*, 239.

Badger, J. H., & Olsen, G. J., (1999). CRITICA: Coding region identification tool invoking comparative analysis. *Molecular Biology and Evolution, 16*(4), 512–524.

Bao, Z., Yang, Z., Huang, Z., Zhou, Y., Cui, Q., & Dong, D., (2019). LncRNADisease 2.0: An updated database of long noncoding RNA-associated diseases. *Nucleic Acids Research, 47*(D1), D1034–D1037.

Bazzini, A. A., Johnstone, T. G., Christiano, R., et al., (2014). Identification of small ORFs in vertebrates using ribosome foot printing and evolutionary conservation. *The EMBO Journal, 33*(9), 981–993.

Bertone, P., Stolc, V., Royce, T. E., et al., (2004). Global identification of human transcribed sequences with genome tiling arrays. *Science, 306*(5705), 2242–2246.

Chacon, D., Beck, D., Perera, D., Wong, J. W. H., & Pimanda, J. E., (2014). BloodChIP: A database of comparative genome-wide transcription factor binding profiles in human blood cells. *Nucleic Acids Research, 42*(D1), D172.

Clough, E., & Barrett, T., (2016). The gene expression omnibus database. *Methods in Molecular Biology, 1418*, 93–110.

Dahariya, S., Paddibhatla, I., Kumar, S., et al., (2019). Long non-coding RNA: Classification, biogenesis and functions in blood cells. *Mol Immunol. 112*, 82–92. doi:10.1016/j.molimm.2019.04.011.

Dinger, M. E., Pang, K. C., Mercer, T. R., & Mattick, J. S., (2008). Differentiating protein-coding and noncoding RNA: Challenges and ambiguities. *PLoS Computational Biology, 4*(11), e1000176.

Dobin, A., Davis, C. A., Schlesinger, F., et al., (2013). STAR: Ultrafast universal RNA-seq aligner. *Bioinformatics, 29*(1), 15–21.

Eddy, S. R., (2011). Accelerated profile HMM searches. *PLoS Computational Biology, 7*(10), e1002195.

ENCODE Project Consortium, Birney, E., Stamatoyannopoulos, J. A., Dutta, A., Guigó, R., Gingeras, T. R., De Jong, P. J., (2007). Identification and analysis of functional elements in 1% of the human genome by the ENCODE pilot project. *Nature, 447*(7146), 799–816.

Engström, P. G., Steijger, T., Sipos, B., et al., (2013). Systematic evaluation of spliced alignment programs for RNA-seq data. *Nature Methods, 10*(12), 1185–1191.

Fernández, J. M., De La Torre, V., Richardson, D., et al., (2016). The BLUEPRINT data analysis portal. *Cell Systems, 3*(5), 491–495.

Frankish, A., Diekhans, M., Ferreira, A. M., et al., (2019). GENCODE reference annotation for the human and mouse genomes. *Nucleic Acids Research, 47*(D1), D766–D773.

Fritah, S., Niclou, S. P., & Azuaje, F., (2014). Databases for lncRNAs: A comparative evaluation of emerging tools. *RNA, 20*(11), 1655–1665.

Frith, M. C., Forrest, A. R., Nourbakhsh, E., et al., (2006). The abundance of short proteins in the mammalian proteome. *PLoS Genetics, 2*(4), e52.

Gish, W., & States, D. J., (1993). Identification of protein coding regions by database similarity search. *Nature Genetics, 3*(3), 266–272.

Glazko, G. V., Zybailov, B. L., & Rogozin, I. B., (2012). Computational prediction of polycomb-associated long noncoding RNAs. *PLoS One, 7*(9), e44878.

Goff, L. A., & Rinn, J. L., (2015). Linking RNA biology to lncRNAs. *Genome Research, 25*(10), 1456–1465).

Goh, S. H., Lee, Y. T., Bouffard, G. G., & Miller, J. L., (2004). Hembase: Browser and genome portal for hematology and erythroid biology. *Nucleic Acids Research, 32*.

Gong, Y., Huang, H. T., Liang, Y., Trimarchi, T., Aifantis, I., & Tsirigos, A., (2017). lncRNA-screen: An interactive platform for computationally screening long noncoding RNAs in large genomics datasets. *BMC Genomics, 18*(1), 434.

Gorodkin, J., & Hofacker, I. L., (2011). From structure prediction to genomic screens for novel noncoding RNAs. *PLoS Computational Biology, 7*(8), e1002100.

Grabherr, M. G., Haas, B. J., Yassour, M., et al., (2011). Full-length transcriptome assembly from RNA-Seq data without a reference genome. *Nature Biotechnology, 29*(7), 644–652.

Gruber, A. R., Findeiß, S., Washietl, S., Hofacker, I. L., & Stadler, P. F., (2010). RNAz 2.0: Improved noncoding RNA detection. *Pacific Symposium on Biocomputing. Pacific Symposium on Biocomputing*, 69–79.

Guh, C. Y., Hsieh, Y. H., & Chu, H. P., (2020). Functions and properties of nuclear lncRNAs-from systematically mapping the interactomes of lncRNAs. *Journal of Biomedical Science, 27*(1), 1–14).

Guttman, M., Garber, M., Levin, J. Z., et al., (2010). Ab initio reconstruction of cell type-specific transcriptomes in mouse reveals the conserved multi-exonic structure of lincRNAs. *Nature Biotechnology, 28*(5), 503–510.

Guttman, M., Russell, P., Ingolia, N. T., Weissman, J. S., & Lander, E. S., (2013). Ribosome profiling provides evidence that large noncoding RNAs do not encode proteins. *Cell, 154*(1), 240–251.

Heng, T. S. P., Painter, M. W., Elpek, K., et al., (2008). The immunological genome project: Networks of gene expression in immune cells. *Nature Immunology, 9*(10), 1091–1094.

Hoeijmakers, W. A., Bártfai, R., & Stunnenberg, H. G., (2013). Transcriptome analysis using RNA-Seq. *Methods in Molecular Biology, 923*, 221–239.

Hofacker, I. L., (2003). Vienna RNA secondary structure server. *Nucleic Acids Research, 31*(13), 3429–3431.

Hon, C. C., Ramilowski, J. A., Harshbarger, J., et al., (2017). An atlas of human long noncoding RNAs with accurate 5' ends. *Nature, 543*(7644), 199–204.

Hong, E. L., Sloan, C. A., Chan, E. T., et al., (2016). Principles of metadata organization at the ENCODE data coordination center. *Database, 2016*, 1.

Hrdlickova, B., Kumar, V., Kanduri, K., Zhernakova, et al., (2014). Expression profiles of long noncoding RNAs located in autoimmune disease-associated regions reveal immune cell-type specificity. *Genome Medicine, 6*(10).

Hrdlickova, R., Toloue, M., & Tian, B., (2017). RNA-Seq methods for transcriptome analysis. *RNA, 8*(1), 10.1002/wrna.1364.

Huang, J. Z., Chen, M., Chen, D., et al., (2017). A peptide encoded by a putative lncRNA HOXB-AS3 suppresses colon cancer growth. *Molecular Cell, 68*(1), 171–184.

Humann, F. C., Tiberio, G. J., & Hartfelder, K., (2013). Sequence and expression characteristics of long noncoding RNAs in honey bee caste development – potential novel regulators for transgressive ovary size. *PLoS One, 8*(10), e78915.

Ingolia, N. T., Lareau, L. F., & Weissman, J. S., (2011). Ribosome profiling of mouse embryonic stem cells reveals the complexity and dynamics of mammalian proteomes. *Cell, 147*(4), 789–802.

Jalali, S., Jayaraj, G. G., & Scaria, V., (2012). Integrative transcriptome analysis suggests processing of a subset of long noncoding RNAs to small RNAs. *Biology Direct, 7*.

Jalali, S., Kapoor, S., Sivadas, A., Bhartiya, D., & Scaria, V., (2015). Computational approaches towards understanding human long noncoding RNA biology. *Bioinformatics, 31*(14), 2241–2251.

Kapranov, P., Cheng, J., Dike, S., et al., (2007). RNA maps reveal new RNA classes and a possible function for pervasive transcription. *Science, 316*(5830), 1484–1488.

Khalil, A. M., Guttman, M., Huarte, M., et al., (2009). Many human large intergenic noncoding RNAs associate with chromatin-modifying complexes and affect gene expression. *Proceedings of the National Academy of Sciences of the United States of America, 106*(28), 11667–11672.

Kim, D., Pertea, G., Trapnell, C., Pimentel, H., Kelley, R., & Salzberg, S. L., (2013). TopHat2: Accurate alignment of transcriptomes in the presence of insertions, deletions and gene fusions. *Genome Biology, 14*(4), R36.

Klomp, J. A., & Furge, K. A., (2012). Genome-wide matching of genes to cellular roles using guilt-by-association models derived from single sample analysis. *BMC Research Notes, 5*, 370.

Lee, H. K., Hsu, A. K., Sajdak, J., Qin, J., & Pavlidis, P., (2004). Coexpression analysis of human genes across many microarray data sets. *Genome Research, 14*(6), 1085–1094.

Li, R., Yu, C., Li, Y., et al., (2009). SOAP2: An improved ultrafast tool for short read alignment. *Bioinformatics, 25*(15), 1966, 1967.

Li, R., Zhu, H., & Luo, Y., (2016). Understanding the functions of long noncoding RNAs through their higher-order structures. *International Journal of Molecular Sciences, 17*(5).

Li, Y., Egranov, S. D., Yang, L., & Lin, C., (2019). Molecular mechanisms of long noncoding RNAs-mediated cancer metastasis. *Genes, Chromosomes and Cancer, 58*(4), 200–207.

Lin, M. F., Jungreis, I., & Kellis, M., (2011). PhyloCSF: A comparative genomics method to distinguish protein coding and noncoding regions. *Bioinformatics, 27*(13), i275–i282.

Lin, Y., Liu, T., Cui, T., et al., (2020). RNAInter in 2020: RNA interactome repository with increased coverage and annotation. *Nucleic Acids Research, 48*(D1), D189–D197.

Liu, H., Luo, J., Luan, S., He, C., & Li, Z., (2019). Long noncoding RNAs involved in cancer metabolic reprogramming. *Cellular and Molecular Life Sciences, 76*(3), 495–504.

Ma, L., Cao, J., Liu, L., et al., (2019). LncBook: A curated knowledgebase of human long noncoding RNAs. *Nucleic Acids Research, 47*(D1), D128–D134.

Mercer, T. R., Dinger, M. E., & Mattick, J. S., (2009). Long noncoding RNAs: Insights into functions. *Nature Reviews Genetics, 10*(3), 155–159.

Mignone, F., Grillo, G., Liuni, S., & Pesole, G., (2003). Computational identification of protein coding potential of conserved sequence tags through cross-species evolutionary analysis. *Nucleic Acids Research, 31*(15), 4639–4645.

Mutz, K. O., Heilkenbrinker, A., Lönne, M., Walter, J. G., & Stahl, F., (2013). Transcriptome analysis using next-generation sequencing. *Current Opinion in Biotechnology, 24*(1), 22–30.

Nesvizhskii, A. I., (2014). Proteogenomics: Concepts, applications and computational strategies. *Nature Methods, 11*(11), 1114–1125.

Pedersen, J. S., Bejerano, G., Siepel, A., et al., (2006). Identification and classification of conserved RNA secondary structures in the human genome. *PLoS Computational Biology, 2*(4), e33.

Ramilowski, J. A., Yip, C. W., & Agrawal, S., (2020). Functional annotation of human long noncoding RNAs via molecular phenotyping. *Genome Research, 30*(7), 1060–1072.

Rice, P., Longden, I., & Bleasby, A., (2000). EMBOSS: The European molecular biology open software suite. *Trends in Genetics, 16*(6), 276–277.

Rivas, E., & Eddy, S. R., (2001). Noncoding RNA gene detection using comparative sequence analysis. *BMC Bioinformatics, 2*, 8.

RNAcentral Consortium, (2020). RNAcentral 2021: Secondary structure integration, improved sequence search and new member databases. *Nucleic Acids Research*. gkaa921.

Sánchez-Castillo, M., Ruau, D., Wilkinson, A. C., et al., (2015). CODEX: A next-generation sequencing experiment database for the haematopoietic and embryonic stem cell communities. *Nucleic Acids Research, 43*(D1), D1117–D1123.

Steijger, T., Abril, J. F., Engström, P. G., et al., (2013). Assessment of transcript reconstruction methods for RNA-seq. *Nature Methods, 10*(12), 1177–1184.

Stuart, J. M., Segal, E., Koller, D., & Kim, S. K., (2003). A gene-coexpression network for global discovery of conserved genetic modules. *Science, 302*(5643), 249–255.

Subramanian, A., Tamayo, P., & Mootha, V. K., (2005). Gene set enrichment analysis: A knowledge-based approach for interpreting genome-wide expression profiles. *Proceedings of the National Academy of Sciences of the United States of America, 102*(43), 15545–15550.

Trapnell, C., Williams, B. A., Pertea, G., et al., (2010). Transcript assembly and quantification by RNA-Seq reveals unannotated transcripts and isoform switching during cell differentiation. *Nature Biotechnology, 28*(5), 511–515.

Volders, P. J., Anckaert, J., Verheggen, K., et al., (2019). LNCipedia 5: Towards a reference set of human long noncoding RNAs. *Nucleic Acids Research, 47*(D1), D135–D139.

Wang, K. C., & Chang, H. Y., (2011). Molecular mechanisms of long noncoding RNAs. *Molecular Cell, 43*(6), 904–914.

Wang, L., Park, H. J., Dasari, S., Wang, S., Kocher, J. P., & Li, W., (2013). CPAT: Coding-potential assessment tool using an alignment-free logistic regression model. *Nucleic Acids Research, 41*(6), e74.

Washietl, S., Findeiss, S., Müller, S. A., Kalkhof, S., Von, B. M., Hofacker, I. L., Stadler, P. F., & Goldman, N., (2011). RNAcode: Robust discrimination of coding and noncoding regions in comparative sequence data. *RNA, 17*(4), 578–594.

Washietl, S., Will, S., Hendrix, D. A., Goff, L. A., Rinn, J. L., Berger, B., & Kellis, M., (2012). Computational analysis of noncoding RNAs. *RNA, 3*(6), 759–778.

Wolfe, C. J., Kohane, I. S., & Butte, A. J., (2005). Systematic survey reveals general applicability of 'guilt-by-association' within gene coexpression networks. *BMC Bioinformatics, 6.*

Wong, N. K., Huang, C. L., Islam, R., & Yip, S. P., (2018). Long noncoding RNAs in hematological malignancies: Translating basic techniques into diagnostic and therapeutic strategies. *Journal of Hematology and Oncology, 11*(1), 131.

Yotsukura, S., Du Verle, D., Hancock, T., Natsume-Kitatani, Y., & Mamitsuka, H., (2017). Computational recognition for long noncoding RNA (lncRNA): Software and databases. *Briefings in Bioinformatics, 18*(1), 9–27.

Zerbino, D. R., & Birney, E., (2008). Velvet: Algorithms for de novo short read assembly using de Bruijn graphs. *Genome Research, 18*(5), 821–829.

Zhu, Y., Mao, D., Gao, W., Han, G., & Hu, H., (2019). Analysis of lncRNA expression in patients with eosinophilic and neutrophilic asthma focusing on LNC-000127. *Frontiers in Genetics, 10.*

CHAPTER 7

Clinical Outcomes of Defective Hematopoiesis

GEETA RAI, DOLI DAS, KHUSHBU PRIYA, and HIRAL THACKER

Department of Molecular and Human Genetics, Institute of Science, Banaras Hindu University, Varanasi, Uttar Pradesh, India

ABSTRACT

The process of production of immune cells through hematopoiesis is an interesting paradigm in stem cell research providing an array of researchable components. Hematopoietic stem cells can differentiate into multiple committed lineages, ensuring the generation of cells with specific functions and also retaining the self-renewing capability. HSCs play a major role in disease biology as irregularities in HSCs generation and differentiation lead to life-threatening diseases. In-depth studies of these defective mechanisms not only reveal the pathophysiology of hematological diseases, including malignancies, immunodeficiencies, bone marrow (BM) failure syndromes, and hemoglobinopathies, but also helps to understand the survival and function of HSCs, thus opening a new door of translational research.

7.1 INTRODUCTION

Hematopoietic stem cells (HSCs) are an atypical population of cells capable of both self-renewal and differentiation into blood cells. HSCs are in charge for the formation of new cells and the maintenance of the immune system (Ivanovs et al., 2017). The generation of immune cells like granulocytes, platelets, monocyte/macrophages, erythrocytes, lymphocytes, mast cells,

Hematopoiesis: Biochemical, Cellular, Molecular, and Genomic Perspectives. Ravi Kumar Gutti (Ed.)

and dendritic cells (DC) from HSCs is termed as hematopoiesis. Various molecules like cytokines, growth factors, and transcription factors regulate the formation of a specific blood cell lineage from HSCs. Due to the relatively short lifespan of many mature types of blood cells, HSCs are necessary to replenish these cells throughout the lifetime of the individual through hematopoiesis (Lee et al., 2019). Long-term HSCs (LT-HSCs) can both – self-renew themselves as well as differentiate into multiple lineages whereas short-term HSCs (ST-HSCs) as also known as progenitors' cells can only differentiate into a specific or limited lineage with a lifespan of merely a few weeks (Pinho & Frenette, 2019).

The hematopoietic road map is an assemblage of several lineage differentiation routes that a stem cell takes to differentiate into different blood cell types. Early in the development, hematopoiesis occurred in yolk sac and embryo proper (Dieterlen-Lievre, 1975); however, in fully developed embryo, HSCs migrate to the fetal liver and proliferate in the bone marrow (BM) for future production in adult life (Lacaud & Kouskoff, 2017). LT-HSC differentiates into ST-HSC, which in turn differentiates into multipotent progenitors (MPP), with reduced self-renewability. CD34 enriched MPP further bifurcates into common lymphoid precursor (CLP) and common myeloid precursor (CMP). CLPs are the earliest lymphoid progenitors from which T lymphocytes, B lymphocytes, DC, and natural killer) cells are produced while CMP further differentiates into megakaryocyte erythroid precursor (MEP) and granulocyte monocyte precursor (GMP). Hao et al. (2001) in his study showed that, CLP activity is found in the $CD7^+$ fraction of $CD7^+$, $CD34^+$, $CD38^-$, $CD45RA^+$ population. However, the same population was absent in the adult BM (Hao et al., 2001). Since IL-7Rα, is an important marker for the murine CLP, is expressed in human $CD10^+$ CLPs in the BM, but not in $CD7^+$ CLPs in the cord blood of human. These findings suggest that the IL7 signaling plays a critical role in human hematopoiesis. The myeloid-erythroid pathway – CMP, MEP, and GMP subsets are isolated within $CD34^+$ and $CD38^+$ fraction. MEP further gives rise to proerythroblast and megakaryocyte that differentiates into erythrocytes and platelets, respectively while the GMP subset ($CD34^+$, $CD38^+$, $CD48^+$, and $CD123^+$) differentiates into granulocyte (neutrophils, eosinophils, and basophils) and agranulocyte (monocytes and mast cell) lineages. The development of B lymphocytes from CLP population is continued in the BM while the maturation of T lymphocyte occurs in the thymus in the case of humans. Different T lymphocyte subset are produced on the basis of its marker and T cell receptor (TCR) development – CD4 T cell ($CD3^+$, $CD4^+$, $CD45RA^+$ and TCR α/β),

CD8 cytotoxic T cell (CD3$^+$, CD8$^+$, and TCR α/β). Similarly, B lymphocytes are also differentiated into – memory B cells (CD27$^+$, CD80$^+$, CD84$^+$, and CD86$^+$) and Plasma cells (CD27$^+$, CD38$^+$, CD78$^+$, and CD138$^+$). All these hematolymphoid progenitors arise from CD34$^+$, CD38$^-$ HSCs while they have no self-renewability (Figure 7.1). Although, the self-renewal capacity HSCs acts as therapeutic cure (BM transplantation and induced pluripotent stem cells-based treatment) for many blood cells malignancies, BM failures syndromes and immunodeficiencies, Improper regulatory mechanisms and modifications of the regulating/controlling genes of hematopoiesis leads to hematological diseases. These disorders include impaired production or lack of important molecules of immune cells, defective maturation of essential components (defective globin chains) and irregular enhanced production of HSC productions (leukemias) (Figure 7.2). We described below various diseases/disorders associated with defective hematopoiesis.

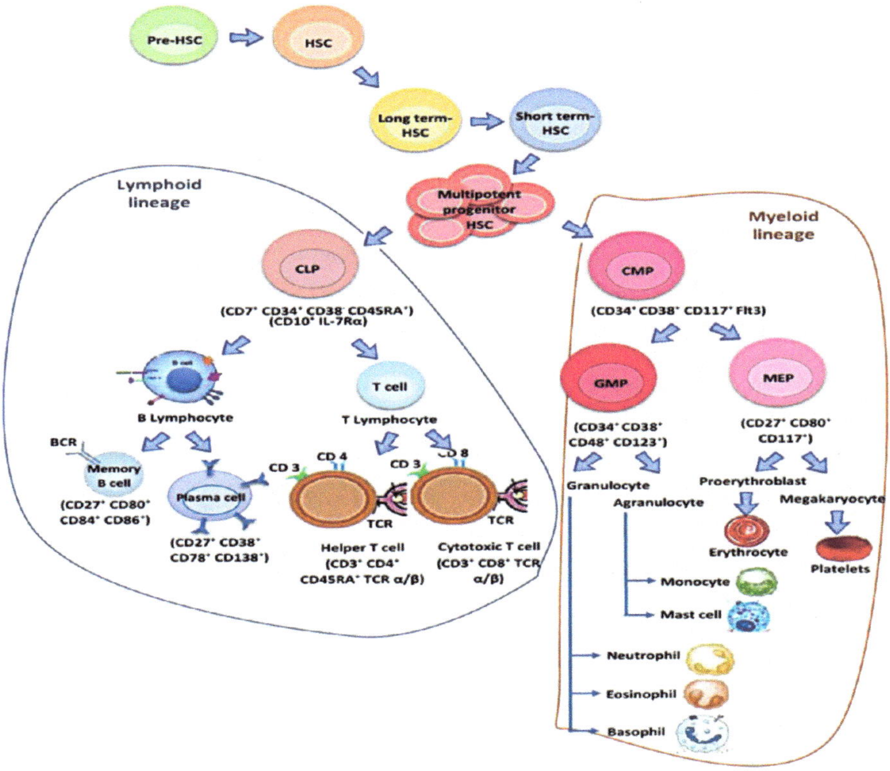

FIGURE 7.1 The hematopoietic road map showing hierarchy of differentiation pathways that a stem cell takes to form different blood cells.

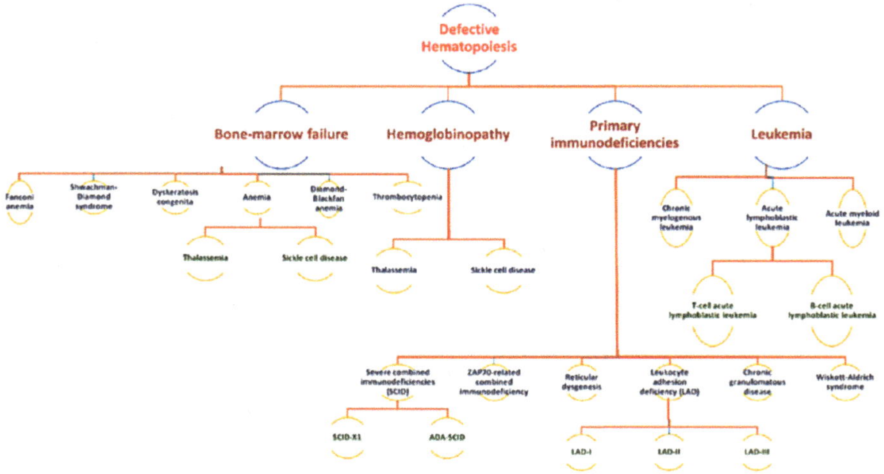

FIGURE 7.2 Disease/disorders associated with defective hematopoiesis. Classification of hematological diseases caused due to aberrant regulatory mechanisms and modifications in different genes of hematopoiesis.

7.2 BONE MARROW FAILURE (BMF)

Bone marrow failure (BMF) syndromes are generally inherited disorders which rarely occur due to defects in developments deficiency hematopoietic segment. Genetic mutation or polymorphism is associated with BMF syndromes. This is a heterogeneous disorder that results into an increased risk of leukemia and cytopenias (Dokal & Vulliamy, 2010). The various types of BMFs can be described in subsections.

7.2.1 FANCONI ANEMIA (FA)

The most common BMF syndrome is Fanconi anemia (FA) which is a recessive autosomal disorder results from mutation in FA pathways genes required for DNA repair mechanism. The complication in FA patients includes anomalies in eyes, skin, ears, kidneys, short stature and endocrine blemishes along with acute myelogenous leukemia and BMF. Two different zebrafish models for FA have been reported for the in-depth study of FA. The first is a fancd2 morphant whose phenotype resembles with FA affected children characterized by microcephaly, shortened body length, and microopthalmia as a result of an escalation in instinctive instability in chromosome architecture (Liu et al., 2003). The second model of zebrafish depicts aloss of function of rad51 a DNA recombination gene, which leads to the development of an FA-like phenotype involving chromosomal instability, shortened body length, and hypocellular kidney marrow (Botthof et al., 2017).

Müller et al. (2012) demonstrate that impaired FA DNA-repair pathway compromises the reprogramming efficiency in primary cells of human and murine. This proves that active DNA repair mechanisms is required for reprogramming of induced pluripotent stem cells (iPSCs) from somatic cells and these genes play a crucial role to conserve multipotency for both BM and iPSCs generated in laboratory. These iPSCs are capable of upholding normal karyotype with HSCs differentiation into other cells. Further lentiviral mediated gene therapy in iPSCs exclusivly for FA patients resulting into differentiation into healthy myeloid and erythroid progenitor cells (Río et al., 2008; Jacome et al., 2009; Raya et al., 2009).

7.2.2 SHWACHMAN-DIAMOND SYNDROME

Shwachman-Diamond syndrome (SDS) is caused by failure of improper maturation of 60S ribosomal subunit which leads to exocrine pancreatic insufficiency and congenital neutropenia (Finch et al., 2011). It is manifested by skeletal abnormalities, hematopoietic defects with exocrine pancreatic insufficiency, where most of the patients suffer from neutropenia and a risk of leukemia. Almost 90% of SDS patients have mutations in the Shwachman-Bodian-Diamond syndrome (SBDS) gene, coding an important ribosome biogenesis protein (Burroughs et al., 2009). Further, mutations in SRP54 results into congenital neutropenia and SDS-like characteristics in individuals (Carapito et al., 2017; Bellanné-Chantelot et al., 2018). Carapita et al. (2017) established a srp54-knockdown zebrafish model where they have restrained the srp54 to promote exocrine pancreas defects neutropenia in embryos of zebrafish (Carapito et al., 2017). Tulpule et al. (2013) show that disease modeling using iPSCs shows enhanced protease levels concerning hematopoietic and pancreatic manifestations of SDS. This study implies that myeloid cells and pancreatic acinar are autodigested by elevated proteases. Hence, anti-proteolytic molecules like aprotinin act as potential therapeutic care.

7.2.3 DYSKERATOSIS CONGENITA (DC)

Dyskeratosis congenita (DC) is a rare inherited disorder that is manifested by BMF susceptibility to cancer, dysfunction of integumentary system mucocutaneous abnormalities and premature aging of the skin, nails, and hair. DC patients' derived reprogramed iPSCs revealed the importance of

DC-associated pathway in pluripotency. Dyskerin gene mutation leads to inhibition of telomerase assembly; further mutations in telomerase reverse transcriptase restricts elongation of telomerase; and WRAP53 mutation triggers improper localization of telomerase leading to consequent shortening of telomerase in BM. Further, these defects result in loss of pluripotency, deteriorating the endurance of iPSCs (Batista et al., 2011).

7.2.4 DIAMOND-BLACKFAN ANEMIA (DBA)

Diamond-Blackfan anemia (DBA) is a heterogeneous erratic disorder. It is manifested as erythrocyte aplasia, normally appearing before one year of age, with broad range of phenotypic abnormalities, such as skeletal deformations and short body stature (Diamond et al., 1961; Ito et al., 2010; Engidaye et al., 2019). Mutations in the ribosomal protein coding genes have been reported in above 50% of DBA patients (Vlachos & Muir, 2010; Taylor & Zon, 2011). Common studies suggest upregulation of the p53 pathway on ribosomal protein deficiency in all zebrafish models. Moreover, involvement of P53 independent mechanisms have been also reported, as knocking down of tp53 leads to inability inrestoring BM defects (Danilova et al., 2008, 2011; Torihara et al., 2011; Zhang et al., 2013, 2014; Yadav et al., 2014; Antunes et al., 2015; Wan et al., 2016; Chakraborty et al., 2018). In addition to this, an exogenous nucleoside when applied to DBA embryo, shows downregulation of tp53 with hematopoiesis restoration and reduction of apoptosis (Danilova et al., 2014). Danilova et al. (2018) reported that overexpression of inflammatory pathways, the complement system, and interferons, with the help of two prototypes – rps19 morphants and rpl11 mutants indicating the involvement of the immune system in the pathogenesis of DBA. Interestingly, two amino acids, L-leucine and L-arginine have shown potential to cure the DBA symptoms by regulating the mTOR signaling mechanism (Payne et al., 2012; Narla et al., 2014; Yadav et al., 2014). This study directed the first clinical pilot phase I/II study of leucine therapy in DBA patents got facilitated by this study (https://clinicaltrials.gov/ct2/show/NCT01362595). Further, a molecule known as SMER28 (6-bromo-N-2-propenyl-4-quinazolinamine), which is an inducer of ATG5-dependent autophagy, has been reported in a screen using DBA induced pluripotent stem cells and confirmed in zebrafish models (Doulatov et al., 2017).

7.2.5 THROMBOCYTOPENIA

Thrombocytopenias, which are heterogeneous disorders, are defined by deficiency in platelet counts ($< 1.5 \times 10^5/\mu l$) (Gauer & Braun, 2012). Lin et al. (2017) developed a thrombocytopenic model in zebrafish by inducing a mutation in the *mpl* gene with transcription activator-like effector nuclease (TALENs). Interestingly, CRISPR/Cas9 method has been used for the restoration of phenotypes by ablating the *ptprj* in zebrafish (Marconi et al., 2019).

7.2.6 ANEMIA

It is defined by deficiency of erythrocytes in the circulation zebrafish models have been reconstructed for the study of hereditary spherocytosis and hereditary elliptocytosis, which are two types of hemolytic anemia caused by abnormal membrane cytoskeleton. Moreover, another form of anemia is Hypochromic anemia which is manifested by pale and small erythrocytes, due to shortage of iron or globin (Iolascon et al., 2009). Hypochromic microcytic anemia occurs due to faulty fabrication of embryonic globin (Brownlie et al., 2003), whereas congenital sideroblastic anemia is caused due to the disruption of heme biosynthesis (Brownlie et al., 1998).

7.3 HEMOGLOBINOPATHY

Hemoglobinopathy, considered as serious hematological disorders which are caused by by irregulation in hemoglobin. Anomaly in globin chains in hemoglobin leads to hemoglobinopathy.

7.3.1 THALASSEMIA

It is characterized by deficiency in healthy hemoglobin resulting in damaged and weak red blood cells. Beta thalassemia is instigated by a splice site mutation affecting the hemoglobin beta globin chain transcription, which leads to insufficient productions of functional $\alpha 2\beta 2$ hemoglobin chains due to the mutation, resulting in anemia. Scientists have shown that iPSCs can be used as a therapeutic cure by reprograming and fixing the genes beta thalassemia affected cells into iPSCs; and then transfusing them beta thalassemia patients (Ye et al., 2009). Researchers have also developed lentiviral mediated practices for improving lentiviral integration sites of the beta globin gene insertion in genomic position as a therapeutic option with high beta globin gene expression without manipulating nearby genes (Papapetrou et al., 2011).

7.3.2 SICKLE CELL DISEASE

Sickle cell disease occurs due to polymorphism in the beta globin chain is considered as a serious hemoglobinopathy. It is characterized by abrupt sickle-shaped red blood cells. Scientists have corrected a hemoglobin A allele (HbS/HbA) into iPSCs derived from sickle cell patients (HbS/HbS) with gene specific plasmids and zinc finger nucleases, expressing 25–40% of the HbA transcript (Zou et al., 2011), showing reliability in the mRNA levels in heterozygous carriers of asymptomatic sickle cell disease. Alteration of gene by zinc finger nuclease increases the homologous recombination opens a door for future gene therapy perspectives modulating the ease of transplantation (Sebastiano et al., 2011).

7.4 PRIMARY IMMUNODEFICIENCIES

Primary Immunodeficiencies (PIDs) are characterized by dysregulated immunity leading to recurrent life-threatening infections and disease conditions (Raje & Dinakar, 2015).

7.4.1 SEVERE COMBINED IMMUNODEFICIENCIES

Severe combined immunodeficiency (SCID) is caused due to impairment in immune system development and function leading to in life-threatening infections in children with SCID (Rai et al., 2020). SCID is a heterogeneous condition as approximately 20 different genetic causes have been identified (Picard et al., 2015). The most common X-linked severe combined immunodeficiency (SCID-X1) and SCID caused by deficiency in adenosine deaminase (ADA) deficiency (ADA-SCID) represents for 40% and 10% of all SCID forms, respectively. It is characterized by defective T cells differentiation. It is a fatal disease in children as it can claim the life of the children if it remains untreated. Fortunately, both SCID-X1 and ADA-SCID have been successfully cured with gene therapy (Cavazzana et al., 2019).

7.4.2 ZAP70-RELATED COMBINED IMMUNODEFICIENCY

ZAP70-related combined immunodeficiency is an unusual form of SCID. It is characterized by a mutation in ZAP70 leads to abruption of TCR signaling, leading to a dearth of peripheral $CD8^+$ and impaired CD4 T^+cells. In addition, the absence of T immunoglobulin production by B cells gets affected due to lack of T-cells (Arpaia et al., 1994; Elder et al., 1994, 1995; Elder, 1996). Using zebrafish model a mutant has been developed by TALENs which shows promising results in reconstituting the defective immunity observed in humans (Moore et al., 2016).

7.4.3 RETICULAR DYSGENESIS

Reticular dysgenesis (RD) with SCID presents in combination with agranulocytosis and sensorineural deafness in patients. RD occurs due to a mutation in the AK2 gene which translates into adenylate kinase 2 enzyme. This enzyme catalyzes the generation of ADP by transferring phosphate molecule from ATP to AMP, leading to ADP production (Dzeja et al., 1998). Recently, transplantation of HSCs has shown promising results to cure RD in patients (Hoenig et al., 2017).

7.4.4 CHRONIC GRANULOMATOUS DISEASE (CGD)

Chronic granulomatous disease (CGD) is the most common inherited PID manifested by dysregulated inflammation, autoimmunity, and life-threatening infections due to defective NADPH oxidase complex in granulocytes and monocytes (Arnold & Heimall, 2017). It has been reported that a useful NADPH oxidase complex is required for reactive oxygen species mediated killing of pathogens engulfed via phagocytosis (Brothers et al., 2011; Yang et al., 2012; Harvie & Huttenlocher, 2015). Further research is going on to improve gene therapy to cure CGD.

7.4.5 WISKOTT-ALDRICH SYNDROME

Wiskott-Aldrich syndrome (WAS) is characterized by mutations in the X-linked WAS gene, which encodes the WAS protein (WASp). HSCs are the place of origin of WASp. It plays a crucial role in communicating signals from cell surface to the actin cytoskeleton. Mutations in the WAS gene causes severe primary immunodeficiency with severe clinical manifestations like microthrombocytopenia, persisting infections with eczema (Massaad et al., 2013). Clinical trial based on LV encoding the human WASP cDNA under the regulation of the human endogenous promoter reconstitutes the T and B cell immunity which makes patients free from recurrent infections with less frequent autoimmune conditions. However, studies are still going on to improve the treatment and poor prognosis also reported in patients (Cavazzana et al., 2019).

7.4.6 LEUKOCYTE ADHESION DEFICIENCY (LAD)

Leukocyte adhesion deficiency (LAD) is defined by malfunctions in leukocyte adhesion. LAD is divided into three forms (LAD I-III). LAD-I is hallmarked by lack or impaired expression of β2 integrins, whereas defective in fucosylation of selectin ligands leads to LAD-II and in LAD-III patients suffer due to defects in integrin signaling (Harris et al., 2013). However, it has been noticed that all LAD patients have enriched circulating neutrophils.

7.5 LEUKEMIA

Leukemias are manifested by unwarranted and uncontrolled augmentation of hematopoietic cells, leading to an excessive amount of immature hematopoietic cells.

7.5.1 CHRONIC MYELOGENOUS LEUKEMIA (CML)

Chronic myelogenous leukemia (CML) is a cancer which affects both BM and blood and can be treated with imatinib, an inhibitor of tyrosine kinase, which aims the BCR-ABL fusion product. The Food and Drug Administration approved Gleevec, also known as imatinib mesylate, an oral drug for the treatment of CML. The new drug exclusively targets the mutant protein, originated exclusively in CML cancer cells. It basically restricts the progenitor cells (white blood cells) proliferation by blocking signaling mechanism (Iqbal & Iqbal, 2014).

7.5.2 ACUTE MYELOID LEUKEMIA (AML)

Acute myeloid leukemia (AML) is described by uncontrolled propagation and aggregation of cancerous cells in the peripheral blood and BM other tissues (Ferrara & Schiffer, 2013; Greim et al., 2014). It is the most familiar type of acute leukemia (AL) among adults. AML shows a deadly disease progression claiming 90% of elderly patient life (Mrózek et al., 2012). Interestingly, HSCs act as a wonder cure for leukemia by matching compatible donor for the HSCs transplantation if treatment starts at an early stage. However, the disease relapses most of the time. To study AML in depth, first zebrafish model was developed by expressing the human AML1 (RUNX1)-ETO fusion oncogene in zebrafish embryos. This destroys regular hematopoiesis by accumulation of early hematopoietic cells in the intermediate cell mass (ICM), and distribution of erythroid cells having dysplastic characteristics (Kalev-Zylinska et al., 2002).

7.5.3 ACUTE LYMPHOID LEUKEMIA

Acute lymphoblastic leukemia (ALL) is a cancerous disease of lymphoid progenitor cells which occurs in both adults and children. There are two types of ALL – T-cell acute lymphoblastic leukemia (T-ALL) and B-cell acute lymphoblastic leukemia (B-ALL). T-ALL is defined by chromosomal abnormalities and genetic alterations involved in differentiation and infiltration of proliferated lymphoid progenitor cells in BM. The cure percentage of multi-agent-based chemotherapy is more than 90% for children and 40% for adults (Dinner & Liedtke, 2018). T-ALL is caused due to recombination and mutations in many genes, such as in HOX genes, histone-modifying genes (e.g., EZH2), RAS signaling genes (e.g., FLT3), transcription-factor tumor suppressors (e.g., AML1, ETV5, or LEF1), mutations in the NOTCH1 pathway, etc. B-ALL B-ALL is a hematologic disorder caused by immature B-cell precursors (BCP) mostly prevalent in children. B-ALL is characterized by recurrent genetic abnormalities like hypodiploidy hyperdiploidy, t(9;22)(q34;q11.2)[BCR-ABL1,t (12;21)(p13;q22)[ETV6-RUNX1], etc. In adults, B-cell ALL accounts for ~ 75% of cases while T-cell ALL represents the remaining cases (Terwilliger & Abdul-Hay, 2017).

The treatment strategy of ALL includes chemotherapy consisting of induction, consolidation and long-term maintenance, with CNS prophylaxis given at intervals throughout therapy. The aim of induction therapy is to achieve complete restoration of normal hematopoiesis with complete reduction from malignancy. The main induction therapy usually involves corticosteroids, vincristine, with an Anthracycline (Scavino et al., 1976; Gottlieb et al., 1984). However, salvage therapy with single agents is more effective in patients with relapses as multiple agents mediated therapy shows toxicity in patients with relapses. Salvage therapy with a single agent mostly includes vinorelbine, clofarabine, nelarabine and topotecan (German, 1997; Bielorai et al., 2013; Shah et al., 2013). Apart from this several monoclonal antibodies based treatment has shown promising results in treating ALL.

7.6 CONCLUSION

As imbalance in hematopoiesis and HSCs function leads to life threatening disease conditions researchers are working tirelessly for understanding and maintaining the normal HSC homeostasis and function. IPSCs disease modeling has helped immensely to understand the need of the hour is to delineate the specific regulating sensors which are important for HSC functions and their ability for modulating cell cycling, self-renewal, homing, and differentiation into lineage potential. With this, the identification of novel molecular targets will help shape better future therapies for hematological disorders (Table 7.1).

TABLE 7.1 Different Hematological Diseases and Their Related Causes

Disorders	Causes/Genes
Bone-marrow failure	Decreased production of one or more major hematopoietic lineages which leads to diminished or absent hematopoietic precursors in the bone marrow.
Fanconi anemia	Impaired FA DNA repair pathway.
Shwachman-diamond syndrome	Mutation in SBDS and SRP54 gene.
Dyskeratosis congenita	Mutation in DKC1, telomerase reverse transcriptase, WRAP53.
Diamond-Blackfan anemia	Mutations in ribosomal proteins coding genes.
Thrombocytopenia	Loss-of-function of variants of PTPRJ.
Anemia	Blood loss, lack of red blood cell production, and high rates of red blood cell destruction.
(i) Hypochromic microcytic anemia	Defect in embryonic globin production.
(ii) Congenital sideroblastic anemia	Disruption of heme biosynthesis.
Hemoglobinopathy	Inherent mutation of genes coded for globin synthesis.
Thalassemia	Mutation in β-globin gene.
Sickle cell disease	Mutation in β-globin gene.
Primary immunodeficiencies	Impaired immunity, resulting in increased susceptibility to infections.
Severe combined immunodeficiencies	Heterogeneous condition, impairment in immune system development.
(i) SCID-X1	X-linked severe combined immunodeficiency.
(ii) ADA-SCID	Deficiency in adenosine deaminase.
ZAP70-related combined immunodeficiency	Mutation in ZAP70 leads to abrupt TCR signaling.
Reticular dysgenesis	Mutations in the adenylate kinase 2 gene.
Chronic granulomatous disease	Defective NADPH oxidase complex.
Wiskott-Aldrich syndrome	Mutation in the X-linked WAS gene.
Leukocyte adhesion deficiency	Adhesion-dependent malfunctions of leukocytes.
(i) LAD-I	Absent or reduced expression of β2 integrin.
(ii) LAD-II	Defects in fucosylation of selectin ligand.
(iii) LAD-III	Defects in integrin signaling.
Leukemia	Uncontrolled augmentation of hematopoietic cells.
Chronic myelogenous leukemia	Caused by translocation between chromosome 22 and 9, creating a BCR-ABL fusion gene.
Acute myeloid leukemia	Mutation in FLT3, NPM1, DNMT3A gene.
Acute lymphoblastic leukemia	Cancerous disease of lymphoid progenitor cells.
(i) T-cell acute lymphoblastic leukemia	Mutation in HOX, FLT3, EZHZ, ETV5 or LEF1 gene.
(ii) B-cell acute lymphoblastic leukemia	Recurrent genetic abnormalities.

KEYWORDS

- **hematopoeitic stem cells**
- **immune deficiency**
- **induced pluripotent stem cells**
- **lentiviral**
- **leukemia**

REFERENCES

Antunes, A. T., Goos, Y. J., Pereboom, T. C., et al., (2015). Ribosomal protein mutations result in constitutive p53 protein degradation through impairment of the AKT pathway. *PLoS Genetics, 11*(7), e1005326.10.1371/journal.pgen.1005326.

Arnold, D. E., & Heimall, J. R., (2017). A review of chronic granulomatous disease. *Advances in Therapy, 34*(12), 2543–2557.

Arpaia, E., Shahar, M., Dadi, H., Cohen, A., & Roifman, C. M., (1994). Defective T cell receptor signaling and CD8+ thymic selection in humans lacking zap-70 kinase. *Cell, 76*(5), 947–958.

Batista, L. F., Pech, M. F., Zhong, F. L., et al., (2011). Telomere shortening and loss of self-renewal in dyskeratosis congenita induced pluripotent stem cells. *Nature, 474*(7351), 399–402.

Bellanné-Chantelot, C., Schmaltz-Panneau, B., Marty, C., et al., (2018). Mutations in the *SRP54* gene cause severe congenital neutropenia as well as Shwachman-Diamond-like syndrome. *Blood, 132*(12), 1318–1331.

Bielorai, B., Fisher, T., Waldman, D., Lerenthal, Y., et al., (2013). Acute lymphoblastic leukemia in early childhood as the presenting sign of ataxia-telangiectasia variant. *Pediatric Hematology and Oncology, 30*(6), 574–582.

Botthof, J. G., Bielczyk-Maczyńska, E., Ferreira, L., & Cvejic, A., (2017). Loss of the homologous recombination gene *rad51* leads to Fanconi anemia-like symptoms in zebrafish. *Proceedings of the National Academy of Sciences of the USA, 114*(22), E4452–E4461.

Brothers, K. M., Newman, Z. R., & Wheeler, R. T., (2011). Live imaging of disseminated candidiasis in zebrafish reveals the role of phagocyte oxidase in limiting filamentous growth. *Eukaryotic Cell, 10*(7), 932–944.

Brownlie, A., Donovan, A., Pratt, S. J., et al., (1998). Positional cloning of the zebrafish sauternes gene: A model for congenital sideroblastic anaemia. *Nature Genetics, 20*(3), 244–250.

Brownlie, A., Hersey, C., Oates, A. C., et al., (2003). Characterization of embryonic globin genes of the zebrafish. *Developmental Biology, 255*(1), 48–61.

Burroughs, L., Woolfrey, A., & Shimamura, A., (2009). Shwachman-Diamond syndrome: A review of the clinical presentation, molecular pathogenesis, diagnosis, and treatment. *Hematology/Oncology Clinics of North America, 23*(2), 233–248.

Carapito, R., Konantz, M., Paillard, C., et al., (2017). Mutations in signal recognition particle SRP54 cause syndromic neutropenia with Shwachman-Diamond-like features. *The Journal of Clinical Investigation, 127*(11), 4090–4103.

Cavazzana, M., Bushman, F. D., Miccio, A., André-Schmutz, I., & Six, E., (2019). Gene therapy targeting haematopoietic stem cells for inherited diseases: Progress and challenges. *Nature Reviews. Drug Discovery, 18*(6), 447–462.

Chakraborty, A., Uechi, T., Nakajima, Y., et al., (2018). Cross talk between TP53 and c-Myc in the pathophysiology of Diamond-Blackfan anemia: Evidence from RPL11-deficient *in vivo* and *in vitro* models. *Biochemical and Biophysical Research Communications, 495*(2), 1839–1845.

Danilova, N., Bibikova, E., Covey, T. M., et al., (2014). The role of the DNA damage response in zebrafish and cellular models of Diamond Blackfan anemia. *Disease Models and Mechanisms, 7*(7), 895–905.

Danilova, N., Sakamoto, K. M., & Lin, S., (2008). Ribosomal protein S19 deficiency in zebrafish leads to developmental abnormalities and defective erythropoiesis through activation of a p53 protein family. *Blood, 112*(13), 5228–5237.

Danilova, N., Sakamoto, K. M., & Lin, S., (2011). Ribosomal protein L11 mutation in zebrafish leads to haematopoietic and metabolic defects. *British Journal of Haematology, 152*(2), 217–228.

Danilova, N., Wilkes, M., Bibikova, E., Youn, M. Y., Sakamoto, K. M., & Lin, S., (2018). Innate immune system activation in zebrafish and cellular models of Diamond Blackfan anemia. *Scientific Reports, 8*(1), (5165). https://doi.org/10.1038/s41598-018-23561-6.

Diamond, L. K., Allen, D. M., & Magill, F. B., (1961). Congenital (erythroid) hypoplastic anemia, A 25-year study. *American Journal of Diseases of Children, 2,* 403–415.

Dieterlen-Lievre, F., (1975). On the origin of haemopoietic stem cells in the avian embryo: An experimental approach. *Journal of Embryology and Experimental Morphology, 33*(3), 607–619.

Dinner, S., & Liedtke, M., (2018). Antibody-based therapies in patients with acute lymphoblastic leukemia. *Hematology, 2018*(1), 9–15.

Dokal, I., & Vulliamy, T., (2010). Inherited bone marrow failure syndromes. *Haematologica, 95*(8), 1236–1240.

Doulatov, S., Vo, L. T., Macari, E. R., et al., (2017). Drug discovery for Diamond-Blackfan anemia using reprogrammed hematopoietic progenitors. *Science Translational Medicine, 9*(376), eaah5645.

Dzeja, P. P., Zeleznikar, R. J., & Goldberg, N. D., (1998). Adenylate kinase: Kinetic behavior in intact cells indicates it is integral to multiple cellular processes. *Molecular and Cellular Biochemistry, 184*(1, 2), 169–182.

Elder, M. E., (1996). Severe combined immunodeficiency due to a defect in the tyrosine kinase ZAP-70. *Pediatric Research, 39*(5), 743–748.

Elder, M. E., Hope, T. J., Parslow, T. G., Umetsu, D. T., Wara, D. W., & Cowan, M. J., (1995). Severe combined immunodeficiency with absence of peripheral blood CD8+ T cells due to ZAP-70 deficiency. *Cellular Immunology, 165*(1), 110–117.

Elder, M. E., Lin, D., Clever, J., Chan, A. C., Hope, T. J., Weiss, A., & Parslow, T. G., (1994). Human severe combined immunodeficiency due to a defect in ZAP-70, a T cell tyrosine kinase. *Science, 264*(5165), 1596–1599.

Engidaye, G., Melku, M., & Enawgaw, B., (2019). Diamond Blackfan anemia: Genetics, pathogenesis, diagnosis and treatment. *Electronic Journal of the International Federation of Clinical Chemistry and Laboratory Medicine, 30*(1), 67–81.

Ferrara, F., & Schiffer, C. A., (2013). Acute myeloid leukaemia in adults. *Lancet, 381*, 484–495.

Finch, A. J., Hilcenko, C., Basse, N., et al., (2011). Uncoupling of GTP hydrolysis from eIF6 release on the ribosome causes Shwachman-Diamond syndrome. *Genes and Development, 25*(9), 917–929.

Gauer, R. L., & Braun, M. M., (2012). Thrombocytopenia. *American Family Physician, 85*(6), 612–622.

German, J., (1997). Bloom's syndrome. XX. The first 100 cancers. *Cancer Genetics and Cytogenetics, 93*(1), 100–106.

Gottlieb, A. J., Weinberg, V., Ellison, R. R., et al., (1984). Efficacy of daunorubicin in the therapy of adult acute lymphocytic leukemia: A prospective randomized trial by cancer and leukemia group B. *Blood, 64*(1), 267–274.

Greim, H., Kaden, D. A., Larson, R. A., et al., (2014). The bone marrow niche, stem cells, and leukemia: Impact of drugs, chemicals, and the environment. *Annals of the New York Academy of Sciences, 1310*(1), 7–31.

Hao, Q. L., Zhu, J., Price, M. A., Payne, K. J., Barsky, L. W., & Crooks, G. M., (2001). Identification of a novel, human multilymphoid progenitor in cord blood. *Blood, 97*(12), 3683–3690.

Harris, E. S., Weyrich, A. S., & Zimmerman, G. A., (2013). Lessons from rare maladies: Leukocyte adhesion deficiency syndromes. *Current Opinion in Hematology, 20*(1), 16–25.

Harvie, E. A., & Huttenlocher, A., (2015). Neutrophils in host defense: New insights from zebrafish. *Journal of Leukocyte Biology, 98*(4), 523–537.

Hoenig, M., Lagresle-Peyrou, C., Pannicke, U., et al., (2017). Reticular dysgenesis: International survey on clinical presentation, transplantation, and outcome. *Blood, 129*(21), 2928–2938.

Iolascon, A., De Falco, L., & Beaumont, C., (2009). Molecular basis of inherited microcytic anemia due to defects in iron acquisition or heme synthesis. *Haematologica, 94*(3), 395–408.

Iqbal, N., & Iqbal, N., (2014). Imatinib: A breakthrough of targeted therapy in cancer. *Chemotherapy Research and Practice, 2014*, 357027.

Ito, E., Konno, Y., Toki, T., & Terui, K., (2010). Molecular pathogenesis in Diamond-Blackfan anemia. *International Journal of Hematology, 92*(3), 413–418.

Ivanovs, A., Rybtsov, S., Ng, E. S., Stanley, E. G., Elefanty, A. G., & Medvinsky, A., (2017). Human haematopoietic stem cell development: From the embryo to the dish. *Development, 144*(13), 2323–2337.

Jacome, A., Navarro, S., Río, P., et al., (2009). Lentiviral-mediated genetic correction of hematopoietic and mesenchymal progenitor cells from Fanconi anemia patients. *Molecular Therapy: The Journal of the American Society of Gene Therapy, 17*(6), 1083–1092.

Kalev-Zylinska, M. L., Horsfield, J. A., Flores, M. V., et al., (2002). Runx1 is required for zebrafish blood and vessel development and expression of a human RUNX1-CBF2T1 transgene advances a model for studies of leukemogenesis. *Development, 129*(8), 2015–2030.

Lacaud, G., & Kouskoff, V., (2017). Hemangioblast, hemogenic endothelium, and primitive versus definitive hematopoiesis. *Experimental Hematology, 49*, 19–24.

Lee, J., Yoon, S. R., Choi, I., & Jung, H., (2019). Causes and mechanisms of hematopoietic stem cell aging. *International Journal of Molecular Sciences, 20*(6), 1272.

Lin, Q., Zhang, Y., Zhou, R., et al., (2017). Establishment of a congenital amegakaryocytic thrombocytopenia model and a thrombocyte-specific reporter line in zebrafish. *Leukemia, 31*(5), 1206–1216.

Liu, T. X., Howlett, N. G., & Deng, M., (2003). Knockdown of zebrafish Fancd2 causes developmental abnormalities via p53-dependent apoptosis. *Developmental Cell, 5*(6), 903–914.

Marconi, C., Di Buduo, C. A., LeVine, K., et al., (2019). Loss-of-function mutations in *PTPRJ* cause a new form of inherited thrombocytopenia. *Blood, 133*(12), 1346–1357.

Massaad, M. J., Ramesh, N., & Geha, R. S., (2013). Wiskott-Aldrich syndrome: A comprehensive review. *Annals of the New York Academy of Sciences, 1285*, 26–43.

Moore, J. C., Mulligan, T. S., Yordán, N. T., et al., (2016). T cell immune deficiency in *zap70* mutant zebrafish. *Molecular and Cellular Biology, 36*(23), 2868–2876.

Mrózek, K., Marcucci, G., Nicolet, D., et al., (2012). Prognostic significance of the European LeukemiaNet standardized system for reporting cytogenetic and molecular alterations in adults with acute myeloid leukemia. *Journal of Clinical Oncology, 30*(36), 4515–4523.

Müller, L. U., Milsom, M. D., Harris, C. E., et al., (2012). Overcoming reprogramming resistance of Fanconi anemia cells. *Blood, 119*(23), 5449–5457.

Narla, A., Payne, E. M., Abayasekara, N., et al., (2014). L-Leucine improves the anaemia in models of Diamond Blackfan anaemia and the, 5q- syndrome in a TP53-independent way. *British Journal of Haematology, 167*(4), 524–528.

Papapetrou, E. P., Lee, G., Malani, N., et al., (2011). Genomic safe harbors permit high β-globin transgene expression in thalassemia induced pluripotent stem cells. *Nature Biotechnology, 29*(1), 73–78.

Payne, E. M., Virgilio, M., Narla, A., et al., (2012). L-Leucine improves the anemia and developmental defects associated with Diamond-Blackfan anemia and del(5q) MDS by activating the mTOR pathway. *Blood, 120*(11), 2214–2224.

Picard, C., Al-Herz, W., Bousfiha, A., et al., (2015). Primary immunodeficiency diseases: An update on the classification from the international union of immunological societies expert committee for primary immunodeficiency. *Journal of Clinical Immunology, 35*(8), 696–726.

Pinho, S., & Frenette, P. S., (2019). Haematopoietic stem cell activity and interactions with the niche. *Nature Reviews. Molecular Cell Biology, 20*(5), 303–320.

Rai, G., Das, D., & Priya, K., (2020). New insights on stem cells modeling and treatment of human diseases. *Frontiers in Bioscience, 25,* 1568–1599.

Raje, N., & Dinakar, C., (2015). Overview of immunodeficiency disorders. *Immunology and Allergy Clinics of North America, 35*(4), 599–623.

Raya, A., Rodríguez-Pizà, I., Guenechea, G., et al., (2009). Disease-corrected haematopoietic progenitors from Fanconi anaemia induced pluripotent stem cells. *Nature, 460*(7251), 53–59.

Río, P., Meza, N. W., & González-Murillo, A., (2008). *In vivo* proliferation advantage of genetically corrected hematopoietic stem cells in a mouse model of Fanconi anemia FA-D1. *Blood, 112*(13), 4853–4861.

Scavino, H. F., George, J. N., & Sears, D. A., (1976). Remission induction in adult acute lymphocytic leukemia. Use of vincristine and prednisone alone. *Cancer, 38*(2), 672–677.

Sebastiano, V., Maeder, M. L., & Angstman, J. F., (2011). In situ genetic correction of the sickle cell anemia mutation in human induced pluripotent stem cells using engineered zinc finger nucleases. *Stem Cells, 29*(11), 1717–1726.

Shah, A., John, B. M., & Sondhi, V., (2013). Acute lymphoblastic leukemia with treatment-naive Fanconi anemia. *Indian Pediatrics, 50*(5), 508–510.

Taylor, A. M., & Zon, L. I., (2011). Modeling Diamond Blackfan anemia in the zebrafish. *Seminars in Hematology, 48*(2), 81–88.

Terwilliger, T., & Abdul-Hay, M., (2017). Acute lymphoblastic leukemia: A comprehensive review and 2017 update. *Blood Cancer Journal, 7*(6), e577.

Torihara, H., Uechi, T., Chakraborty, A., Shinya, M., Sakai, N., & Kenmochi, N., (2011). Erythropoiesis failure due to RPS19 deficiency is independent of an activated Tp53 response in a zebrafish model of Diamond-Blackfan anaemia. *British Journal of Haematology, 152*(5), 648–654.

Tulpule, A., Kelley, J. M., Lensch, M. W., et al., (2013). Pluripotent stem cell models of Shwachman-Diamond syndrome reveal a common mechanism for pancreatic and hematopoietic dysfunction. *Cell Stem Cell, 12*(6), 727–736.

Vlachos, A., & Muir, E., (2010). How I treat Diamond-Blackfan anemia. *Blood, 116*, 3715–3723.

Wan, Y., Zhang, Q., Zhang, Z., et al., (2016). Transcriptome analysis reveals a ribosome constituents disorder involved in the RPL5 downregulated zebrafish model of Diamond-Blackfan anemia. *BMC Medical Genomics, 9*, 13.

Yadav, G. V., Chakraborty, A., Uechi, T., & Kenmochi, N., (2014). Ribosomal protein deficiency causes Tp53-independent erythropoiesis failure in zebrafish. *The International Journal of Biochemistry and Cell Biology, 49*, 1–7.

Yang, C. T., Cambier, C. J., Davis, J. M., Hall, C. J., Crosier, P. S., & Ramakrishnan, L., (2012). Neutrophils exert protection in the early tuberculous granuloma by oxidative killing of mycobacteria phagocytosed from infected macrophages. *Cell Host and Microbe, 12*(3), 301–312.

Ye, L., Chang, J. C., Lin, C., Sun, X., Yu, J., & Kan, Y. W., (2009). Induced pluripotent stem cells offer a new approach to therapy in thalassemia and sickle cell anemia and option in prenatal diagnosis in genetic diseases. *Proceedings of the National Academy of Sciences of the United States of America, 106*(24), 9826–9830.

Zhang, Y., Ear, J., Yang, Z., Morimoto, K., Zhang, B., & Lin, S., (2014). Defects of protein production in erythroid cells revealed in a zebrafish Diamond-Blackfan anemia model for mutation in RPS19. *Cell Death and Disease, 5*(7), e1352.

Zhang, Z., Jia, H., Zhang, Q., et al., (2013). Assessment of hematopoietic failure due to Rpl11 deficiency in a zebrafish model of Diamond-Blackfan anemia by deep sequencing. *BMC Genomics, 14*, 896.

Zou, J., Mali, P., Huang, X., Dowey, S. N., & Cheng, L., (2011). Site-specific gene correction of a point mutation in human iPS cells derived from an adult patient with sickle cell disease. *Blood, 118*(17), 4599–4608.

CHAPTER 8

Circulating Messengers of Blood and Their Clinical Applications

ANULA DIVYASH SINGH[1] and RASMITA SAMAL[2]

[1]*Apollo Hospitals Educational and Research Foundation, Hyderabad, Telangana, India*

[2]*Department of Life Sciences, Central University of Karnataka, Gulbarga, Karnataka, India*

ABSTRACT

Blood cell-derived microvesicles are an indispensable component of the systemic circulation, mediating long distant communication between cells and tissues. These nano-to-micro-sized vesicles are abundantly secreted by various blood cells, particularly those derived from red blood cells and platelets are extensively studied. The molecular cargo packaged inside the vesicles largely depends on the environmental stimulus conferring MVs as an excellent source of biomarkers. Owing to their easy accessibility, these minimally invasive tool is of immense clinical relevance for early diagnosis of diseases such as cancers, diabetes, cardiovascular diseases, etc. Besides, they serve as delivery vehicles of therapeutic molecules such as drugs or miRNAs, which when encapsulated inside the lipid bilayer are protected from degradation in the harsh pathological environment. Engineered MVs used to deliver therapeutic cargo to the targeted region, offers an off-the-shelf alternative for therapeutic interventions. This chapter summarizes the characteristic properties of various MVs present in circulation and the significant role they play in regulating normal homeostasis or adaptive response in diseased conditions. We further discuss the multifaceted clinical

Hematopoiesis: Biochemical, Cellular, Molecular, and Genomic Perspectives. Ravi Kumar Gutti (Ed.)

accreditations of these circulating nanosystems focusing on the diagnostic, immunomodulatory and drug delivering potential implicating MVs as novel platform for cell-free therapeutic development for precision medicine.

8.1 INTRODUCTION

Circulating microvesicles (MVs) released from blood cells into the extracellular milieu play a crucial role in orchestrating intercellular communication by shuttling their bioactive contents, such as proteins, metabolites, and nucleic acids either locally or systemically. Horder reported the existence of extracellular like vesicles for the first time (Horder, 1899). Later Wolf described tiny vesicles as platelet dust that contained procoagulant derivatives of platelets (Wolf, 1967). The origin of vesicles was demonstrated in sheep reticulocytes, where the authors identified reticulocytes selectively released transferrin receptors within extracellular vesicles (EVs) during the maturation of red blood cells (Pan et al., 1983). Although vesicles were considered as means of disposal of cellular waste, their importance came to the limelight only in the last decade when their active role in intercellular communication and cell signaling was revealed. Since then, studies on microvesicles have gained momentum in unraveling their multifaceted role in health and disease (Shah et al., 2018; Kalluri et al., 2020).

MVs represent a heterogeneous group of populations broadly classified into subgroups based on their size, origin, and sedimentation properties (Muller et al., 2014; Butler et al., 2018). Ectosomes are vesicles that shed off from the surface of the plasma membrane following outward budding, including microvesicles with a size range of 50–500 nm, while exosomes are nanovesicles of endocytic origin with a size range of 40–160 nm, formed by reverse budding of the peripheral membrane of late endosome. Previously, it was suggested that phospholipids present on MV membranes lose their asymmetric distribution, exposing the negatively charged Phosphatidylserine (PS) on the outer surface of the membrane, which further leads to the shedding of MVs (Morel et al., 2011). Alternatively, MV biogenesis occurs due to Ca^{+2} ion influx into the cytosol, which is RBC senescence mechanism, leading to the externalization of PS and vesiculation. These tiny spherical membrane-bound vesicles are secreted by almost all types of cells and exist in different biofluids such as saliva, urine, cerebrospinal fluid, and breast milk, but blood is the richest source (Pisitkun et al., 2004; Caby et al., 2005; Kapsogeorgou et al., 2005; Admyre et al., 2007; Vella et al., 2008). The molecular cargo packed in the vesicles mostly resembles the parental cells of origin consisting of surface receptors, integral membrane proteins, cytosolic proteins, mRNAs, and miRNAs. However, packaged material inside the vesicles varies depending on the cellular origin and the environmental stimulus acting on cells.

Peripheral blood contains trillions of MVs released from various blood cells, such as megakaryocytes, platelets, erythrocytes, leukocytes, and endothelial cells (Arraud et al., 2014). Several efforts were made to understand the potential function of MVs and their role in disease progression and therapeutics. In this context, considerable evidence has demonstrated that bloodborne MVs are primarily responsible for thrombosis and hemostasis (Burnouf et al., 2015; Weisel et al., 2019). So far, most of the studies on circulating MVs were preferably done on plasma/serum samples because of their easy access as liquid biopsies. However, clinical translation of such non-invasive technology for disease monitoring is far from the reality due to technical challenges such as small sample volume, preanalytical conditions, viscosity, and concentrations of proteins and lipoproteins in serum/plasma samples that overlap in size and density of MVs and that finally impedes the yield and purity. The tendency of MVs to form aggregates further aggravates the complexity of the isolation process. Though several methods were developed to isolate such micro- to nano-sized vesicles based on different principles, a standard procedure for effective isolation is yet to be established, which becomes even trickier with blood samples. This chapter discusses the biology of various blood-derived microvesicles and exosomes; for ease of our understanding, we refer collectively as circulating microvesicles (MVs). Here we provide a comprehensive overview of the characteristic biological properties of various types of MVs, and on the current techniques implemented to isolate such tiny vesicles. We further provide insight on the clinical applications of circulating MVs rendering to their non-invasive diagnostic potential and therapeutic drug delivery.

8.2 TYPES OF CIRCULATING MICROVESICLES

Circulating MVs constitute a group of heterogeneous subpopulations, including platelet-derived microvesicles (PMV), RBC-derived microvesicles (RBC-MV), leukocyte-derived microvesicles (LMV), and Endothelium-derived microvesicles (EC-MV). Each subgroup expresses a specific set of markers, which mark their identity and improve our understanding of their potential functions. An overview of different types of circulating MVs and their biological functions is shown in Figure 8.1. Previous enumeration studies indicated that a large proportion of plasma MVs have originated from platelets and erythrocytes, but only a minor proportion from leukocytes and endothelial cells (Arraud et al., 2014). Hence, in this section, we discuss the biological properties and functional implications of different blood-derived MVs focusing on PMVs and RBC-MVs.

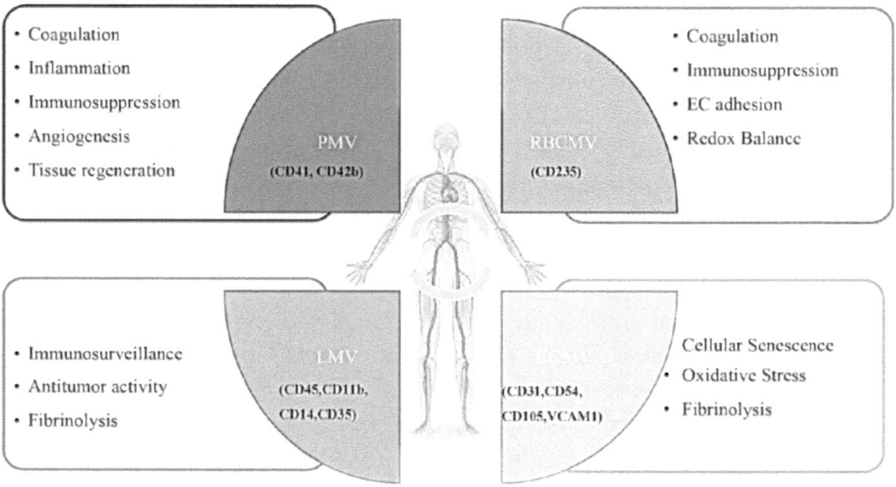

FIGURE 8.1 Overview of different types of MVs present in circulation: platelet-derived microvesicles (PMVs), red blood cell-derived microvesicles (RBC-MVs), leukocyte-derived microvesicles (LMVs), endothelial cell-derived microvesicles (EC-MVs). Diverse biological functions exhibited by each subgroup of MVs are displayed, and specific markers corresponding to each subgroup are listed in parenthesis.

8.2.1 RBC-DERIVED MICROVESICLES

The red blood cell population is the highest (~5 billion/ml blood) among other blood cells present in circulation. RBC-MVs are secreted during erythropoiesis, aging, environmental stimulus, and under pathological perturbations. Most of the studies on RBC-MVs, in the recent past, are focused on vesicles derived from stored RBC units to investigate blood transfusion-related complications (Kim et al., 2018; Thangaraju et al., 2020).

RBC-MVs are composed of a large number of smaller-sized vesicles of about 0.06 μm (Allan et al., 1980) expressing exosomal marker CD63, while large-sized MVs >200 nm do not express CD63 (Noubouossie et al., 2020).

RBC-MVs typically express CD235a markers on their cell surface. These MVs are composed of a diverse array of proteins, including cytoskeletal proteins (actin, band3), denatured HbA, membrane organizers tetraspanins (CD9, CD63, CD81), chaperones heat shock protein 70, 90, biogenesis factor (Alix, TSG10), but they are devoid of Spectrin. Apart from proteins, RBC secretes several miRNAs into the recipient cells and regulates the gene expression via targeted mRNA degradation or translation repression. About 78 different types of miRNA are secreted by RBC in stored blood (Huang et al., 2019), of which miR-125b-5p was the most abundant and reportedly target the inflammatory genes via TRAF6/MAPK/NFkB pathways in chondrocytes (Rasheed et al., 2019).

The earlier conception of RBC playing a passive role in thrombosis had recently changed when Noubouossie et al. (2020) identified a novel mechanism in which MVs from stored RBC units activate coagulation by direct and independent activation of Factor XII and Prekallikrein. Previous studies have reported the accumulation of MVs during blood storage, which may be due to increased cytosolic Ca^{+2}, ATP depletion, and increased potassium leakage (Thangaraju et al., 2020). Although RBC-MVs lack tissue factor (TF), which is the major initiator of coagulation, an alternative mechanism has been identified to initiate thrombin formation via the Factor XII-dependent pathway. PS levels from high RBC counts enhance the negative charge on the surface and induce a procoagulant effect, further aggravating thrombotic complications. On the other hand, free hemoglobin released during RBC hemolysis in stored blood leads to sequestration of NO. It induces platelet aggregation and activation of endothelial cells driving towards prothrombotic events (Weisel et al., 2019). The potential of RBC-MVs as NO scavengers is high, which reacts 1,000-fold faster than RBC bound Hb (Donadee et al., 2011). A recent study reported that RBC-MVs and PMVs derived from blood bank units induced procoagulant activity in a Factor XII-dependent manner, while monocyte-derived MVs did not exhibit any thrombogenic effect (Tripisciano et al., 2017).

Besides the physiological coagulant effect, RBC-MVs impart immunomodulatory properties. Earlier the immunosuppressive potential of RBC-MVs was reported after blood transfusion. RBC-MVs exhibit anti-inflammatory effects when incubated with activated macrophages (Sadallah et al., 2011). The author reported that after RBC-MVs uptake by macrophages inhibited its activation by zymosan A or lipopolysaccharide (LPS) and reduced the release of TNFα and IL-8 (Sadallah et al., 2008). Moreover, RBCs are enriched in antioxidant enzymes such as thioredoxin, superoxide dismutase, catalases, glutathione peroxidase, etc., that help to maintain prooxidant and antioxidant levels in circulation. Following transfusion, the RBC-MVs released due to oxidative processes leads to neutrophil activation and rapid release of ROS, suggesting a role of RBC-MVs in transfusion-related lung injury (Jank et al., 2011). All these findings reveal the clinical relevance of RBC-MVs; since most of the results reported are from stored blood units, further investigations of fresh blood samples will reveal novel mechanistic insight into their clinical potential.

8.2.2 PLATELET-DERIVED MICROVESICLES (PMV)

Platelet-derived microvesicles (PMVs) represent the most abundant microvesicles constituting about 70–90% of total circulating MVs (Aatonen et al., 2012). PMVs were first reported by Wolf as lipid-rich particles called 'platelet dust' by electron microscopy studies (Tao et al., 2017). PMVs are released into the circulation upon activation and in response to the pathological stimulus. PMVs express specific surface markers CD41 and CD42a, specific miRNAs including miR-21, miR-223, miR339, and glycoproteins GPIb and GPVI (Zarà et al., 2019), that help to track the origin among the heterogeneous groups of MVs present in circulation. PMVs are known to adhere to recipient cells via specific lipid and ligand receptors on their cell surface and stimulate the recipient cell function (Del Conde et al., 2005). Several mechanisms were proposed to internalize PMVs into the recipient cells, including phagocytosis, endocytosis, and micropinocytosis via lipid drafts, clathrin, caveolae, or via direct fusion to the membranes (Costa Verdera et al., 2017).

Like RBC-MVs, PMVs exhibit a multifaceted role in various biological processes such as homeostasis, angiogenesis, coagulation, tissue repair, and immunity (Zarà et al., 2019). They transport a wide array of cargo consisting of receptors, chemokines, enzymes, micro-RNA, and other bioactive molecules. PMVs exhibit coagulant activity by exposing the phosphatidylserine (PS) that provides a catalytic surface for the formation of tenase and prothrombinase complexes. Also, PS aids in the conversion of TF from its inactive to biologically active state, which further forms complex with factors VII and VIIa leading to the initiation of the coagulation cascade (Tripisciano et al., 2017). Previously, PMVs were reported to modulate the behavior of leukocytes after P-selectin present on the surface of PMVs bind to PSGL-1 (glycoprotein ligand-1) present on the leukocyte, resulting in leukocyte aggregation and accumulation (Forlow et al., 2000). Moreover, PMVs are recognized as angiogenic inducers, known to induce sprouting and angiogenic activity of endothelial progenitor cells (Prokopi et al., 2009). PMVs actively regulate the proliferation, adhesion, survival, and chemotaxis of hematopoietic cells (Baj-Krzyworzeka et al., 2002). The authors identified that PMVs might transfer platelet antigens to hematopoietic stem progenitor cells and stimulate their proliferation. In another study, PMVs were demonstrated to induce proliferation and migration of endothelial cells and fibroblast to improve angiogenesis and aid in re-epithelization of chronic wounds (Guo et al., 2017). All these findings indicate PMVs regulative actions on various biological functions of cells.

PMVs released from platelets play a major role in thrombogenesis and inflammation (Goetzl et al., 2016). CD36 is downregulated in platelets and macrophages by PMVs via ubiquitination and proteasomal degradation, which further inhibits platelet activation and prevents thrombus formation (Srikanthan et al., 2014). PMV levels are elevated in diseases associated with increased thromboembolism, such as in cancer, sepsis, pre-eclampsia, atherosclerosis, etc. Besides, the immunomodulatory potential of PMVs is also well known. PMV-containing miRNA (miR126-3p) are internalized by macrophages and increase the phagocytic ability by suppressing the expression of genes of immune cells. PMVs are known to regulate the phenotypic modification of vascular smooth muscle cells by modulating platelet-derived growth factor receptor-beta, suggesting the potential role PMVs in immune modulation (Zarà et al., 2019). It is now clear that tiny circulating microvesicles act as a storehouse of regulatory molecules, understanding the role of individual subpopulations of PMVs would shed more light on their specific actions in health and disease.

8.2.3 LEUKOCYTE-DERIVED MICROVESICLES (LMV)

LMVs are released from the cells of innate and adaptive immune systems, constituting a minor proportion of total Evs derived from plasma. Various cell types of leukocyte subpopulation, including T cells, B cells, dendritic cells (DC), macrophages, monocytes, polymorphonucleate cells, mast cells, and natural killer cells (NKs), release MVs, which differ in their plasma membrane and surface protein composition. LMVs are typically marked by expression of surface proteins, including ICAM-1, TF, PSGL-1, CD62L, C3, MMPs, inflammatory cytokines, and contain a wide array of microRNAs, namely, miR-222, miR-155, miR-146a, miR-146b, and miR-125a-5p, miR-21-5p (Zarà et al., 2019). LMVs exhibit immunoregulatory functions based on their molecular compositions. Further, it was demonstrated that MVs released from NK cells marked by CD56 and perforin-induced cytotoxicity in immune cells (Lugini et al., 2012). In another study the immunosuppressive role of LMVs secreted from regulatory T cells was observed. The authors reported that CD8+25+ T cells inhibit cytotoxic T lymphocyte response, which was induced by DC in B16 melanoma (Xie et al., 2013). LMVs derived from immature DCs have also exhibited inhibitory roles in transplant tolerance and allograft rejection (Pugholm et al., 2016). The role of LMVs in cellular regulation is slowly being understood. Further investigation on the individual categories of LMVs would provide a better understanding of their specific functions and implications in clinical settings.

8.2.4 ENDOTHELIAL CELL-DERIVED MICROVESICLES (EC-MV)

EC-MVs exhibit a dual role in vascular homeostasis, showing both deleterious and beneficial effects. Under physiological conditions, they are released at a concentration of 10^3–10^4 per ml blood, while pathophysiological concentrations 10^5/ml were found to be toxic to EC, impairing the angiogenesis, proliferation, and increases apoptosis (Mezentsev et al., 2005). *In vivo* studies have shown that EC-MVs induce endothelial dysfunction, inflammation, and promote atherosclerosis under high glucose conditions (Jansen et al., 2013). Similar to other MVs present in circulation, EC-MV also imparts procoagulant activity due to the presence of negative charge from PS and tissue factor on the surface (Dignat-George et al., 2011). Earlier, it was reported that EC stimulation with TNFα, induces TF-mediated coagulation cascade (Combes et al., 1999). Apart from their damaging effect, EC-MVs promote EC survival and pro-angiogenic actions *in vitro* due to the presence of MMPs in EC-MV (Taraboletti et al., 2002). As endothelial cells play a crucial role in vascular processes, a mechanistic insight into EC-MV would provide deeper knowledge of their functional potential.

8.3 CURRENT TRENDS IN ISOLATION PROCEDURE

MVs isolated from diverse biological fluids vary in their viscosity, fat, and protein content, determining their yield and purity (Ramirez et al., 2018). The choice of isolation method depends on several factors, including source, biological application, amount of starting material, and available resources. Blood being a complex body fluid, the isolation procedure should be chosen carefully to minimize co-precipitants and contaminants, thereby minimizing artifacts during application and analysis. Blood collection method is also an essential factor as shear forces can lead to hemolysis and affect the MV content. In this aspect, usage of anticoagulants reportedly reduce blood cell aggregates during MV isolation (Ramirez et al., 2018). Isolation of PMVs requires platelet-rich plasma (PRP) extracted from whole blood in the presence of an anticoagulant. The choice of anticoagulant is crucial for PMV characterization and quantification, as anticoagulants could activate platelets during blood collection and storage, triggering the release of MVs. Sodium citrate is well known for minimizing *in vitro* platelet activation and release (Ramirez et al., 2018; Taus et al., 2019).

Ultracentrifugation is one of the conventional methods used for the optimal separation of exosomes from almost all biological fluids, including blood plasma and serum. However, there are many limitations associated with this technique. Apart from high cost and time consuming, it separates microvesicles, including lipoproteins and protein aggregates that might interfere with their functional implications (Aatonen et al., 2014). To further improve the efficiency, density gradient centrifugation is used as an alternative. The components in the sample are separated according to their respective gradient. In this context, ultrafiltration was employed for the separation of MVs based on their size. Although it is a convenient option, clogging of filters leads to lower yield. Recently, size-exclusion chromatography (SEC) has been identified to minimize the vesicular fraction's structural damage. MVs pass through a column made of porous material like dextran polymer, agarose, polyacrylamide, etc. This technique is highly recommended as exosomes' biological activity remains preserved while maintaining the vesicle integrity and structure. Additionally, minimal volumes of samples, less time and labor make them ideal choice for isolation of circulating MVs. Further SEC technique coupled with ultracentrifugation aids in eliminating co-precipitants with exosomes that generally arise due to overlapping sizes.

Polymer-induced precipitation is another strategy used, where highly hydrophilic polymers like polyethylene glycol (PEG) interact with water molecules creating a hydrophobic environment for the precipitation of exosomes. However, the existence of nonspecific protein contaminants still remains a major problem. Alternatively, immunoaffinity technique is employed to obtain highly purified exosomes based on surface antigen-antibody reaction. Various exosomal marker proteins like CD63, CD82,

CD81, CD9, Rab5, Alix, and Annexin are substantially utilized for isolation. Furthermore, microfabrication technology is implemented to isolate and characterize MVs for in situ non-invasive disease diagnosis. The exosome-containing fluid is mixed with sheath fluid, and particles in the medium containing the exosomes get separated according to their physio-chemical properties. The apparatus is facilitated with signal detection platforms that allow fast detection and real-time exosome characterization (Yang et al., 2020). Microfluidics technology is another approach that specifically allows isolation and evaluation of Evs and requires minimum working volumes. There are on-chip devices and Exochip, which combine microfluidics with immunoaffinity and quantification by staining with fluorescent dyes (Ramirez et al., 2018). A brief overview of the current methods implemented in isolation is shown in Figure 8.2.

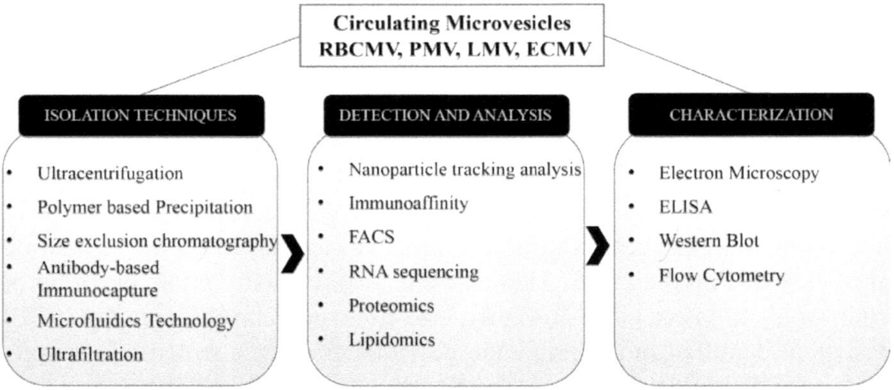

FIGURE 8.2 Schematic representation of various techniques implemented in isolation, detection, and analysis of circulating MVs.

With extensive studies and developing technologies, there are a plethora of choices available to isolate MVs, but none of the methods work independently, giving a pure MV population. In most of the studies, a mixed vesicle population is obtained. Hence, the ideal method is chosen by considering the analytical sample properties and its area of application (Konoshenko et al., 2018). An international consortium of 92 researchers from 12 countries has recently launched a database that provides information on isolation and characterization procedures for vesicles. This database, EV-TRACK, gave scope to several groups to compare methodologies and improve isolation and characterization techniques (Ramirez et al., 2018).

8.4 MOLECULAR CHARACTERISTICS OF CIRCULATING MVs

Apart from isolation, the characterization of MVs is another critical area to focus on due to overlapping sizes, morphology, and origin of different subgroups of MVs; as a result, identifying true MVs becomes a daunting task. MVs carry diverse cargoes of proteins, lipids, and RNA that are potent regulators of biological functions and modulators of disease environment. It is now widely accepted that MVs' protein content tends to change when transitioning from a healthy environment to a disease milieu (Shah et al., 2018). Moreover, MVs-derived directly from patient samples are of high clinical relevance as they represent an enriched source of biomarkers and may serve as a promising non-invasive diagnostic tool.

Conventional techniques used for detection and analysis of microvesicles are not enough for a complete analysis of MVs; hence it is suggested to use a combination of techniques (Taus et al., 2019). Visualization and morphological quantitation of isolated MVs is obtained by transmission electron microscopy (TEM) or cryo-electron microscopy (Cryo-EM) technique (Issman et al., 2013). Particle size is quantified using dynamic light scattering (DLS) technique and nanoparticle tracking analysis (NTA); both technique measures the fluctuation in the light scattering of particles in suspension, which relates Brownian motion to the particle size. NTA is commonly used to evaluate the size of blood exosomes (Kapogiannis et al., 2019). In clinical settings, flow cytometry is the preferred choice for the enumeration of MVs using various fluorescent-labeled immunoglobulins against CD62P, CD61, lactadherin, and Annexin V (Xiong et al., 2012). A summary of methods implemented in MVs characterization is illustrated in Figure 8.2. Although these techniques provide an understanding of the biophysical properties, size, and purity of MVs, a deeper insight into their molecular profiles is required to understand the potential regulation in biological processes that is acquired by using high throughput technologies.

In this context, mass spectrometry-based proteome analysis has been widely employed to understand the MVs cargo. One of the major challenges with circulating MVs is the sample preparation that must be free of any contaminants arising from blood cell components. As blood is a complex mixture of proteins and lipoproteins with a large dynamic range, their presence may mask the identification of low abundant effector molecules present in MVs. Therefore, it is highly recommended to examine the MVs sample enrichment and confirm the purity using TEM or atomic force microscopy-based platforms prior to proteome analysis. In addition, the low yield is another limiting factor for such studies. Although initially PMVs were quantified with high sensitivity using 2-dimensional gel electrophoreses (Jin et al., 2005), this approach was not appreciated due to the demand of large sample volume. Earlier, the importance and challenges of proteomics-based studies in characterization RBC-MVs were discussed (Rubin et al., 2010).

Over the last decades, label-free approaches have attracted major attention in EV research, particularly in understanding protein-protein interaction, protein abundances, and possible post-translational modifications. Label-free approaches are extensively used as the method is simple and cost-effective, with a possibility to analyze unlimited proteins with a minimal sample amount. In many recent studies, potential biomarkers were identified for Alzheimer's diseases, breast cancer, and cardiovascular diseases (Shah et al., 2018).

Circulating MVs also harbor conserved microRNAs of parental origin, which are tiny noncoding RNAs (22 nucleotides long) playing a crucial role in post-translational regulation of RNA either by mRNA degradation or translational repression (Ambros, 2004; van Rooij et al., 2007). Although bodily fluids contain cell-free miRNAs present in circulation, the long-term prognostic value of such biomarker is a matter of debate. On the contrary, MVs contain selective miRNAs protected from circulating ribonuclease (Rnase) and serve as stable prognostic biomarkers (Grasedieck et al., 2012). This was evident from a previous study by Jansen et al. where the authors identified miRNA-126 and miRNA-199a in circulating MVs obtained from coronary artery disease patients, which was absent in plasma-derived miRNA, indicating selective packaging of miRNA in MVs and their potential usage in risk stratification (Jansen et al., 2013). Another study by (Xiao et al., 2012) has provided the first in-depth screening of miRNA profile from tumor-derived melanoma exosomes and identified I-miR-31, miR-185, and miR-34b involved in melanoma progression. Overall, these findings suggest that MVs-derived miRNA could serve as a surrogate for biopsy/circulating miRNA profiling.

8.5 CLINICAL APPLICATIONS OF BLOOD-DERIVED MVs

Circulating MVs are ubiquitously released into the bloodstream in physiological conditions or in response to various pathological stimuli. Among other circulating MVs, RBC-MVs and PMVs are the preferred choices as they are anucleated and lack nuclear or mitochondrial DNA, avoiding the risk of horizontal gene transfer (Wahlgren et al., 2012). Unlike the parental blood cells, these natural carriers impart low immunogenicity, non-toxicity, easy availability, and a longer half-life, making them an excellent source for biomedical applications (Kalluri et al., 2020). A schematic representation on the clinical applications of MVs is shown in Figure 8.3. Thus, MVs are currently under active investigation for their therapeutic potential and diagnostic implications.

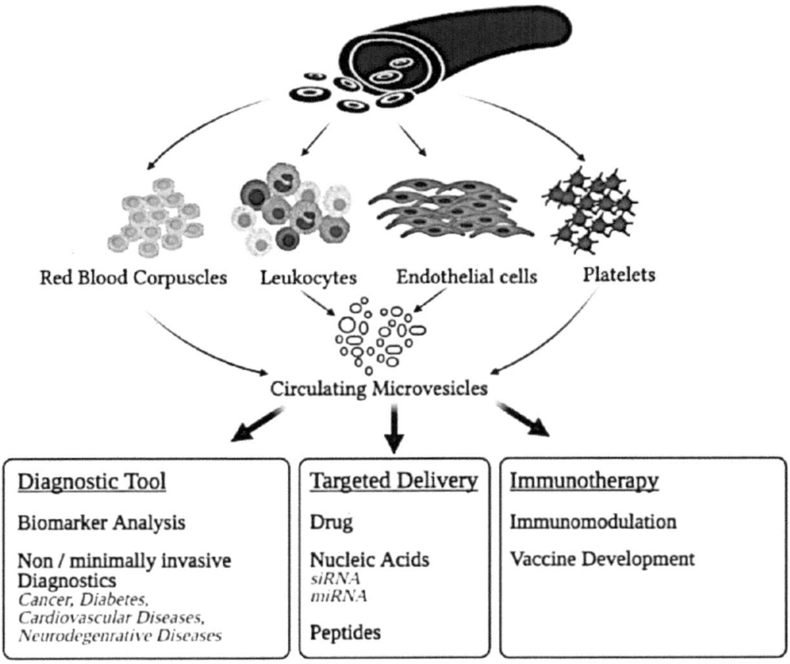

FIGURE 8.3 Clinical applications of blood cells-derived circulating microvesicles.

Source: Created in BioRender.com.

8.5.1 DIAGNOSTIC POTENTIAL

Exosome-based diagnostics are gaining importance as liquid biopsies that aid in monitoring disease severity, risk stratification, and prognosis prediction. MVs are released from various biological fluids, including blood, saliva, urine, breast milk, etc. Apart from their easy accessibility, they display the characteristic surface markers of cellular origin, carry disease-specific cargoes, and are known to be highly stable in biological fluids. In addition, the release of MVs tends to increase in response to the diseased environment, and accordingly, the vesicular contents, including proteins, lipids, and nucleic acids (DNA, mRNA, microRNA, circulating RNA, etc.), are altered. Analysis of circulating MVs in such conditions will provide direct insight into the disease state and aid in early predicting disease in a minimally invasive manner.

Several strategies have been implemented in the development of MV-based diagnostic tools, particularly explored in the area of cancer diagnostics. The first blood-based exosomal diagnostic, ExoDx for prostate cancer, was commercially available in 2016, which employed both exosomal RNA and cell-free DNA quantification in a single-step (Brinkmann et al., 2018). In the recent past, researchers had proposed a MV-based test for the detection of EGFRT790M mutations in the plasma of non-small cell lung carcinoma patients (Castellanos-Rizaldos et al., 2018). Further, ELISA-based test ExoTest was developed using anti-Rab5 antibody for the detection of exosomal marker antigen CD63 and Caveolin 1. Caveolin-1 is known to play a role in tumor metastasis and was detectable in plasma-derived exosomes of cancer patients (Urbanelli et al., 2015). In this context, exosomal DNA was also implemented to identify tumor-specific gene mutation such as BRAFV600E, which is present universally in multiple cancer cells, including breast cancer, melanoma, lung, pancreas, and prostate cancer. Rupp et al. have identified CD24, derived from circulating exosomes, as a biomarker for breast cancer (Rupp et al., 2011). Glypican-1 (Gpc-1) was identified for early detection of pancreatic cancer (Melo et al., 2015).

Apart from cancer, MV-based biomarker analysis has been implicated in other diseases such as diabetes, neurodegenerative disease, cardiovascular diseases as well. Recently, thrombotic mechanism mediated by elevated levels of PMVs and RBC-MVs under hyperglycemia in Type 2 diabetes patients was reported (Gkaliagkousi et al., 2019). Earlier studies have shown PMVs and EC-MVs levels correlated that with myocardial risk in patients suffering from myocardial infarction (Jung et al., 2012). Researchers have implemented blood-derived exosomal markers to characterize bone marrow (BM) conditions, including post-transplant recovery (Mitsuhashi, 2013). Although a number of biomarkers related to multiple diseases have been identified so far, the translation process of biomarker profiles from bench to bedside is yet to gain full momentum. Studies in this direction must validate the reliable biomarkers identified on a large cohort with better specificity and sensitivity that may aid in the development of new diagnostic panels.

8.5.2 THERAPEUTIC POTENTIAL

MVs are well known for their role as drug delivery vehicles because of their high stability and bioavailability. Striking evidence on MVs carrying drugs through the blood-brain barrier (Banks et al., 2020) indicates that MVs can be an attractive choice for therapeutic applications as they can reach distal tissues without any physiological hindrance. Most of the drugs or RNA, which are directly introduced into our body, are exposed to degrading enzymes present in the extracellular space and get cleared up prior to reaching the target site. While encapsulated MVs bypass this degradative environment due to the protective membranous cover and prolong their circulation time for effective targeting. Drug loading onto MVs is usually performed in two ways, either via preloading or post-loading mechanisms. Preloading utilizes the cellular machinery for the incorporation of siRNA or miRNA, which occurs during MV biogenesis. On the other hand, post-loading is done in a passive way using a physical model of electroporation or sonication techniques or using chemical agents such as saponin and transfection agents (Walker et al., 2019). Hence, MVs can be exploited for autologous delivery of drug from patients that offers new hope for therapeutic interventions and personalized medicine.

In recent times, circulating MVs are engineered for the targeted delivery of therapeutic agents. Several strategies have been implemented to improve the cellular uptake of drugs, preventing their rapid clearance as well as trafficking to the site of action while avoiding the off-target effects (Luan et al., 2017). Selective targeting of MVs is largely dependent on their surface protein expressions such as tetraspanins (Rana et al., 2012), lipid contents such as phosphotidylserine (Flannagan et al., 2014), and surface glycan levels, which are known to direct MVs to specific sites (Hood, 2016). In a previous study, engineered PMVs were used for loading anticancer drugs against leukemia and authors demonstrated targeted delivery of drugs with high efficiency and minimal off-target toxicity (Kailashiya et al., 2019). Recently the prospectus and challenges of delivery of MVs in heart injury was discussed, focusing on challenges of rapid clearance of MVs once injected into the patients (Chen et al., 2021).

Besides, circulating MVs are enriched in proteins with immunomodulatory functions, suggesting their immunotherapy application against human cancer. Accumulating evidence revealed an immunosuppressive role of cancer-derived exosomes (Li et al., 2006; Yang et al., 2020). A recent study has reported that melanoma-derived exosomes inhibited the proliferation and activation of CD8+ T cells and induced the T regulatory cells to exhibit an immunosuppressive role (Sharma et al., 2020). Moreover, it was shown that RBC-MVs stimulate T-cell response through antigen-presenting cells (Danesh et al., 2014). Recently, PMVs were employed for atherosclerotic plaque-targeted therapy (Ma et al., 2021). The authors have demonstrated that PMVs encapsulated inflammasome inhibitor lowers inflammation and inhibited the proliferation of macrophages and T cells at the plaque region.

Cell-free vaccine development is another area where EVs have garnered recent attention. Although the first evidence on the release of exosomes from DC (referred as dexosomes) was provided by Zitvogel two decades ago suggesting their role in immune suppression in malignancy (Zitvogel et al., 1998), efforts to understand the property and potential function is slowly being explored. Dexosomes exhibits Class I and Class II MHC molecules contributing to antigen specific immune response, hence being considered as a good cell free system for vaccine development. Earlier studies suggested engineering the microvesicles to deliver peptide vaccines at targeted sites (Geuze and Melief, 1997). In addition, monocyte derived DC in peripheral blood can be induced to release human dexosomes, and used for antitumor vaccine delivery, which is safe to administer and elicits both innate (NK cells) and adaptive (T cell) immune response (Pitt et al., 2016; Nikfarjam et al., 2020). It is noteworthy to observe the immense therapeutic potential of microvesicles, yet clinical translation hasn't reached its full potential due to challenges associated with bioavailability and efficacy, which needs further clarification.

8.6 CONCLUSION

Circulating MVs, released from cellular components of blood, are of great clinical interest for therapeutic delivery and diagnostic potential. This chapter discusses the current understanding on the biological properties of circulating MVs, focusing on the functional implications of the abundantly present ones in circulation. This chapter provides a comprehensive overview of the optimal methods in usage for isolation and characterization of these nano- to micro-sized vesicles. Though several methods have emerged, and new methods are still evolving to identify the ideal gold standard technique for isolation, we believe it is not very far from reality.

Earlier studies on MVs are aligned towards the effect of intercellular communication on target cells suggesting a major connection with health and disease. In this regard, MVs are well-studied for their implications in the areas of therapeutic drug delivery and as non-invasive means for early disease diagnosis. Nevertheless, many specific disease biomarkers have been identified so far, but only a few have received clinical accreditation. It is crucial to validate such potential markers in large cohorts to identify reliable and reproducible candidates for disease diagnosis. Moreover, with respect to their applications as drug delivery vehicles, bioavailability, and efficacy need to be better understood to achieve the full potential of these tiny yet powerful nanocarriers packed with potential therapeutic targets.

KEYWORDS

- **biomarkers**
- **circulating microvesicles**
- **diagnostics**
- **drug delivery**
- **platelets**
- **red blood cells**

REFERENCES

Aatonen, M. T., Ohman, T., Nyman, T. A., Laitinen, S., Grönholm, M., & Siljander, P. R., (2014). Isolation and characterization of platelet-derived extracellular vesicles. *Journal of Extracellular Vesicles, 3*. doi: 10.3402/jev.v3.24692.

Aatonen, M., Grönholm, M., & Siljander, P. R., (2012). Platelet-derived microvesicles: Multitalented participants in intercellular communication. *Seminars in Thrombosis and Hemostasis, 38*(1), 102–113. doi: 10.1055/s-0031-1300956.

Admyre, C., Johansson, S. M., Qazi, K. R., et al., (2007). Exosomes with immune modulatory features are present in human breast milk. *Journal of Immunology, 179*(3), 1969–1978. doi: 10.4049/jimmunol.179.3.1969.

Allan, D., Thomas, P., & Limbrick, A. R., (1980). The isolation and characterization of 60 nm vesicles ('nanovesicles') produced during ionophore A23187-induced budding of human erythrocytes. *The Biochemical Journal, 188*(3), 881–887. doi: 10.1042/bj1880881.

Ambros, V., (2004). The functions of animal microRNAs. *Nature, 431*(7006), 350–355. doi: 10.1038/nature02871.

Arraud, N., Linares, R., Tan, S., et al., (2014). Extracellular vesicles from blood plasma: Determination of their morphology, size, phenotype and concentration. *Journal of Thrombosis and Haemostasis, 12*(5), 614–627. doi: 10.1111/jth.12554.

Baj-Krzyworzeka, M., Majka, M., Pratico, D., et al., (2002). Platelet-derived microparticles stimulate proliferation, survival, adhesion, and chemotaxis of hematopoietic cells. *Experimental Hematology, 30*(5), 450–459. doi: 10.1016/s0301-472x(02)00791-9.

Banks, W. A., Sharma, P., Bullock, K. M., Hansen, K. M., & Ludwig, N., (2020). Transport of extracellular vesicles across the blood-brain barrier: Brain pharmacokinetics and effects of inflammation. *International Journal of Molecular Sciences, 21*(12). doi: 10.3390/ijms21124407.

Brinkmann, K., Enderle, D., Flinspach, C., Meyer, L., Skog, J., & Noerholm, M., (2018). Exosome liquid biopsies of NSCLC patients for longitudinal monitoring of ALK fusions and resistance mutations. *Journal of Clinical Oncology, 36*(15_suppl), e24090-e90. doi: 10.1200/JCO.2018.36.15_suppl.e24090.

Burnouf, T., Chou, M. L., Goubran, H., Cognasse, F., Garraud, O., & Seghatchian, J., (2015). An overview of the role of microparticles/microvesicles in blood components: Are they clinically beneficial or harmful? *Transfusion and Apheresis Science, 53*(2), 137–145. doi: 10.1016/j.transci.2015.10.010.

Butler, J. T., Abdelhamed, S., & Kurre, P., (2018). Extracellular vesicles in the hematopoietic microenvironment. *Haematologica, 103*(3), 382–394. doi: 10.3324/haematol.2017.183335.

Caby, M. P., Lankar, D., Vincendeau-Scherrer, C., Raposo, G., & Bonnerot, C., (2005). Exosomal-like vesicles are present in human blood plasma. *International Immunology, 17*(7), 879–887. doi: 10.1093/intimm/dxh267.

Castellanos-Rizaldos, E., Grimm, D. G., Tadigotla, V., et al., (2018). Exosome-based detection of EGFR T790M in plasma from non-small cell lung cancer patients. *Clinical Cancer Research, 24*(12), 2944–2950. doi: 10.1158/1078-0432.ccr-17-3369.

Chen, P., Wang, L., Fan, X., et al., (2021). Targeted delivery of extracellular vesicles in heart injury. *Theranostics, 11*(5), 2263–2277. doi: 10.7150/thno.51571.

Combes, V., Simon, A. C., Grau, G. E., et al., (1999). *In vitro* generation of endothelial microparticles and possible prothrombotic activity in patients with lupus anticoagulant. *The Journal of Clinical Investigation, 104*(1), 93–102. doi: 10.1172/jci4985.

Costa, V. H., Gitz-Francois, J. J., Schiffelers, R. M., & Vader, P., (2017). Cellular uptake of extracellular vesicles is mediated by clathrin-independent endocytosis and macropinocytosis. *Journal of Controlled Release, 266*, 100–108. doi: 10.1016/j.jconrel.2017.09.019.

Danesh, A., Inglis, H. C., Jackman, R. P., et al., (2014). Exosomes from red blood cell units bind to monocytes and induce proinflammatory cytokines, boosting T-cell responses *in vitro*. *Blood, 123*(5), 687–696. doi: 10.1182/blood-2013-10-530469.

Del, C. I., Shrimpton, C. N., Thiagarajan, P., & López, J. A., (2005). Tissue-factor-bearing microvesicles arise from lipid rafts and fuse with activated platelets to initiate coagulation. *Blood, 106*(5), 1604–1611. doi: 10.1182/blood-2004-03-1095.

Dignat-George, F., & Boulanger, C. M., (2011). The many faces of endothelial microparticles. *Arteriosclerosis, Thrombosis, and Vascular Biology, 31*(1), 27–33. doi: 10.1161/atvbaha.110.218123.

Donadee, C., Raat, N. J., Kanias, T., et al., (2011). Nitric oxide scavenging by red blood cell microparticles and cell-free hemoglobin as a mechanism for the red cell storage lesion. *Circulation, 124*(4), 465–476. doi: 10.1161/circulationaha.110.008698.

Flannagan, R. S., Canton, J., Furuya, W., Glogauer, M., & Grinstein, S., (2014). The phosphatidylserine receptor TIM4 utilizes integrins as coreceptors to effect phagocytosis. *Molecular Biology of the Cell, 25*(9), 1511–1522. doi: 10.1091/mbc.E13-04-0212.

Forlow, S. B., McEver, R. P., & Nollert, M. U., (2000). Leukocyte-leukocyte interactions mediated by platelet microparticles under flow. *Blood, 95*(4), 1317–1323.

Geuze, J. J. & Melief, C. J. .M., (1997). *Cell Derived Antigen Presenting Vesicles*. WO1997005900A1.

Gkaliagkousi, E., Nikolaidou, B., Gavriilaki, E., et al., (2019). Increased erythrocyte- and platelet-derived microvesicles in newly diagnosed type 2 diabetes mellitus. *Diabetes and Vascular Disease Research, 16*(5), 458–465. doi: 10.1177/1479164119844691.

Goetzl, E. J., Goetzl, L., Karliner, J. S., Tang, N., & Pulliam, L., (2016). Human plasma platelet-derived exosomes: Effects of aspirin. *FASEB Journal, 30*(5), 2058–2063. doi: 10.1096/fj.201500150R.

Grasedieck, S., Schöler, N., Bommer, M., et al., (2012). Impact of serum storage conditions on microRNA stability. *Leukemia, 26*(11), 2414–2416. doi: 10.1038/leu.2012.106.

Guo, S. C., Tao, S. C., Yin, W. J., Qi, X., Yuan, T., & Zhang, C. Q., (2017). Exosomes derived from platelet-rich plasma promote the re-epithelization of chronic cutaneous wounds via activation of YAP in a diabetic rat model. *Theranostics, 7*(1), 81–96. doi: 10.7150/thno.16803.

Hood, J. L., (2016). Post isolation modification of exosomes for nanomedicine applications. *Nanomedicine, 11*(13), 1745–1756. doi: 10.2217/nnm-2016-0102.

Horder, E., (1899). Blood dust or blood granules: A new constituent of the blood? *The Lancet, 154*, 1015. doi: 10.1016/S0140-6736(01)59201-6.

Huang, H., Zhu, J., & Fan, L., (2019). MicroRNA profiling of exosomes derived from red blood cell units: Implications in transfusion-related immunomodulation. *BioMed Research International, 2019*, 2045915. doi: 10.1155/2019/2045915.

Issman, L., Brenner, B., Talmon, Y., & Aharon, A., (2013). Cryogenic transmission electron microscopy nanostructural study of shed microparticles. *PLoS One, 8*(12), e83680. doi: 10.1371/journal.pone.0083680.

Jank, H., & Salzer, U., (2011). Vesicles generated during storage of red blood cells enhance the generation of radical oxygen species in activated neutrophils. *The Scientific World Journal, 11*, 173–185. doi: 10.1100/tsw.2011.25.

Jansen, F., Yang, X., Franklin, B. S., et al., (2013). High glucose condition increases NADPH oxidase activity in endothelial microparticles that promote vascular inflammation. *Cardiovascular Research, 98*(1), 94–106. doi: 10.1093/cvr/cvt013.

Jin, M., Drwal, G., Bourgeois, T., Saltz, J., & Wu, H. M., (2005). Distinct proteome features of plasma microparticles. *Proteomics, 5*(7), 1940–1952. doi: 10.1002/pmic.200401057.

Jung, C., Sörensson, P., Saleh, N., Arheden, H., Rydén, L., & Pernow, J., (2012). Circulating endothelial and platelet derived microparticles reflect the size of myocardium at risk in patients with ST-elevation myocardial infarction. *Atherosclerosis, 221*(1), 226–231. doi: 10.1016/j.atherosclerosis.2011.12.025.

Kailashiya, J., Gupta, V., & Dash, D., (2019). Engineered human platelet-derived microparticles as natural vectors for targeted drug delivery. *Oncotarget, 10*(56), 5835–5846. doi: 10.18632/oncotarget.27223.

Kalluri, R., & LeBleu, V. S., (2020). The biology, function, and biomedical applications of exosomes. *Science, 367*(6478). doi: 10.1126/science.aau6977.

Kapogiannis, D., Mustapic, M., Shardell, M. D., et al., (2019). Association of extracellular vesicle biomarkers with Alzheimer disease in the baltimore longitudinal study of aging. *JAMA Neurology, 76*(11), 1340–1351. doi: 10.1001/jamaneurol.2019.2462.

Kapsogeorgou, E. K., Abu-Helu, R. F., Moutsopoulos, H. M., & Manoussakis, M. N., (2005). Salivary gland epithelial cell exosomes: A source of autoantigenic ribonucleoproteins. *Arthritis and Rheumatism, 52*(5), 1517–1521. doi: 10.1002/art.21005.

Kim, Y., Xia, B. T., Jung, A. D., et al., (2018). Microparticles from stored red blood cells promote a hypercoagulable state in a murine model of transfusion. *Surgery, 163*(2), 423–429. doi: 10.1016/j.surg.2017.09.028.

Konoshenko, M. Y., & Lekchnov, E. A., (2018). Isolation of extracellular vesicles: General methodologies and latest trends. *BioMed Research International, 2018*, 8545347. doi: 10.1155/2018/8545347.

Li, X. B., Zhang, Z. R., Schluesener, H. J., & Xu, S. Q., (2006). Role of exosomes in immune regulation. *Journal of Cellular and Molecular Medicine, 10*(2), 364–375. doi: 10.1111/j.1582-4934.2006.tb00405.x.

Luan, X., Sansanaphongpricha, K., Myers, I., Chen, H., Yuan, H., & Sun, D., (2017). Engineering exosomes as refined biological nanoplatforms for drug delivery. *Acta Pharmacologica Sinica, 38*(6), 754–763. doi: 10.1038/aps.2017.12.

Lugini, L., Cecchetti, S., Huber, V., et al., (2012). Immune surveillance properties of human NK cell-derived exosomes. *Journal of Immunology, 189*(6), 2833–2842. doi: 10.4049/jimmunol.1101988.

Ma, Q., Fan, Q., Han, X., et al., (2021). Platelet-derived extracellular vesicles to target plaque inflammation for effective anti-atherosclerotic therapy. *Journal of Controlled Release, 329*, 445–453. doi: 10.1016/j.jconrel.2020.11.064.

Melo, S. A., Luecke, L. B., Kahlert, C., et al., (2015). Glypican-1 identifies cancer exosomes and detects early pancreatic cancer. *Nature, 523*(7559), 177–182. doi: 10.1038/nature14581.

Mezentsev, A., Merks, R. M., O'Riordan, E., et al., (2005). Endothelial microparticles affect angiogenesis *in vitro*: Role of oxidative stress. American journal of physiology. *Heart and Circulatory Physiology, 289*(3), H1106–1114. doi: 10.1152/ajpheart.00265.2005.

Mitsuhashi, M., (2013). *Assessment of Bone Marrow Recovery by Measuring Plasma Exosome mRNAs*. US20130172208A1.

Morel, O., Jesel, L., Freyssinet, J. M., & Toti, F., (2011). Cellular mechanisms underlying the formation of circulating microparticles. *Arteriosclerosis, Thrombosis, and Vascular Biology, 31*(1), 15–26. doi: 10.1161/atvbaha.109.200956.

Muller, L., Hong, C. S., Stolz, D. B., Watkins, S. C., & Whiteside, T. L., (2014). Isolation of biologically-active exosomes from human plasma. *Journal of Immunological Methods, 411*, 55–65. doi: 10.1016/j.jim.2014.06.007.

Nikfarjam, S., Rezaie, J., Kashanchi, F., & Jafari, R., (2020). Dexosomes as a cell-free vaccine for cancer immunotherapy. *Journal of Experimental and Clinical Cancer Research, 39*(1), 258. doi: 10.1186/s13046-020-01781-x.

Noubouossie, D. F., Henderson, M. W., Mooberry, M., et al., (2020). Red blood cell microvesicles activate the contact system, leading to factor IX activation via 2 independent pathways. *Blood, 135*(10), 755–765. doi: 10.1182/blood.2019001643.

Pan, B. T., & Johnstone, R. M., (1983). Fate of the transferrin receptor during maturation of sheep reticulocytes *in vitro*: Selective externalization of the receptor. *Cell, 33*(3), 967–978. doi: 10.1016/0092-8674(83)90040-5.

Pisitkun, T., Shen, R. F., & Knepper, M. A., (2004). Identification and proteomic profiling of exosomes in human urine. *Proceedings of the National Academy of Sciences of the United States of America, 101*(36), 13368–13373. doi: 10.1073/pnas.0403453101.

Pitt, J. M., André, F., Amigorena, S., et al., (2016). Dendritic cell-derived exosomes for cancer therapy. *The Journal of Clinical Investigation, 126*(4), 1224–1232. doi: 10.1172/jci81137.

Prokopi, M., Pula, G., Mayr, U., et al., (2009). Proteomic analysis reveals the presence of platelet microparticles in endothelial progenitor cell cultures. *Blood, 114*(3), 723–732. doi: 10.1182/blood-2009-02-205930.

Pugholm, L. H., Bæk, R., Søndergaard, E. K., Revenfeld, A. L., Jørgensen, M. M., & Varming, K., (2016). Phenotyping of leukocytes and leukocyte-derived extracellular vesicles. *Journal of Immunology Research, 2016*, 6391264. doi: 10.1155/2016/6391264.

Ramirez, M. I., Amorim, M. G., Gadelha, C., & Milic, I., (2018). Technical challenges of working with extracellular vesicles. *Nanoscale, 10*(3), 881–906. doi: 10.1039/c7nr08360b.

Rana, S., Yue, S., Stadel, D., & Zöller, M., (2012). Toward tailored exosomes: The exosomal tetraspanin web contributes to target cell selection. *The International Journal of Biochemistry & Cell Biology, 44*(9), 1574–1584. doi: 10.1016/j.biocel.2012.06.018.

Rasheed, Z., Rasheed, N., Abdulmonem, W. A., & Khan, M. I., (2019). Author correction: MicroRNA-125b-5p regulates IL-1β induced inflammatory genes via targeting TRAF6-mediated MAPKs and NF-κB signaling in human osteoarthritic chondrocytes. *Scientific Reports, 9*(1), 14729. doi: 10.1038/s41598-019-50844-3.

Rubin, O., Crettaz, D., Tissot, J. D., & Lion, N., (2010). Pre-analytical and methodological challenges in red blood cell microparticle proteomics. *Talanta, 82*(1), 1–8. doi: 10.1016/j.talanta.2010.04.025.

Rupp, A. K., Rupp, C., Keller, S., et al., (2011). Loss of EpCAM expression in breast cancer derived serum exosomes: Role of proteolytic cleavage. *Gynecologic Oncology, 122*(2), 437–446. doi: 10.1016/j.ygyno.2011.04.035.

Sadallah, S., Eken, C., & Schifferli, J. A., (2008). Erythrocyte-derived ectosomes have immunosuppressive properties. *Journal of Leukocyte Biology, 84*(5), 1316–1325. doi: 10.1189/jlb.0108013.

Sadallah, S., Eken, C., & Schifferli, J. A., (2011). Ectosomes as modulators of inflammation and immunity. *Clinical and Experimental Immunology, 163*(1), 26–32. doi: 10.1111/j.1365-2249.2010.04271.x.

Shah, R., Patel, T., & Freedman, J. E., (2018). Circulating extracellular vesicles in human disease. *The New England Journal of Medicine, 379*(10), 958–966. doi: 10.1056/NEJMra1704286.

Sharma, P., Diergaarde, B., Ferrone, S., Kirkwood, J. M., & Whiteside, T. L., (2020). Melanoma cell-derived exosomes in plasma of melanoma patients suppress functions of immune effector cells. *Scientific Reports, 10*(1), 92. doi: 10.1038/s41598-019-56542-4.

Srikanthan, S., Li, W., Silverstein, R. L., & McIntyre, T. M., (2014). Exosome poly-ubiquitin inhibits platelet activation, downregulates CD36 and inhibits pro-atherothombotic cellular functions. *Journal of Thrombosis and Haemostasis, 12*(11), 1906–1917. doi: 10.1111/jth.12712.

Tao, S. C., Guo, S. C., & Zhang, C. Q., (2017). Platelet-derived extracellular vesicles: An emerging therapeutic approach. *International Journal of Biological Sciences, 13*(7), 828–834. doi: 10.7150/ijbs.19776.

Taraboletti, G., D'Ascenzo, S., Borsotti, P., Giavazzi, R., Pavan, A., & Dolo, V., (2002). Shedding of the matrix metalloproteinases MMP-2, MMP-9, and MT1-MMP as membrane vesicle-associated components by endothelial cells. *The American Journal of Pathology, 160*(2), 673–680. doi: 10.1016/s0002-9440(10)64887-0.

Taus, F., Meneguzzi, A., Castelli, M., & Minuz, P., (2019). Platelet-derived extracellular vesicles as target of antiplatelet agents. What is the evidence? *Frontiers in Pharmacology, 10*, 1256. doi: 10.3389/fphar.2019.01256.

Thangaraju, K., & Neerukonda, S. N., (2020). Extracellular vesicles from red blood cells and their evolving roles in health, coagulopathy and therapy. *International Journal of Molecular Sciences, 22*(1). doi: 10.3390/ijms22010153.

Tripisciano, C., Weiss, R., Eichhorn, T., et al., (2017). Different potential of extracellular vesicles to support thrombin generation: Contributions of phosphatidylserine, tissue factor, and cellular origin. *Scientific Reports, 7*(1), (6522). doi: 10.1038/s41598-017-03262-2.

Urbanelli, L., Buratta, S., Sagini, K., Ferrara, G., Lanni, M., & Emiliani, C., (2015). Exosome-based strategies for diagnosis and therapy. *Recent Patents on CNS Drug Discovery, 10*(1), 10–27. doi: 10.2174/1574889810666150702124059.

Van, R. E., & Olson, E. N., (2007). MicroRNAs: Powerful new regulators of heart disease and provocative therapeutic targets. *The Journal of Clinical Investigation, 117*(9), 2369–2376. doi: 10.1172/jci33099.

Vella, L. J., Greenwood, D. L., Cappai, R., Scheerlinck, J. P., & Hill, A. F., (2008). Enrichment of prion protein in exosomes derived from ovine cerebral spinal fluid. *Veterinary Immunology and Immunopathology, 124*(3, 4), 385–393. doi: 10.1016/j.vetimm.2008.04.002.

Wahlgren, J., De, L. K. T., Brisslert, M., et al., (2012). Plasma exosomes can deliver exogenous short interfering RNA to monocytes and lymphocytes. *Nucleic Acids Research, 40*(17), e130. doi: 10.1093/nar/gks463.

Walker, S., Busatto, S., Pham, A., et al., (2019). Extracellular vesicle-based drug delivery systems for cancer treatment. *Theranostics, 9*(26), 8001–8017. doi: 10.7150/thno.37097.

Weisel, J. W., & Litvinov, R. I., (2019). Red blood cells: The forgotten player in hemostasis and thrombosis. *Journal of Thrombosis and Haemostasis, 17*(2), 271–282. doi: 10.1111/jth.14360.

Wolf, P., (1967). The nature and significance of platelet products in human plasma. *British Journal of Haematology, 13*(3), 269–288. doi: 10.1111/j.1365-2141.1967.tb08741.x.

Xiao, D., Ohlendorf, J., Chen, Y., et al., (2012). Identifying mRNA, microRNA and protein profiles of melanoma exosomes. *PloS One, 7*(10), e46874. doi: 10.1371/journal.pone.0046874.

Xie, Y., Zhang, X., Zhao, T., Li, W., & Xiang, J., (2013). Natural $CD8^{+}25^{+}$ regulatory T cell-secreted exosomes capable of suppressing cytotoxic T lymphocyte-mediated immunity against B16 melanoma. *Biochemical and Biophysical Research Communications, 438*(1), 152–155. doi: 10.1016/j.bbrc.2013.07.044.

Xiong, Z., Oriss, T. B., Cavaretta, J. P., Rosengart, M. R., & Lee, J. S., (2012). Red cell microparticle enumeration: Validation of a flow cytometric approach. *Vox Sanguinis, 103*(1), 42–48. doi: 10.1111/j.1423-0410.2011.01577.x.

Yang, D., Zhang, W., Zhang, H., et al., (2020). Progress, opportunity, and perspective on exosome isolation - efforts for efficient exosome-based theranostics. *Theranostics, 10*(8), 3684–3707. doi: 10.7150/thno.41580.

Zarà, M., Guidetti, G. F., Camera, M., et al., (2019). Biology and role of extracellular vesicles (EVs) in the pathogenesis of thrombosis. *International Journal of Molecular Sciences, 20*(11). doi: 10.3390/ijms20112840.

Zitvogel, L., Regnault, A., Lozier, A., et al., (1998). Eradication of established murine tumors using a novel cell-free vaccine: Dendritic cell-derived exosomes. *Nature Medicine, 4*(5), 594–600. doi: 10.1038/nm0598-594.

CHAPTER 9

Hematopoiesis and Cancer Stem Cells: The Seed and the Soil Crosstalk

AKHILESH PRAJAPATI

Department of Biotechnology, School of Science, GSFC University, Vadodara, Gujarat, India

ABSTRACT

Hematopoiesis is a balanced cellular mechanism to maintain blood lineage cellular turnover throughout life. In adults, bone marrow (BM) is the major site for hematopoiesis and provides an appropriate cellular and molecular microenvironment consisting of a heterogeneous stem cell population for proliferation, differentiation, and survival characteristics of the blood system. Neoplastic alteration in the hematopoietic stem cell and cancer stem cells (CSC) is primarily associated with all hematological malignancies. Since HSCs are a major contributor to the hematopoietic system, MSCs actively support hematopoiesis. In hematological cancers, neoplastic cells and CSCs trigger the normal development of hematopoiesis by usurping HSCs and MSCs functions. This altered function reprogramed MSCs to generate a favorable microenvironment to support neoplastic cells, leading to hematological malignancies.

9.1 INTRODUCTION

Hematopoiesis is a continuous and regulated cellular process of blood and its component production and turnover of the blood system as per demand in the body throughout life since embryonic development. However, the formation of blood varies in individuals with the advancement of age and onset of the disease condition. Moreover, understanding the process of

Hematopoiesis: Biochemical, Cellular, Molecular, and Genomic Perspectives. Ravi Kumar Gutti (Ed.)

hematopoiesis and its regulation further helps researchers and clinicians to gauge the pathologies associated with it.

Over the last few decades, regenerative medicine and modern biotechnological approaches have opened up new avenues to study oncology and other life-threatening diseases in humans. Stem cell biology is always a promising area to study such life-threatening disorders for better cures. Hematopoiesis is regulated by specific types of cells; collectively called hematopoietic stem cells (HSCs). HSCs have the ability to self-renewal and differentiation capability into all blood lineage cells, which comprises monocytes/MNC, erythrocytes/RBC, platelets, lymphocytes, granulocytes, and macrophages/WBC. After fertilization till zygote formation in mammals, including humans, the hematopoiesis process is being regulated by the yolk sac and the liver, respectively, later the process transferred to the bone marrow (BM) and thymus, where the HSCs finally reside. In adulthood, BM takes over the complete charges to produce blood cells in the body through the HSCs (Lee-Six et al., 2018).

Cancers have increased dramatically across the globe in the last decade with the topmost death-causing factor among others. Many recent reports and studies showed the incidence of cancer pathology and deaths due to cancer (Moliner et al., 2019). Hematopoietic cancers (HC) still cover approximately 10% of all the total diagnosed cancer with an equal death ratio in the world.

Increasing incidences of HC and unique properties of HSCs in the cancer progression further attract work on this aspect. This chapter will focus more on transforming HSCs' niche and their contribution in the formation of blood cancers.

Hematopoietic stem cells: the major site for blood formation in the human body is BM; a spongy and soft tissue in long bones. The proximal portion of the trabeculae bone region is majorly site for the HSC niche in the long bones (Wasnik et al., 2016). Initially, these niche cells do not express any lineage specific marks. In mice studies, at the early stage of hematopoiesis, mice HSCs do not show any blood cells lineage markers. These cells are collectively called Lin negative cells (Weissman & Shizuru, 2008). Interestingly, these cells have the ability to generate all types of lineage-specific progenitor cells. However, as the development proceeds in mice, hematopoietic stem cells start expressing Sca-1 cell markers and cover almost 25–30% of the total Lin negative cells. These Sca-1 cells also express cell surface antigen CD90 markers. Report showed high repopulation capacity during the developmental stages. Morrison et al. (1994) showed low CD90 cells express high repopulation capacity in the BM cells of mice (Morrison & Weissman, 1994). Moreover, these stem cells also express c-kit (tyrosine kinase receptor) and its ligand (KL) significantly. These combinations are further used by many researchers to isolate these HSCs (Noda et al., 2008).

Another important cell surface marker CD34 has also been reported in murine research. However, the expression of $CD34^{high}$ has reflected controversy in available reports. Data on mouse fetal HSCs showed $CD^{34\ high}$ expression but adult cells lack its presence and considered $CD34^{-}$ (Osawa et al., 1996). Similar investigations by Ito et al. (2000) demonstrated large population of cells are seen CD34- and showed less expression of $CD34^{high}$ in mice (Ito, Tajima, & Ogawa, 2000). Interestingly, Sato et al. (1999) report says these $CD34^{high}$ cells expression has been activated by one or more cytokines and their combinations (Sato, Laver, & Ogawa, 1999).

In the context of human HSC development, early-stage HSC has been identified by expression of cell surface markers which includes $CD34^{high}$ $CD90^{high}Lin^{-}$ $CD38^{low}$ $CD45RA^{-(negative)}$ (Baum et al., 1992; Bhatia et al., 1997; Kondo et al., 2003; Majeti, Park, & Weissman, 2007). Additionally, various combinations of cell surface marks are being used in FACS analysis. CMP (erythrocytes, granulocytes, megakaryocytes, and macrophages) and CLP (B cell, T cell, NK cell) are identified by cellular markers including $CD34^{high}CD38^{high}CD10^{high}$ and $Lin^{-(negative)}$ $CD34^{high}CD38^{high}$ $CD45RA^{-(negative)}$, respectively.

Cytokines play a very crucial role in HSC activation and tissue specific localization. Cytokine CXCL12, also known as SDF-1 expression, has been reported high in BM cells. SDF-1 has been known for localization of the endosteal niche. An investigation by Sugiyama et al. (2006) has revealed abundant expression of CXCL-12 shown by a number of HSC's.

Interestingly, shifting of HSC niche from fetal liver during embryonic stage to adult BM is facilitated by CXCL12 expression and its receptor (CXC4). Any alteration in the CXCL12 and CXC4 action severely affects the migration and translocation activity of HSCs from fetal liver to BM and leads to severe pathologies followed by death of the organism (Sugiyama et al., 2006). Moreover, apart from the migration of HSCs, CXCL12 also involves in the regulation of self-renewal, differentiation capability, and other developmental processes of HSCs (Nie et al., 2008).

9.2 PRODUCTION OF CANCER STEM CELLS (CSC)

Self-renewal and differentiation capability are the main characteristics of stem cells for the production of the required number of differentiated specialized cells in the tissue system. These cells undergo several transition phases also called transit-amplifying cells (TA) from progenitor to highly specialized cells types during normal development. HSCs as described above in the chapter, having capability to differentiate into the different cell lineages of the blood system. Moreover, during the developmental stages, if any cell acquires a number of mutations and accumulation of the same, it leads to a progression of cancer. However, to combat such faulty mechanisms many protective mechanisms evolved (Figure 9.1). To maintain tissue homeostasis, a subset of stem cells divides and maintains appropriate cell numbers in the tissue. Notably, normal tissue stem cells share developmental and proliferation properties with cancer stem cells (CSC), including self-renewal and differentiation capabilities. It has been shown in mouse models that a cascade of regulatory genes is actively involved in HSCs genesis (Morrison & Spradling, 2008). Self-renewal capability genes are considered to be oncogenes and participate in cancer development and progression rather than another subset of genes. Studies have shown *Bmi1*, one of the factors of silencing machinery for polycomb development regulates target genes by transcriptional epigenetic repression mechanisms (Jacobs et al., 1999).

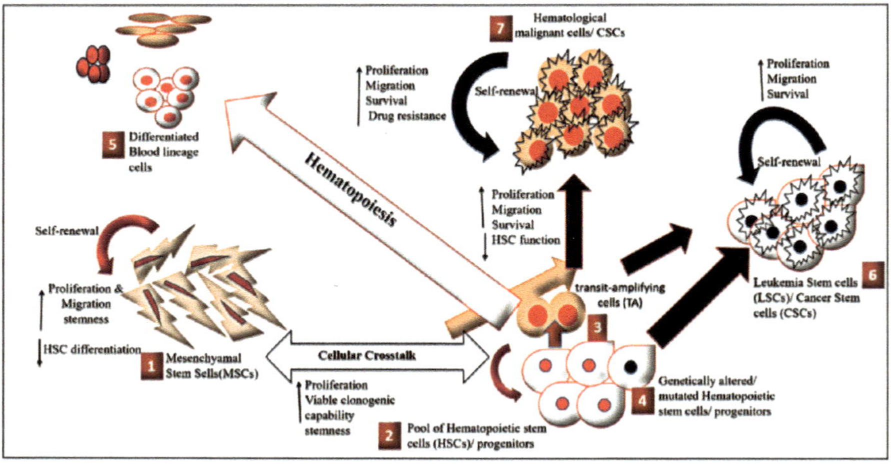

FIGURE 9.1 A pictorial representation of hematopoiesis and hematological malignancy. (1) Modified MSCs and their crosstalk between HSCs; (2) HSC niche in normal hematopoiesis and HMs; (3) TA cells in hematopoiesis and HMs; (4) genetic alterations in HSCs and undefined malignant cells; (5) hematopoiesis and differentiated blood system cells; (6) aggressive leukemic stem cells; and (7) chemotherapy resistant advanced hematological malignant cells.

Additionally, *Bmi1* is also found actively involved in self-renewal and maintenance of neural stem cells and HSC pool in mouse studies (Molofsky et al., 2003; Park et al., 2003). It believes that CSC in the blood cancer malignancies follows the same molecular pathways and events that regulate normal stem cells. These CSC proliferate and migrate uncontrollably to produce tumors. Moreover. In the absence of *Bmi1* in mouse leukemia models showed failure of self-renewal capability further strengthening its role in cancer progression through stem cell niche (Lessard & Sauvageau, 2003). Many embryonic developmental pathways also revealed up-regulated expression of *Bmi1* in mouse HSC and progenitor cells (Leung et al., 2004). Likewise, tumor suppressor oncogenes, including *p19Arf*, *p16Ink4a*, and *Tp53* negatively regulate *Bmi* expression in normal stem cells, which indicates its pivotal role in oncogenic progression.

9.3 MOLECULAR AND CELLULAR EVENTS

Genetic transformation/modification in HSCs during proliferation/division or its progenitors are associated with several oncological events in the hematopoietic tissue and its microenvironment, leads to give rise to the birth of various hematologic malignancies (HMs), which includes myelodysplastic syndrome (MDS), myeloproliferative neoplasm (MPNs), acute myeloid leukemia (AML), chronic myeloid leukemia (CML), chronic lymphocytic leukemia (CLL), and acute lymphoblastic leukemia (ALL) (Sanchez-Aguilera et al., 2017). These genetic alterations, HSCs or progenitors are transformed into leukemia stem cells (LSCs). Isolated LSCs demonstrated stem cell characteristics, self-renewal capability, uncontrolled differentiation into leukemic blasts and oncologic progression (Schepers, Campbell, & Passegue, 2015).

Oncogenic development through the CSC would be a series of by various cellular and molecular events, which includes activation of oncogenic mutation in the normal stem cell niche or acquire mutation/genetic alteration in TA or progenitor cells stage during the series of developmental process and, abrupt normal proliferation mechanism. Scientific investigations showed leukemia arises at the level of HSC and acquires several mutations required to become malignant (Dash & Gilliland, 2001; Lowenberg, Downing, & Burnett, 1999). Bonnet et al. and colleagues have shown a stem cell population expressing $CD34^{+}/CD38^{-(\text{negative})}$ cell surface markers in AML in NOD/SCID mouse model system (Bonnet & Dick, 1997).

Studying normal HSCs stem cells and using the same strategies to characterize CSC have revealed categorized structure development of LSCs populations in AML. Obtaining a detailed understanding of LSC-specific molecular characteristics is a high priority for the future development of leukemia drugs. Importantly, while some phenotypic features of LSCs are similar to HSCs, recent studies have also described features that are unique to AML. These findings help to design treatment strategies to combat leukemia and other HMs (Jordan, 2002; Sutherland et al., 1996).

9.4 CROSS TALK BETWEEN NORMAL AND CANCER STEM CELLS (CSC): A TWO-FACE EVE

As discussed above, LSC contains a significant population of self-renewal and metastatic characteristic cells in the normal HSC niche. Both the cell types are believed to interact with each other in the BM cellular environment and said to be responsible for blood malignancies (Lapidot et al., 1994; Wang & Dick, 2005). Additionally, the communication between the LSCs and the HSCs niche in tissue maintain their quiescent state, which is believed to be a kind of protection of the cancer cells from the cytotoxic effect of conventional chemotherapy agents (Zagozdzon & Golab, 2015; Bonnet & Dick, 1997).

The BM tissue is considered to be a rich source of MSCs and HSCs (Murray & Peault, 2015). The crosstalk between the cell types is required for normal hematopoiesis. These MSCs help in retaining long-term, quiescent HSCs cell population by secreting several cellular factors and cytokines which include leukemia inhibitor factor and IL-6 which further help in maintaining stemness characteristics of HSC (Calvi et al., 2003; Arai et al., 2004).

However, isolated MSCs from leukemia patients showed an inhibitory effect on expansion and differentiation properties of hematopoietic cells when allowed to grow in a laboratory environment. P. Chandran *et al.* has shown an impaired expansion of $CD34^{+}$ hematopoietic cells in AML patients by MSCs and reduced their differentiation capability by maintaining the cellular marker, including $CD34^{+}$ $CD38^{-}$ quiescent precursor cells in the immature state. However, the same event has not been observed with healthy individuals (Chandran et al., 2015).

Moreover, the role of MSCs has been extensively studied in terms of BM microenvironment modulation by triggering angiogenesis and immunomodulation to support hematopoiesis (Zhao et al., 2010; Anthony & Link, 2014; Kfoury & Scadden, 2015). The characteristics of MSCs in hematopoiesis modulation and HSC pool maintenance initiated their usage in allogeneic stem cell transplantation in the medical field (Hoogduijn & Dor, 2011; Zhao & Liu, 2016).

9.5 CONCLUSION

Studying the crosstalk at molecular and cellular level between HSCs, LSCs, and MSCs niche in the BM tissue microenvironment will further enhance our deep understanding about the tissue development and oncogenic progression of various HMs. Such findings further open up new avenues in the discovery of potential therapies and medical interventions to treat the pathologies.

KEYWORDS

- **hematological malignancy**
- **hematopoiesis**
- **hematopoietic stem cells**
- **leukemia stem cells**

REFERENCES

Anthony, B. A., & Link, D. C., (2014). Regulation of hematopoietic stem cells by bone marrow stromal cells. *Trends in Immunology, 35*(1), 32–37. doi: 10.1016/j.it.2013.10.002.

Arai, F., Hirao, A. M. O., et al., (2004). Tie2/angiopoietin-1 signaling regulates hematopoietic stem cell quiescence in the bone marrow niche. *Cell, 118*(2), 149–161. doi: 10.1016/j.cell.2004.07.004.

Baum, C. M., Weissman, I. L., Tsukamoto, A. S., Buckle, A. M., & Peault, B., (1992). Isolation of a candidate human hematopoietic stem-cell population. *Proceedings of the National Academy of Sciences of the United States of America, 89*(7), 2804–2808. doi: 10.1073/pnas.89.7.2804.

Bhatia, M., Wang, J. C., Kapp, U., Bonnet, D., & Dick, J. E., (1997). Purification of primitive human hematopoietic cells capable of repopulating immune-deficient mice. *Proceedings of the National Academy of Sciences of the United States of America, 94*(10), 5320–5325. doi: 10.1073/pnas.94.10.5320.

Bonnet, D., & Dick, J. E., (1997). Human acute myeloid leukemia is organized as a hierarchy that originates from a primitive hematopoietic cell. *Nature Medcine, 3*(7), 730–737. doi: 10.1038/nm0797-730.

Calvi, L. M., Adams, G. B., et al., (2003). Osteoblastic cells regulate the haematopoietic stem cell niche. *Nature, 425*(6960), 841–846. doi: 10.1038/nature02040.

Chandran, P., Le, Y., Li, Y., et al., (2015). Mesenchymal stromal cells from patients with acute myeloid leukemia have altered capacity to expand differentiated hematopoietic progenitors. *Leukemia Research, 39*(4), 4864–4893. doi: 10.1016/j.leukres.2015.01.013.

Dash, A., & Gilliland, D. G., (2001). Molecular genetics of acute myeloid leukaemia. *Best Practice & Research Clinical Haematology, 14*(1), 49–64. doi: 10.1053/beha.2000.0115.

Hoogduijn, M. J., & Dor, F. J., (2011). Mesenchymal stem cells in transplantation and tissue regeneration. *Frontiers in Immunology, 2*, 84. doi: 10.3389/fimmu.2011.00084.

Ito, T., Tajima, F., & Ogawa, M., (2000). Developmental changes of CD34 expression by murine hematopoietic stem cells. *Experimental Hematology, 28*(11), 1269–1273. doi: 10.1016/s0301-472x(00)00535-x.

Jacobs, J. J., Kieboom, K., Marino, S., DePinho, R. A., & Van, L. M., (1999). The oncogene and polycomb-group gene bmi-1 regulates cell proliferation and senescence through the ink4a locus. *Nature, 397*(6715), 164–168. doi: 10.1038/16476.

Jordan, T. C., (2002). Unique molecular and cellular features of acute myelogenous leukemia stem cells. *Leukemia, 16*(4), 559–562. doi: 10.1038/sj.leu.2402446.

Kfoury, Y., & Scadden, D. T., (2015). Mesenchymal cell contributions to the stem cell niche. *Cell Stem Cell, 16*(3), 239–253. doi: 10.1016/j.stem.2015.02.019.

Kondo, M., Wagers, A. J., Manz, M. G., et al., (2003). Biology of hematopoietic stem cells and progenitors: Implications for clinical application. *Annual Review of Immunology, 21*, 759–806. doi: 10.1146/annurev.immunol.21.120601.141007.

Lapidot, T., Sirard, C., Vormoor, J., Murdoch, B., et al., (1994). A cell initiating human acute myeloid leukaemia after transplantation into SCID mice. *Nature, 367*(6464), 645–648. doi: 10.1038/367645a0.

Lee-Six, H., Obro, N. F., Shepherd, M. S., et al., (2018). Population dynamics of normal human blood inferred from somatic mutations. *Nature, 561*(7724), 473–478. doi: 10.1038/s41586-018-0497-0.

Lessard, J., & Sauvageau, G., (2003). Bmi-1 determines the proliferative capacity of normal and leukaemic stem cells. *Nature, 423*(6937), 255–260. doi: 10.1038/nature01572.

Leung, C., Lingbeek, M., Shakhova, O., et al., (2004). Bmi1 is essential for cerebellar development and is overexpressed in human medulloblastomas. *Nature, 428*(6980), 337–341. doi: 10.1038/nature02385.

Lowenberg, B., Downing, J. R., & Burnett, A., (1999). Acute myeloid leukemia. *The New England Journal of Medicine, 341*(14), 1051–1062. doi: 10.1056/NEJM199909303411407.

Majeti, R., Park, C. Y., & Weissman, I. L., (2007). Identification of a hierarchy of multipotent hematopoietic progenitors in human cord blood. *Cell Stem Cell, 1*(6), 635–645. doi: 10.1016/j.stem.2007.10.001.

Moliner, P., Lupon, J., De Antonio, M., et al., (2019). Trends in modes of death in heart failure over the last two decades: Less sudden death but cancer deaths on the rise. *European Journal of Heart Failure, 21*(10), 1259–1266. doi: 10.1002/ejhf.1569.

Molofsky, A. V., Pardal, R., Iwashita, T., Park, I. K., Clarke, M. F., & Morrison, S. J., (2003). Bmi-1 dependence distinguishes neural stem cell self-renewal from progenitor proliferation. *Nature, 425*(6961), 962–697. doi: 10.1038/nature02060.

Morrison, S. J., & Spradling, A. C., (2008). Stem cells and niches: Mechanisms that promote stem cell maintenance throughout life. *Cell, 132*(4), 598–611. doi: 10.1016/j.cell.2008.01.038.

Morrison, S. J., & Weissman, I. L., (1994). The long-term repopulating subset of hematopoietic stem cells is deterministic and isolatable by phenotype. *Immunity, 1*(8), 661–673. doi: 10.1016/1074-7613(94)90037-x.

Murray, I. R., & Peault, B., (2015). Q&A: Mesenchymal stem cells – where do they come from and is it important? *BMC Biology, 13*, 99. doi: 10.1186/s12915-015-0212-7.

Nie, Y., Han, Y. C., & Zou, Y. R., (2008). CXCR4 is required for the quiescence of primitive hematopoietic cells. *Journal of Experimental Medicine, 205*(4), 777–783. doi: 10.1084/jem.20072513.

Noda, S., Horiguchi, K., Ichikawa, H., & Miyoshi, H., (2008). Repopulating activity of ex vivo-expanded murine hematopoietic stem cells resides in the CD48-c-Kit+Sca-1+lineage marker- cell population. *Stem Cells, 26*(3), 646–655. doi: 10.1634/stemcells.2007-0623.

Osawa, M., Hanada, K., Hamada, H., & Nakauchi, H., (1996). Long-term lymphohematopoietic reconstitution by a single CD34-low/negative hematopoietic stem cell. *Science, 273*(5272), 242–245. doi: 10.1126/science.273.5272.242.

Park, I. K., Qian, D., Kiel, M., et al., (2003). Bmi-1 is required for maintenance of adult self-renewing haematopoietic stem cells. *Nature, 423*(6937), 302–305. doi: 10.1038/nature01587.

Sanchez-Aguilera, A., & Mendez-Ferrer, S., (2017). The hematopoietic stem-cell niche in health and leukemia. *Cellular and Molecular Life Sciences, 74*(4), 579–590. doi: 10.1007/s00018-016-2306-y.

Sato, T., Laver, J. H., & Ogawa, M., (1999). Reversible expression of CD34 by murine hematopoietic stem cells. *Blood, 94*(8), 2548–2554.

Schepers, K., Campbell, T. B., & Passegue, E., (2015). Normal and leukemic stem cell niches: Insights and therapeutic opportunities. *Cell Stem Cell, 16*(3), 254–267. doi: 10.1016/j.stem.2015.02.014.

Sugiyama, T., Kohara, H., Noda, M., & Nagasawa, T., (2006). Maintenance of the hematopoietic stem cell pool by CXCL12-CXCR4 chemokine signaling in bone marrow stromal cell niches. *Immunity, 25*(6), 977–988. doi: 10.1016/j.immuni.2006.10.016.

Sutherland, H. J., Blair, A., & Zapf, R. W., (1996). Characterization of a hierarchy in human acute myeloid leukemia progenitor cells. *Blood, 87*(11), 4754–4761.

Wang, J. C., & Dick, J. E., (2005). Cancer stem cells: Lessons from leukemia. *Trends in Cell Biology, 15*(9), 494–501. doi: 10.1016/j.tcb.2005.07.004.

Wasnik, S., Kantipudi, S., Kirkland, M. A., & Pande, G., (2016). Enhanced ex vivo expansion of human hematopoietic progenitors on native and spin coated acellular matrices prepared from bone marrow stromal cells. *Stem Cells International, 2016*, 7231567. doi: 10.1155/2016/7231567.

Weissman, I. L., & Shizuru, J. A., (2008). The origins of the identification and isolation of hematopoietic stem cells, and their capability to induce donor-specific transplantation tolerance and treat autoimmune diseases. *Blood, 112*(9), 3543–3553. doi: 10.1182/blood-2008-08-078220.

Zagozdzon, R., & Golab, J., (2015). Cancer stem cells in haematological malignancies. *Contemporary Oncology, 19*(1A), A1–6. doi: 10.5114/wo.2014.47127.

Zhao, K., & Liu, Q., (2016). The clinical application of mesenchymal stromal cells in hematopoietic stem cell transplantation. *Journal of Hematology and Oncology, 9*(1), 46. doi: 10.1186/s13045-016-0276-z.

Zhao, S., Wehner, R., Bornhauser, M., Wassmuth, R., Bachmann, M., & Schmitz, M., (2010). Immunomodulatory properties of mesenchymal stromal cells and their therapeutic consequences for immune-mediated disorders. *Stem Cells and Development, 19*(5), 607–614. doi: 10.1089/scd.2009.0345.

CHAPTER 10

lncRNAs in Leukemia

SWATI DAHARIYA,[1] ADITYA VELIDANDI,[2] DUSHYANT KUMAR GAUTAM,[1] RAVI KUMAR GUTTI,[1] and INDIRA PADDIBHATLA[1]

[1]*Department of Biochemistry, School of Life Sciences, University of Hyderabad, Hyderabad, Telangana, India*

[2]*Department of Biotechnology, National Institute of Technology, Warangal, Telangana, India*

ABSTRACT

Mammals display a genetic similarity with other organisms in the coding genes yet exhibit diversity and complexity in the long noncoding RNAs (lncRNAs). The past two decades of research on lncRNAs shed light on their functional significance in various pathological conditions. lncRNAs belong to a diverse class of noncoding RNA (ncRNAs) molecules. They are defined as the new paradigm for epigenetic regulation in normal and disease development, especially in leukemia. Functional studies in recent years support the role of lncRNA in cell transformation, tumorigenesis, and metastasis. Such scientific evidences unraveled the need for utilizing a particular lncRNA, serving as a biomarker and/or a target, as a therapeutic for leukemia. Therefore, understanding the mechanistic significance of lncRNAs in leukemia is vital as it provides valuable insights into the transcriptional deregulation in leukemogenesis. Here, we focus on a brief overview of the current knowledge on lncRNAs expression that contributes to leukemogenesis associated with specific blood cells. Furthermore, we discuss lncRNAs mediated molecular mechanisms involved in epigenetic regulation. This chapter underscores the importance of research on lncRNAs regulatory mechanism and lncRNA therapeutic-based approaches in leukemia.

Hematopoiesis: Biochemical, Cellular, Molecular, and Genomic Perspectives. Ravi Kumar Gutti (Ed.)

10.1 INTRODUCTION

The lncRNAs are a new class of ncRNA molecules with more than 200 nucleotides in length identified in eukaryotic cells (Costa, 2005; Mattick & Makunin, 2006). These type of transcripts don't possess the ability to translate into proteins (Dahariya et al., 2019; Delás & Hannon, 2017; Wang et al., 2018). While the expression pattern of lncRNAs is studied extensively in disease condition, the exact role of lncRNAs and the mechanism involved in various diseases, specifically in cancer is not completely deciphered. In recent years, lncRNAs have gained widespread attention due to their remarkable expression patterns associated with the regulation of several genes at both transcriptional- and post-transcriptional levels (Spitale et al., 2011; Long et al., 2017; Wierzbicki, 2012; Delás & Hannon, 2017; Kornienko et al., 2013; Rinn & Chang, 2012).

Recent literature indicates the regulation of gene expression by lncRNAs at multiple levels through their interaction with several biological molecules, such as proteins, RNA and DNA, by a variety of mechanisms, including promoter, scaffold, decoy, and guide (Dahariya et al., 2019; Kornienko et al., 2013). Thereby, lncRNAs play a significant role in multiple biological processes vital for processes such as gene expression, genetic imprinting, histone modification, and chromatin regulation and in diseases including pathogenesis and prognosis of several types of cancers. These lncRNAs have a crucial role at the cellular level in proliferation, differentiation, and apoptosis (Schwarzer et al., 2017; Tian et al., 2016; Wang et al., 2018). It is no surprise that accumulating evidences underscore the decisive role of lncRNAs in all the facets of malignancies and tumorigenesis. There exists literature suggesting a link between the lncRNAs and different types of leukemia with defective hematopoietic system (Delás & Hannon, 2017; Li et al., 2013; Kazemzadeh et al., 2015; Yuan et al., 2017). Given the ability of a single lncRNA to regulate the expression of several genes, it is conceptually attractive to propose that lncRNAs modulate the process of leukemopoiesis. Previous research has shown the efficacy of anti-cancerous drugs on the leukemic pathways in different model system (Yadav et al., 2021). One possible mechanistic approach targeted by the lncRNAs could be similar to the chemotherapeutics targeting inflammatory pathways in leukemia.

In the current chapter, we discuss the characteristics and role of lncRNAs related to Leukemia. We provide a succinct overview of the current understanding of lncRNA expression profiles in various leukemia types, processes or mechanisms which might contribute to leukemogenesis, and their role as a biomarker. Identifying the functional significance of diverse lncRNAs can enhance the understanding of the underlying biological events. This will further lead to the identification of some of the lncRNAs as novel therapeutic targets, eventually leading to novel prospects for prognostication, diagnosis, and treatment for various types of leukemia.

10.2 TYPES OF LEUKEMIA

Leukemogensis is a process which involves several intricate and independent epigenetic and genetic events (He et al., 2019). Leukemia is an oligo clonal blood disease that arises in the bone marrow (BM) with defective blood cells that are abnormal in growth and number. The hematopoietic precursor cells in leukemia carry genetic material that is altered, mutated or damaged. These abnormal cells are not capable of differentiation into matured blood cells and remain as blasts or leukemia cells exhibiting uncontrolled division. Therefore, such changes can drive the immature cells to multiply and can overburden the BM resulting in the suppression of normal and healthy blood cells. Decades of research shed light on the ability of leukemogenic cells to invade other extra-medullary tissues such as liver, spleen, lungs, kidney, central nervous system (CNS) and testicles, etc. (Ohtake, 2007). Leukemia is observed in both children and adults (Campo et al., 2011). Clinically and pathologically, leukemia transforms from the blast cells mainly affecting two types of cell lineages, i.e., myeloid and lymphoid (Figure 10.1).

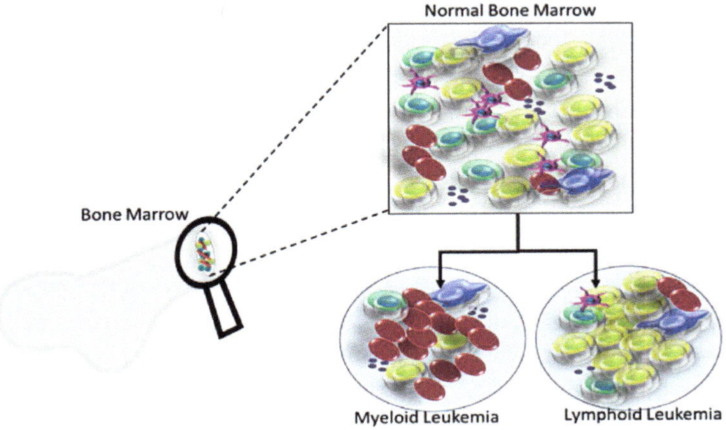

FIGURE 10.1 There are primarily two types of leukemia based on the type of cell lineages affected (a) myeloid; and (b) lymphoid leukemia. In myeloid leukemia, the abnormal myeloid lineage cells are produced as immature blood cells (blasts), which cannot carry out their regular functions and rapid multiplication quickly, worsen the disease. In lymphoid leukemia, particularly lymphoid lineages cells get affected and produce too many abnormal lymphoid cells which affect the lymphoid cells (lymphocytes), which play an important role in maintaining our immune system.

Acute leukemia (AL) results in an exponential increase in the population of immature blood cells, and as a consequence, the BM won't be able to generate healthy blood cells. Immediate treatment is needed for AL patients because of the quick accumulation and progression of the malignant cells. These cancerous cells metastasize in the body by spreading to various organs by spilling over into the bloodstream (Ohtake, 2007). AL is the most common type of leukemia diagnosed in children. On the other hand, chronic leukemia (CL) is frequently observed in older individuals, but it can also be found in any age group (Ciccone & Ferrajoli, 2014). CL is characterized by the occurrence of excessive relatively mature blood cells, but still abnormal. While CL is slow to progress, the rate of cell production is at a much higher resulting in many abnormal blood cells (Delgado et al., 2020). AL must be treated immediately. However, careful monitoring of patients diagnosed with CL is necessary before treatment to make sure maximum therapeutic efficacy.

Based on the rate of progression it is broadly divided into AL and CL with either abnormal lymphoid or myeloid lineages. Malignancies are therefore classified as Acute Lymphoid Leukemia (ALL), Acute Myeloid Leukemia (AML), Chronic Lymphoid Leukemia (CLL), and Chronic Myeloid Leukemia (CML) (Figure 10.2) (Bennett et al., 1976; Dorantes-Acosta & Pelayo, 2012; Pelayo et al., 2005). ALL progresses rapidly replacing functional lymphocytes with leukemia cells with defective differentiation into matured cells (Terwilliger & Abdul-Hay, 2017). It can affect the entire BM and easily spread to lymph nodes, liver, thymus, tonsils, adenoids, and spleen, scattered throughout the digestive and respiratory systems finally to the entire body. While AML is a type of fast-growing blood cancer within the soft inner parts of BM specially starts in myeloid cells and interferes with the production of normal myeloid lineages like granulocytes, red blood cells, and platelets (Bach & Slany, 2009; Ohtake, 2007). It is called by many other names, like acute myelocytic leukemia, acute myelogenous leukemia, acute granulocytic leukemia, and acute non-lymphocytic leukemia. As AML develops in other types of blood-forming cells, it quickly moves into the blood and blowout into entire body parts.

Whereas CLL is a slow growing clinically heterogeneous disease that initially starts in the lymphocytes present in the BM which then extends into the blood. It can also metastasize in lymph nodes and organs (like the liver and spleen). The clinical heterogeneity of CLL reflects the differences in the chromosomal alterations (del17p, trisomy 12, del11q and del13q) as well as the physiology of the disease, which can affect the outcome of the

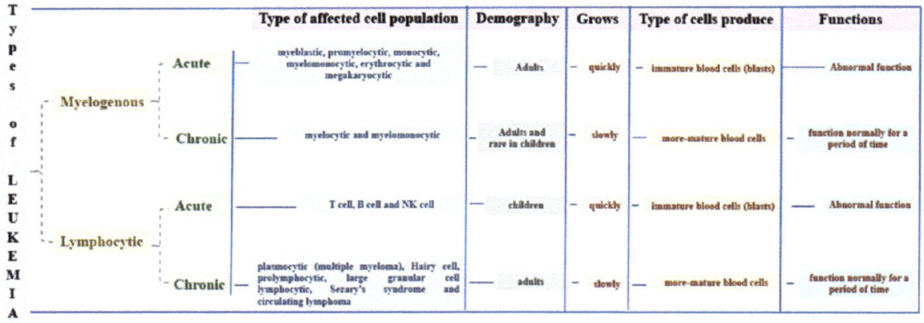

FIGURE 10.2 Four common types of leukemia.

Note: There are four main types of leukemia, based on progress and lineage whether they are acute or chronic, and myeloid or lymphocytic: acute myeloid (or myelogenous) leukemia (AML), chronic myeloid (or myelogenous) leukemia (CML), acute lymphocytic (or lymphoblastic) leukemia (ALL), and chronic lymphocytic (or lymphoblastic) leukemia (CLL).

disease (Fabris et al., 2020). CLL develops due to the high rate of production of lymphocytes, resulting in too many abnormal lymphocytes, which then outnumber the normal blood cells in turn making it really hard for the body to fight against the infection (Tsiodras et al., 2000). CML also known as chronic myelogenous leukemia is rare, only 10% of all leukemia cases. It begins in the BM cells, then spreads slowly into the complete blood and eventually spreads to others parts of the body (Kantarjian & Cortes, 2006). It has been found that CML is associated with an abnormal chromosome known as Philadelphia (Ph) chromosome; this is an abnormality that occurs when a broken piece of chromosome 9 attaches itself to chromosome 22 and also vice-versa. This causes breakpoint cluster region (*BCR*) and Abelson murine leukemia gene (*ABL*) genes to combine and work as a cancer gene (Kang et al., 2016).

In all types of leukemia, highly regulated division of healthy cells can be impaired by downregulating the tumor suppressor genes that results abnormal increase in the defective BM cells as well as other types of blood cells and hinder the production of functionally healthy blood cells. Therefore, patients diagnosed with leukemia not only suffers from anemia, clotting disorders but also with reduced immune ability to fight against infections that resulting in bruising of the body easily (Troup et al., 1960). Several epigenetic and genetic events take place altering the cancer cells during their transformation. Hence, for decades the research was focussed on the underlying molecular processes in which epigenetically altered noncoding and coding tumor suppressor genes mediating different molecular pathways involved in acquisition of mesenchymal attributes leading to tumor plasticity and metastasis as well as tumor plasticity, metabolic shift, genomic instability and unlimited capacity of cell division in leukemia (Bach & Slany, 2009; Wang, 2019).

10.3 DEVIANT EXPRESSION OF lncRNAs IN LEUKEMIA

Several studies have reported that aberrant lncRNA profiling in leukemic condition have demonstrated that relatively few lncRNAs' expression patterns, thus suggesting that lncRNA signature within the leukemia (Garzon et al., 2014; Schwarzer et al., 2017; Wang et al., 2018; Wang, 2019). Interestingly in late years, the scientific community has become aware of the remarkable dysregulated expression and mutational spectrum of genes which are closely related with the expression of lncRNAs (Li et al., 2018; Gomes et al., 2017; Garzon et al., 2014b). LncRNAs can contribute to tumorigenesis either as a promoter for oncogenes by targeting tumor-suppressor genes or as tumor suppressors' genes against oncogenes or onco-miRNAs (Garzon et al., 2014; Nitzan et al., 2014; Pan et al., 2017). Furthermore, a number of studies have documented that certain lncRNAs have a causal role in the regulatory mechanism of vital oncogenic processes like apoptosis, angiogenesis, migration, proliferation, and differentiation of healthy cells of different leukemia etiologies (Zheng et al., 2019; Pouyanrad et al., 2019; Wang et al., 2016; Shi et al., 2019; Schwarzer et al., 2017; Gallipoli et al., 2015; Garzon et al., 2014; Gioia et al., 2017).

In the past few years, there has been a significant expansion in the pathogenesis of all types of leukemia research. To enhance the life expectancy of cancer patients, oncologists believe that newly discovered RNA molecule "lncRNAs" have an aberrant expression in various leukemia. Taking advantage of the genome-wide profiling technologies such as next-generation sequencing, chip sequencing, and microarray, etc., several studies revealed that many lncRNAs were transcribed and exhibit distinct expression profiles in various leukemic malignancies. Studies on lncRNAs indicate that aberrant expression of certain lncRNAs contributes to leukemogenesis (Figure 10.3). Nonetheless, we summarize the lncRNAs that have specifically been expressed in major types of leukemic pathogenesis, which are discussed in detail in the following.

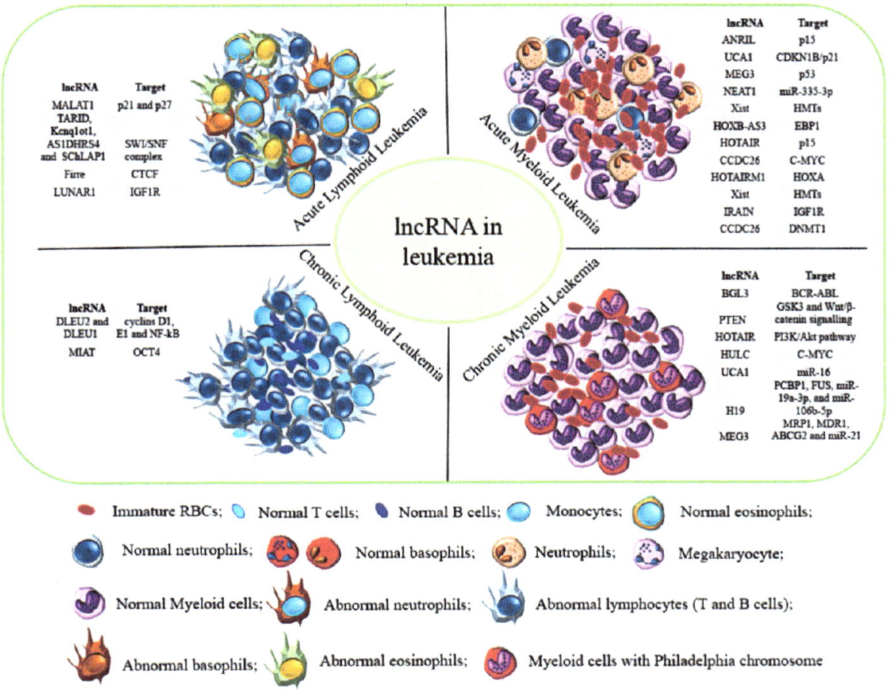

FIGURE 10.3 LncRNAs involved in acute myeloid (or myelogenous) leukemia (AML), chronic myeloid (or myelogenous) leukemia (CML), acute lymphocytic (or lymphoblastic) leukemia (ALL), and chronic lymphocytic (or lymphoblastic) leukemia (CLL).

10.3.1 LncRNAs IN ALL

ALL is clinically and biologically a heterogeneous childhood cancer recurrently affected due to chromosomal aberrations, including aneuploidies, amplifications, and translocations. The ALL onset is indicated by seizure of T or B cell progenitor's differentiation and abnormal proliferation. Overwhelming evidence indicates with the advent of NGS technologies, and overwhelming progress was achieved in the overall understanding of the lncRNAs expression signatures in leukemogenesis of B-lineage ALL (B-ALL). In recent years, several genome sequencing-based studies have been performed to assess the differential expression of various lncRNAs among different subtypes of ALL leukemogenesis, which attribute to tumor-suppressive or oncogenic attributes performing several biochemical and molecular functions in the cells. Cuadros M et al. performed comparative profiling to determine the lncRNA expression in children diagnosed with

B-ALL without and with ETV6-RUNX1 fusion. Microarray profiling was done on human samples diagnosed with B-ALL, which in turn observed the expression of a subset of lncRNAs, termed as B-ALL associated long RNAs (BALRs), which correspond to particular abnormalities in B-ALL cytogenetics. A functional assay indicated that the knockdown of BALR-2 (lncRNA among BALRs) has resulted in the apoptosis of B-ALL cell lines specifically and upregulation of BALR-2 was associated with reduced patients' response to prednisone and to make matters worse, overall survival was shown to be at 85%. LncPath Human Cancer Array (Arraystar Inc.) was done to analyze the expression profiles of messenger RNA (mRNA) and lncRNA which included about 1906 protein-coding genes and 2829 lncRNAs (Cuadros et al., 2020). Because lncRNA functions are poorly explored, their pathway-focused approach has made it possible to predict the functions of lncRNAs by determining the functions of their associated mRNAs. Based on the predicted lncRNA/mRNA associations, they have determined a total of 19 mRNA/lncRNA pairs which were expressed differentially amongst the pediatric B-ALL patients with ETV6-RUNX1-negative and ETV6-RUNX1-positive. Further research results showed that the high expression levels of TCL6 and TCL1B lncRNAs were highly linked with ETV6-RUNX1-positive B-ALL. Similarly, Ghazavi et al. confirmed the essential role played by lncRNAs in the molecular pathogenesis of pediatric B-cell precursor (BCP) ALL by defining an ETV6/RUNX1 specific expression signature with 596 lncRNA transcripts in ETV6/RUNX1-positive BCP-ALL and revealed that lnc-TIMM21-5, lnc-NKX2-3-1, lnc-RTN4R-1, and lnc-ASTN1-1 are regulated by the oncogenic fusion protein and are functionally involved in the biology of this human leukemia subtype. Additionally, an epigenetic mark known as H3K27ac (which is linked with enhancers) was observed to be present on the nine loci of the lncRNAs as well as their adjoining coding genes. Furthermore, the occurrence of this specific expression signature associated with these coding genes in ETV6/RUNX1 pre-B ALL, has in turn revealed a cis interaction amongst the lncRNAs and their adjoining coding genes (Ghazavi et al., 2016). In the recent past, two such studies reporting profiling expression has shed some light on how these transcripts were associated with the hematopoietic malignancy. Fang and colleagues have performed an expression profiling of the lncRNA of the entire genome of patients diagnosed with Mixed-lineage leukemia-rearranged (MLL-r) ALL. This led to the discovery of total 111 lncRNAs which were expressed differentially in MLL-r ALL patients in comparison to bnoe marrow samples from healthy individuals. Furthermore, specific expression patterns amongst the different MLL translocations were reported. Functional studies with siRNAs demonstrated that particular lncRNAs could affect cellular proliferation and apoptosis (Zeng et al., 2020). Affinito et al. provided dysregulated lncRNA-mRNA pairs to discover strongly correlated lncRNAs with the most aberrantly expressed protein-coding genes potentially underlying B-ALL pathogenesis. They constructed a co-expression network of the differentially expressed lncRNAs (30) and protein-coding genes (754) (Affinito et al., 2020). They have highlighted the role of 24 essential lncRNAs and their associated mRNAs which might have a pivotal role in B-ALL pathogenesis using the guilt-by-association approach. Bárcenas-López et al., are curious to know the role lncRNAs in early relapse and early mortality had done a microarray-based gene expression analysis in children with B-ALL. From cox regression and hazard ratios analysis, it was found that LINC01013 and LINC00152 lncRNAs were the most differentially expressed in patients with early relapse and early mortality, and potentially regulate genes involved in cell-substrate adhesion and peptidyl-tyrosine auto-phosphorylation processes associated with the development of early relapses and early deaths in children with B-ALL (Barcenas-Lopez et al., 2020). Another study reported that the expression inhibition of the lncRNAs *RP11–624C23.1*, *RP11-203E8*, and *RP11-446E9* could grant advantages to ALL cells by influencing biological processes such as DNA damage response, cell migration, proliferation, and apoptosis as well as resistance to genotoxic stress by performing the whole transcriptome sequencing of 56 ALL patients (Gioia et al., 2017). Gioia et al., studying 56 ALL children by microarray, noted that lncRNA transcripts were specifically silenced in tumoral cells. They could pinpoint the roles of 3 lncRNAs RP11-624C23.1, RP11-446E9, and RP11-203E8, which have an affect on cancer hallmarks cellular phenotypes such as DNA damage response, cell migration, proliferation and apoptosis (Gioia et al., 2017). So far, the majority of lncRNA profiling studies explored lncRNAs have a role in the regulation of gene expression at the epigenetic and post-transcription levels. Genome-wide association studies involving cancer have revealed that about 80% of cancer-related Single-Nucleotide Polymorphisms (SNPs) are present in noncoding regions, which particularly include lncRNAs, indicating that a significant part of the genetic etiology of ALL can be related to lncRNAs. Alva Rani James et al. explored the DNA methylation profiles associated with the specific lncRNAs expression that agrees with the already established BCP-ALL subtypes by performing the DNA methylation and RNA sequencing of 42 relapses and 40 diagnosis samples from 45 BCP-ALL patients via IlluminaInfinium microarray technique. From three major BCP-ALL subtypes they described a set of relapse-specific and subtype-specific lncRNAs by using integrated bioinformatics analysis. Then determined potential functions of the subtype-specific lncRNAs like DUX4-specific lncRNAs and Ph-like-specific lncRNAs, which correlated with the genes associated with the regulation of signaling and metabolic processes such as JAK-STAT, mTOR, PI3K-AKT, Hippo and TGF-signaling pathways (James et al., 2019). A large-scale transcriptome sequencing and DNA methylation array study by James and colleagues investigated lncRNA expression in a cohort of 82 BCP-ALL samples and reported 1,235 specific lncRNAs as being aberrantly dysregulated BCP-ALL subtypes (James et al., 2019). Ouimet et al. reported a study by performing total transcriptome analysis of 56 pre-B ALL pediatric patients to find five lncRNAs, which were specifically overexpressed in pre-B ALL (Ouimet et al., 2017). These genes might have a part in cancer hallmarks such as treatment response, cell apoptosis, migration, and proliferation. In particular, RP11-137H2.4 lncRNA has a significant role to play in cell migration, proliferation, and apoptosis; therefore, silencing of that particular lncRNA in glucocorticoid (GC)-resistant pre-B ALL cells was adequate to restore NR3C1-independent cellular respose to GC in turn leading to GC-induced apoptosis. Additionally, Garitano-Trojaola et al. reported that about 43 lncRNAs were found to be expressed abnormally in ALL patients by analyzing the peripheral blood samples of ALL patients and healthy donors and they have also reported that Linc-PINT was found to be downregulated in both B- and T-ALL patients (Garitano-Trojaola et al., 2018). Furthermore, Fernando et al. undertook a microarray study to define patterns of lncRNAs expression in three different subsets of B-ALL finding that lncRNAs expression segregated with cytogenetic subtypes. Among the four most differentially regulated lncRNAs (BALR-1, -2, -6, and LINC00958), high expression of BALR-2 (B-ALL associated long RNA-2) was encoded by a locus on chromosome 7q21.2, correlated with worse overall survival and a reduced response to treatment using prednisone in a huge group of B-ALL cases (Fernando et al., 2015). The advent of NGS analyzes generated pieces of evidence on lncRNA expression data in ALL, which might have correlated with overall survival and treatment response in ALL patients.

10.3.2 LncRNAs IN CLL

Subhash et al. identified 5,800 hyper-methylated and 12,570 hypo-methylated CLL-specific differentially methylated genes via methyl-CpG-binding domain protein-enriched genome-wide sequencing technique, among which hypo-methylated *AC012065.7* and hyper-methylated colorectal neoplasia differentially expressed (*CRNDE*) were two novel lncRNAs validated in CLL samples (Subhash et al., 2016). Notably, survival analysis has in turn determined that hypo-methylation of *AC012065.7* and hyper-methylation of *CRNDE* were associated with poor outcomes in patients with CLL. In addition, lncRNA myocardial infarction associated transcript (*MIAT*) is an lncRNA whose expression level was considered to be positively associated with the aggressive CLL forms carrying either trisomy 12, 11q-deletion or 17p-deletion over indolent form carrying 13p-deletion (Ishii et al., 2006). Ronchetti et al. studied the expression of lncRNAs in 26 different subpopulations of healthy B-cells and 217 early-stage Binet A CLL patients by employing microarray data using custom annotation pipeline. Microarray data analyzes unraveled 24-lncRNAs signature especially dysregulated in CLL in comparison to healthy B-cells. Among the 24 lncRNAs, the most downregulated lnc-*TOMM7-1*, mapped to chromosome 7p antisense to the interleukin-6 gene, and top upregulated lncRNAs includes lnc-*SNX29P2-3*, lnc-*SEL1L3-6,* lnc-*AC004696.1-1*, also known as *ZNF667-AS1* which in CLL samples, which promotes B-cell lineage proliferation and differentiation (Ronchetti et al., 2016).

10.3.3 LncRNAs IN AML

Feng and team using iron deficiency anemia as control samples by microarray, has analyzed the samples from AML patients to determine the mRNAs and lncRNAs expression profile in order to understand the role of lncRNAs in AML (Feng et al., 2018; Wang, 2019). Further they revealed large numbers of differentially expressed lncRNAs were clinically significant. Mainly eight lncRNAs (*RP11-222K16.2*, *AC092580.4*, *LINC00899*, *CTB-83J4.1*, *RP1-90G24.11*, *HOXB-AS3*, *RP6-109B7.5*, and *RP11-305O.6*) showing extensive involvement in AML development, by playing an indispensable role like affecting the immune system and hematopoietic cell differentiation (Wang, 2019). Similarly, Cao L and his group reported for differentially expressed lncRNAs in samples secured from the BM of three pediatric patients diagnosed with AML, whereas healthy children were used as controls (Cao et al., 2016). They found 372 dysregulated lncRNAs (difference ≥10-fold) and furthermore confirmed expression of 24 lncRNAs was altered in pediatric AML patients. Furthermore, the study reported the lncRNA *ENST00000435695* as the most upregulated, while lncRNA *ENST00000415964* was the most downregulated lncRNA in pediatric AML. According to a study by Zeng et al., they identified a total of 66 lncRNAs between healthy and diseased samples. Among them, lncRNA *RP11-834C111*, *RP1-137K24*, *CTD-2234N14*, and *LINC01003* were observed to be independent predictors of prognosis. Furthermore, based on the expression of these 4 lncRNAs, a risk scoring method was formed (Zeng et al., 2020). From Kaplan-Meier survival analysis it was found that the risk score system can classify patients into low-risk and high-risk categories with considerably different survival outcomes. Function analysis of the lncRNAs showed that in these lncRNA-mRNA networks the mRNAs were extensively associated with T cell receptor signaling pathway, natural killer cell mediated cytotoxicity, B cell receptor signaling pathway, Fc epsilon RI signaling pathway, apoptosis and mTOR signaling pathway. It implies that these four signature lncRNAs are closely associated with survival and prognosis of leukemic cells. To explore the detailed mechanism of lncRNAs in AML, Jia-Qi Pan et al. constructed lncRNA co-expression network based on Kaplan-Meier survival analysis that results 8 lncRNA co-expression modules related to AML prognosis (Pan et al., 2017). Another study reported the association of upregualted lncRNAs with low degree of DNA methylation and a higher ratio of being bound by transcription factors like ELK-1, ATF-2, STAT4, and SP1 in comparison with those downregulated in AML samples by performing the transcriptome-wide expression profile of lncRNAs. Furthermore, they have identified a novel lncRNA *LOC285758*, which was associated with poor prognosis in AML patients and also found to be regulating the AML cell line proliferation through increasing the histone deacetylase 2 (HDAC2) expression, a prospective therapeutic target in AML blasts (Bug et al., 2005; Lei et al., 2018). The prognosis-module pathway network illustrates potential lncRNA-mRNA co-expression associations in the AML pathway, WNT signaling pathway and chemokine signaling pathway. The results of detailed study on core prognostic modules explored the molecular mechanism of an lncRNA *ZNF571-AS1* in AML via acting as the Janus kinase (JAK) – signal transducer and activator of transcription proteins (STAT) signaling pathway by regulating KIT and STAT5 (Pan et al., 2017). To explore the comprehensive lncRNA expression profiles in AML, Wang (2019) subjected BM mononuclear cell samples to NGS. RNA sequencing results showed 412 downregulated and 216 upregualted lncRNAs in patients diagnosed with AML in comparison with controls. Enrichment analyses indicated that these lncRNAs which were expressed differentially were associated with neoplastic signaling pathways such as MAPK and cAMP signaling pathways. Furthermore, lncRNA *RP11-342M1.7* was associated with high CR correlated with AML development and progression by promoting cell proliferation (Wang, 2019). Zheng et al. by sequencing clearly revealed that lncRNA expression profile was closely associated with the progression and development of AML. 621 downregulated and 630 upregulated lncRNAs were recognized in patients diagnosed with AML in comparison to controls, which were primarily enriched in AML oncogene-related biological processes and mechanisms such as hematopoietic cell lineage, leukocyte trans-endothelial migration and neutrophil degranulation. Furthermore, the *in vitro* experiments have revealed that the lncRNA *RP4-576H24.2* could be potential oncogene to promote proliferation in AML pathogenesis (Zheng et al., 2019). Garzon et al. determined lncRNAs expression were closely associated with clinical features and recurrent mutations such as mutations in RUNX1, ASXL1, IDH2, CEBPA, and NPM1 genes and internal tandem duplications in the FLT3 gene (FLT3-ITD) in AML by performing custom microarray (Garzon et al., 2014). To explore subtype specific lncRNAs in AML, Wang et al. suggests lncRNA *LRRC75A-AS1* as a potential molecular biomarker which showed strong association with the target genes of transcription factors ETS variant 6 and tumor protein 53 by using the three lncRNA expression based risk score (Wang et al., 2018). De Clara et al. showed evidence on the importance of lncRNAs in AML initiation, development and progression through quantifying 11,036 lncRNAs expression from transcriptomic study performed in cytogenetically normal AML patients (De Clara et al., 2017). Schwarzer et al. reconstructed a high-density model of the human noncoding and coding hematopoietic landscape to discover lncRNA fingerprint associated with lineage specification, cellular differentiation and hematopoietic stem progenitor cells maintenance (Schwarzer et al., 2017). They define a core lncRNA stem cell signature in normal hematopoietic stem cells and AML blast that could function as a prognosis marker in various groups of AML patient's and might result in new therapeutic interventions which can target the noncoding transcritptome. Kotake et al. showed that Antisense Noncoding RNA in the INK4 Locus (*ANRIL*) exerts an oncogenic function via modulating expression of $p15^{INK4b}$ and $p16^{INK4a}$. In the development of BCP in both types of acute leukemia (ALL and AML), lncRNA *ANRIL* was overexpressed, thereby aggravating the inhibition of $p15^{INK4b}$ locus (Badalà et al., 2008).

10.3.4 LncRNAs IN CML

According to the present shreds of evidence, unique lncRNA profiles were reported to associate with the abnormal interference of lncRNAs in human CML malignancies have been identified. Chuang et al. profiled and validated an array of miRNA and lncRNAs at different stages of CML to achieve a clear understanding of the disease and help to develop miRNA and lncRNA targeted therapy intervention for CML patients. They identified the different patterns of miRNAs and lncRNAs expression at different stages of CML. They found a high number of aberrantly expressed lncRNAs tend to be more downregulated, including lncRNAs *Zfas1*, *H19*, etc., and others such as *HOTAIR*, *IPW*, *NTT*, etc., were upregulated and were serving as tumor oncogenes (Chuang & Khorram, 2018). Xu group investigated the dysregulated mRNA-lncRNA pairs from whole-genome lncRNA and mRNA expression profile CML data. By screening a module in the dysregulated lncRNA-associated competitive endogenous RNA (ceRNA) networks, they uncovered the three potential lncRNAs (*SNHG5*, *SNHG3*, and *DLEU2*) functionally involve in the pathogenesis of CML as ceRNAs (Xu et al., 2020). As little is reports being available about lncRNAs profiles related in CML. In the future, it is hope to explore more about lncRNAs deviate expression in CML that has not been fully elucidated yet. Several lncRNAs might offer insight into CML pathogenesis and novel treatment options.

10.4 LncRNAs IN REGULATION OF LEUKEMIA

Rapid developments in cell and molecular biology fields have been crucial to the identification of dysregulated molecules interrelated with leukemia which in turn suggests that leukemia could be linked to the heterogeneity of molecular and cellular genetics. Empowered by recent advances in sequencing performances, transcriptome-wide studies have revealed spatial and temporal control of lncRNA expression is fundamental in the course of tumor aggressiveness, cellular and histological heterogeneity, as well as the complexity of molecular and epigenetically mechanisms. Accumulating evidence has demonstrated that the tremendous numbers of lncRNAs are considered to be dynamic regulators which are significantly associated with leukemia types which are widely studied in both lymphoid and myeloid sub-types of leukemia in acute and chronic status (Table 10.1). Generally, lncRNAs are directly or indirectly interacting with DNA, RNA and proteins, by a variety of mechanisms like promoter, scaffold, decoy, guide and signal (Figure 10.4) to induce chromatin remodeling, to regulate varieties of splicing modes and protein activity, to modify and alter nucleosome and chromatin structure, and to change the protein localization, etc.

TABLE 10.1 Involvement of lncRNAs in Myeloid and Lymphoid Leukemia Development

Types of Leukemia		LncRNAs Involved	Expression of lncRNAs		Mechanism	References
			Up-Regulated	Down-Regulated		
Myeloid	AML	XIST	√	–	Chromosome inactivation	Wutz et al. (2002); Llinàs-Arias & Esteller (2017)
		HOXB-AS3	√	–	Guide	–
		HOTAIR	√	–	Decoy	Neff et al. (2012)
		CCDC26	√	–	Interact with DMNT1	Jones et al. (2020)
		RUNXOR	√	–	scaffold	Wang & Wei (2015)
		ANRIL	√	–	Decoy	Yu et al. (2009)
		HOTAIRM1	√	–	Scaffold	Wang & Dostie (2017)
	–	BGL3	√	–	Sponge	Zan & Casali (2015); Carabia et al. (2017)
		PLIN2	√	–	Signal	Sun et al. (2017)
		HOTAIR	√	–	Signal	Wang et al. (2017)
		HULC	√	–	Decoy	Lu et al. (2017)
		UCA1	√	–	Sponge	Xiao et al. (2017)
		MEG3	–	√	Decoy and Sponge	Zhou et al. (2017)
		H19	–	√	Decoy	Yang et al. (2020)
Lymphoid	ALL	MALAT1	√	–	Scaffold	Wang et al. (2016)
		TARID, Kcnq1ot1, AS1DHRS4, DACOR1 and SChLAP1	√	–	Scaffold	Lee & Roberts (2013); Arab et al. (2014); Merry et al. (2015)
		Firre	√	–	Decoy	Yang et al. (2015)
		LUNAR1	√	–	Decoy	Koch & Radtke (2011); Trimarchi et al. (2014)
	CLL	DLEU1 and DLEU2	√	–	Decoy	Ulitsky & Bartel (2013); Klein et al. (2010); Mercer et al. (2009)
		MIAT	√	–	Signal	Mercer et al. (2009); Sattari et al. (2016)

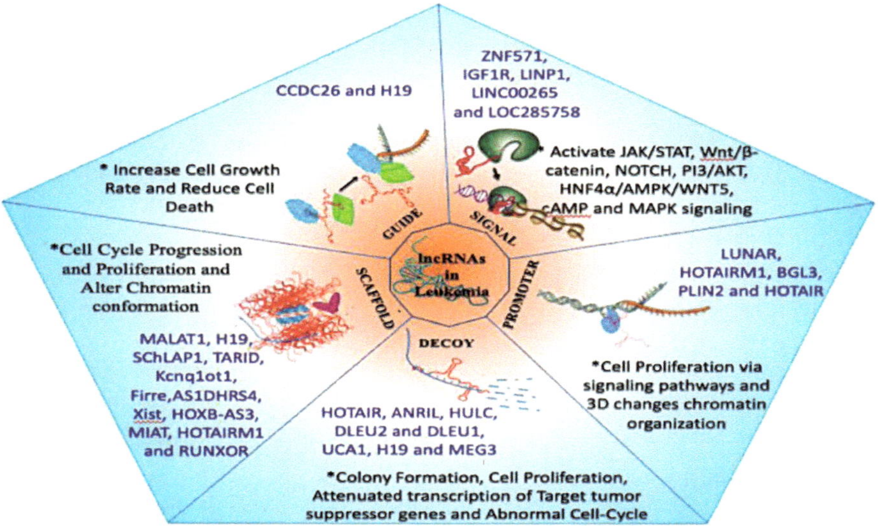

FIGURE 10.4 Genetic susceptible mechanisms of lncRNAs involved in leukemia.

Note: Recurrently several numbers of lncRNAs involved in leukemia. The main molecular mechanism governed by lncRNAs are promoter, scaffold, decoy as well as guide are associated in the progression of leukemia via playing a role in several biological processes.

10.4.1 lncRNAs IN ALL

In the recent past, the role of the lncRNAs in various cancers have been widely explored, however, the role of lncRNAs in the ALL pathogenesis is yet to be fully explored. Genome-wide association studies revealed few lncRNAs are attributed to ALL. The majority of research studies suggest that a significant portion of lncRNAs could regulate genes as ALL suppressors or oncogenes that play a role in exerting various cellular functions. The histone methyl-transferase of Polycomb Repressive Complex 2 (PRC2) is a well-studied protein that is regulated by ncRNAs and is also known as an intermediary target of lncRNAs (Davidovich & Cech, 2015). PRC2 is considered to take part in the inactivation of chromatin via instigating the inhibition of H3K427me3 histone marks. To unravel the role of lncRNAs associated with epigenetic regulation of BCP-ALL, Wang X group found lncRNA metastasis-associated lung adenocarcinoma transcript 1 (MALAT1) has participated in epigenetic modifications through interacting with the EZH2 and induce H3K27me3 modification of its target genes *p21* and *p27* expression resulting in cell cycle progression at G1/S transition in cell lymphoma (Wang et al., 2016). It is clearly evident from the recent literature that few lncRNAs, such as *AS1DHRS4*, *Kcnq1ot1*, and *TARID* tend to take action to modify the nucleosome position via the SWI/SNH complex, which was found in lncRNA SWI/SNF Complex Antagonist Associated with Prostate Cancer 1 (*SChLAP1*) or employ DNA methyl-transferases to change structural conformation of chromatin (Arab et al., 2014; Merry et al., 2015; Lee & Roberts, 2013). The lncRNA Functional intergenic repeating RNA element (Firre) was determined to be essential to maintain the inactivation of the X chromosome. X-linked lncRNA Firre, in association with chromatin remodelers, CTCF, and attachment, is one of the fundamental events in the process, and this also includes altering the confirmation of chromatin during the inactivation process of X chromosome. Furthermore, the inactivated X chromosome will be placed nearer to the nucleolus, thus maintaining the methylation of H3K27me3 (Yang et al., 2015). Leukemia-induced noncoding activator RNA (*LUNAR1*) is a crucial lncRNA which is associated with T-cell leukemia, an aggressive hematological neoplasm derived from malignant T-lymphocyte progenitors with abnormal NOTCH1 signaling (Trimarchi et al., 2014; Koch & Radtke, 2011). Trimarchi et al. has reported that the lncRNA *LUNAR1* expression was upregulated in the samples of primary T-cell leukemia particularly in the samples with Notch mutation, whereas it was downregulated upon inhibition of Notch (Trimarchi et al., 2014). Furthermore, the results has revealed that the lncRNA *LUNAR1* was highly associated with *IGF1R*, its neighbor coding gene, which has suggested earlier that the high expression of *IGF1R* supported Notch signaling to play a role in leukemia-initiating cell activity in T-cell leukemia (Trimarchi et al., 2014; Medyouf et al., 2011).

10.4.2 lncRNAs IN AML

LncRNAs are central players in the unique epigenetic programming of AML. From the molecular point of view, it will be relevant to understand lncRNAs' potential contribution in epigenetic dysfunction is central to the leukemogenesis in AML. The success achieved in the recent past in understanding the AML genetic landscape has highlighted that, regardless of its heterogeneity, several AML cases consist of recurrent mutations in the genes encoding for epigenetic regulators. Recurrently mutated important epigenetic regulators in AML include Histone acetyl-transferases (HATs) – namely EP300/CBP and members of the MYST/MOZ family, through point mutation or chromosomal translocation (Gallipoli et al., 2015). Histone methyl-transferases (HMTs)

– namely KMT2B/MLL1, a chromosomal translocation hotspot leading to a uniquely aggressive disease of bad prognosis in leukemia (Bach & Slany, 2009). DNA methyl-transferases and demethylases – respectively DNMT3A and TET2 or the metabolic enzymes IDH1/2, that interfere with target-specific methylation (Metzeler et al., 2011). It can also reorganize DNA methylation patterns genome-wide (Wurm & Pina, 2019). One of the best-studied interactions of lncRNAs with the epigenetic machinery is provided by X-inactive specific transcript *(Xist)*, which mediates X chromosome inactivation via interaction with and under the guidance of HMTs (Wutz et al., 2002; Llinàs-Arias & Esteller, 2017). A breakthrough observation in AML proliferation positioned *HOXB-AS3* lncRNAs as epigenetic regulators via interacting with the ErbB3-binding protein 1 (EBP1) and guiding EBP1 to ribosomal DNA (rDNA) locus which allows transcription of ribosomal RNA (rRNA) and de novo protein synthesis in Nucleophosmin (NPM1)-mutated leukemic blasts (Papaioannou et al., 2019). The most abundantly expressed *HOTAIR* lncRNA plays a role in leukemogenesis by enhancing the self-renewal of LSC in solid cancer and AML. *HOTAIR* inhibited $p15^{INK4b}$ expression through Zeste Homolog 2 (EZH2)-enrolled tri-methylation of Lys 27 of histone H3 (H3K27me3) in $p15^{INK4b}$ promoter. In addition, $p15^{INK4b}$ partially reversed the decrease of colony and proliferation induced by *HOTAIR* knockdown, suggesting that $p15^{INK4b}$ plays a significant part in the leukemogenesis by *HOTAIR*. A growing body of evidence suggests that PRC2 has a role in cancer development and progression. Genome-wide analysis of PRC2-mediated trimethylation of histone 3 determined the locus-specific perseverance of H3K27me3 in spite of its enhancer *EZH2* inactivation which suggests the biased compensation by *EZH1*. Contrastingly, the inactivation of the much needed PRC2 gene, Eed has resulted in the total removal of the PRC2 function, which was not compatible with the growth of the leukemia (Neff et al., 2012). Evidence suggests that among the epigenetic modifications required for the transcriptional regulation of genes, DNA methylation is one of the most significant. DNA methyltransferases such as DNMT1, DNMT3b, and DNMT3a are much needed in the maintenance and establishment of DNA methylation in cells. DNA methylation is important for several crucial cellular processes, such as genomic stability, X-chromosomal inactivation and heterochromatin formation. Abnormal modifications were observed in the activity of DNMT1 in several developmental defects and diseases. *CCDC26,* a nuclear lincRNA is transcribed from a 328-kilobase gene on chromosome 8, from a region (8q24.21) neighboring the proto-oncogene *c-MYC* and frequently mutated in AML, directly interacts with DNMT1. The absence of *CCDC26* lncRNA causes dislocation of DNMT1 into the cytoplasm resulting in the considerable hypo-methylation of genomic DNA which is responsible for enhanced cell death and minimal cell growth rate in myeloid leukemia cells (Jones et al., 2020). *RUNX1* overlapping promoter-derived noncoding RNA *(RUNXOR)* is a novel intragenic 216 kb lncRNA within the *RUNX1* locus transcribed from an upstream promoter that overlaps all of the *RUNX1* gene's exons and introns. Researches have shown evidence of overexpression of lncRNA *RUNXOR* in BM of AML pateitns' and AML cell lines. Furthermore, researches identified epigenetic molecular mechanisms underlying *RUNX1*-associated translocation promoted by lncRNA *RUNXOR* as a scaffold in AML (Hong & Wei, 2015). An initial study showed that *ANRIL* expression inversely correlates with $p15^{INK4b}$ expression in ALL and AML. It was observed that the silencing of the tumor suppressor gene $p15^{INK4b}$ was mediated by lncRNA *ANRIL* through heterochromatin formation and DNA methylation in a Dicer-independent manner which in turn excludes the association of endogenous small RNAs in the process (Yu et al., 2009). Wang and Dostie found that HOXA Transcript Antisense RNA, Myeloid-Specific 1 (*HOTAIRM1)* is a promising biomarker for various leukemia through presenting unique HOTAIRM1 variants contributing to three-dimensional in changes chromatin organization involve in the *HOXA* genes temporal collinear activation. The variants of *HOTAIRM1* preferentially associate with either UTX/MLL or PRC2 complexes to modulate the levels of activating and silencing marks at the proximal end of cluster in turn delaying the transcription of central *HOXA* genes and recruitment of the UTX, a histone demethylase. Overall, *HOTAIRM1* has been found to interact with different comnplexes needed for chromatin modifications in AML condition (Wang & Dostie, 2017). In AML cell lines, one of the highly phosphorylated receptor tyrosine kinases is IGF1R, which in turn promotes cell growth via PI3K/Akt signaling pathway. J. Sun et al. found a novel intragenic lncRNA IRAIN, which is transcribed in antisense direction from intronic promoter inside the IGF1R locus. LncRNA IRAIN directly interacts with the IGF1R promoter and enhancer as well as involves in the orchestration of an intra-chromosomal loop linking the IGF1R enhancer and promoter (Sun et al., 2014).

10.4.3 LncRNAs IN CLL

In the recent past, several studies have reported few newly identified lncRNAs which are involved in CLL. In B-CLL, over 50% of patients have 13q14.3 deficiencies. Garding et al. found 13q deletions associated with minimally deleted region encompassing Deleted in leukemia 1 and 2 (*DLEU2* and

DLEU1). LncRNAs *DLEU1* and *DLEU2* were transcribed from head to head in the minimally deleted region of chromosome 13. Furthermore, it was also observed that lncRNAs *DLEU1* and *DLEU2* were considerably demethylated at the 5′ end in most of the patients diagnosed with CLL, leading to the attenuation in the transcription of a series of adjoining sequences which encode tumor-suppressing genes. Cyclins E1 and D1 were negatively regulated by lncRNA *DLEU2* through miR-16-1 and miR-15a, which play an significant role in the pathogenesis of CLL (Ulitsky & Bartel, 2013). Klien et al. found that homozygous loss has considerable impact on the control and regulation of healthy CD5^{+}B lymphocytes and their homeostasis of lncRNAs *DLEU1* and *DLEU2*. This was partly because of the loss of miR-16-1 and miR-15a which are situated inside the intron 4 of *DLEU2*, has a correlation with clonal lymphoproliferative disorders, development of indolent B cell-autonomous, abnormal cell-cycle progression and recapitulating the spectrum of CLL-associated phenotypes which were observed in humans (Klein et al., 2010). Furthermore, recent observations reported has shown that the lncRNAs *DLEU1* and *DLEU2* could regulate transcription of theri adjoining tumor suppressor genes, which might serve as a positive regulators for NF-kB activity, affecting leukemia cell survival (Mercer et al., 2009; Garding et al., 2013). LncRNA *MIAT* has a significant functional role in the proliferation of CLL. LncRNA *MIAT* was first discovered as a susceptible locus for myocardial infarction is situated at 22q12.1 with 30,051 bp in length (Ishii et al., 2006). LncRNA *MIAT* is retained by nucles which has demonstrated no involvement with the chromatin and developed a distinguishing nuclear structure to escape nuclear export, in contrast to lncRNA *Xist*, which in turn has a significant structural role in the formation of heterochromatin during the inactivation of X chromosome. LncRNA *MIAT* was observed to be overexpressed in malignant mature B cells, which also includes primary leukemic cells secured from patients diagnosed with CLL, cell lines of non-Hodgkin lymphoma origin and established CLL cell lines (Sattari et al., 2016). Expression of lncRNA *MIAT* directly associates with the aggressiveness of the diseases and was controlled directly by *OCT4*, and positively regulates transcription of *MIAT* (Gomafu) resulting in the consitution of regulatory feedback loop in the modulation of mouse embryonic stem cells pluripotency (Zan & Casali, 2015). Likewise, lncRNA *MIAT* has demostrated to form a regulatory loop with *OCT4*, in tumoral mature B-cell and both *MIAT* and *OCT4* were found to significant for the survival of cells. Overall, these findings suggest lncRNA *MIAT* upregulation supports proliferation of the monoclonal malignant B cell with naive B-cell phenotype by protecting from apoptosis, which in turn contributes to the progression of CLL (Sattari et al., 2016). Overall, these observations demonstrate the association of lncRNA *MIAT* to support the proliferation of the malignant mature B-cells. Some examples of lncRNAs that are involved in the therapeutic application are compiled in Table 10.2. While the list is not exhaustive, it summarizes the best examples of lncRNAs reported specifically in types of leukemia, which are discussed in detail.

TABLE 10.2 Examples of lncRNAs That are Involved in the Therapeutic Application

LncRNAs Applicable in Therapeutics	Name of LncRNAs	References
Biomarker	AL133346.1	Cuadros et al. (2020)
	CCDC26	Chen et al. (2019)
	SNHG5	Li & Sun (2018)
	LINC00899	Wang et al. (2018)
	LINC00265	Wang et al. (2018)
	FBXL19	Sheng et al. (2021)
	CRNDE	Wang et al. (2017)
Onco-lncRNAs	HOTTIP	Zhuang et al. (2019)
	CRNDE	Ma et al. (2020)
	LINC00641	Wang et al. (2019)
	UCA1	Li et al. (2019); Sun et al. (2018)
	ANRIL	Sun et al. (2018)
	HOTAIR	Xing et al. (2015)
	CCAT1	Chen et al. (2016)
	PVT1	Zeng et al. (2015); Colombo et al. (2015); Cui et al. (2016); Salehi & Sharifi (2018)
	TUG1	–
	MONC and MIR100HG	Emmrich et al. (2014)
Onco-suppressor	linc00460	Zhaung et al. (2021)
	HOTAIRM1	Chen et al. (2017)
	IRAIN	Sun et al. (2014)
	linc-223	Mangiavacchi et al. (2016)
	NEAT1	Zhao et al. (2019)
	CCDC26	Hirano et al. (2015)
	MEG3	Yang et al. (2007)

10.4.4 LncRNAs IN CML

CML is the only myeloproliferative disease that is caused by reciprocal translocation of *ABL* on chromosome 9 and *BCR* on chromosome 22 creating a new chromosome called Ph chromosome. A unique lncRNA *BGL3* was identified which might function as a primary regulator in BCR-ABKL-mediated cellular transformation through performing comprehensive analysis of human CML cells for lncRNAs. In addition, functional assays have demonstrated that the lncRNA *BGL3* was induced greatly in reply to the inhibition of *BCR-ABL* kinase activity or due to the *BCR-ABL* expression disruption in CML patients and cell lines. Remarkably, lncRNA *BGL3* might act as a ceRNA that binds with a subset of microRNAs to cross-regulate Phosphatase and Tensin homolog *(PTEN)*, a crucial tumor suppressor gene expression (Zan & Casali, 2015; Carabia et al., 2017). Another study reported the probable functions of *CEBPA*-related lncRNA *PLIN2* during CML. It was clearly evident from the results, during the CML process both *PLIN2* and *CEBPA* were upregualted, and it was observed that they have a positive association among the *PLIN2* and *CEBPA* in CML patients. Furthermore, it was reported that *CEBPA*-mediated upregulation of expression of *PLIN2* can promote the CML development through Wnt/β-catenin and GSK3 signaling (Sun et al., 2017). For instance, Wang investigated the mechanisms of lncRNA *HOTAIR* could play a vital part in the development of multidrug resistance in CML to imatinib through PI3K/Akt-dependent mechanism (Wang et al., 2017). Another study reported the positive correlation between the imatinib-induced apoptosis in CML cells and lncRNA *HULC* which in turn leads to the PI3K/Akt pathway activity inhibition and *c-Myc* expression reduction (Lu et al., 2017). Another study identified LncRNA urothelial carcinoma-associated 1 (*UCA1*) a crucial regulator of multidrug resistance protein-1, which was initially reported as a ceRNA. LncRNA *UCA1* inhibit miR-16 expression by competitively binding which was regarded as the primary cause for imatinib resistance in CML cells (Xiao et al., 2017). *H19* is an lncRNA that was lowly expressed in CML. They found that colony formation and cell viability was considerably inhibited due to the overexpression of *H19* and also extended the survival in three xenografted mouse models and CML cell lines. Furthermore, by using a combination of RNA pull-down and computational predictions, target miRNAs and proteins of *H19* were recognized, which include miR-106b-5p, miR-19a-3p, FUS protein and PCBP1. Colony formation and cell growth of CML cell lines can be considerably inhibited by targeting miR-106b-5p, miR-19a-3p, FUS protein and PCBP1. Inhibitory impact of

H19 in CML cell lines can be considerably reduced by the co-overexpression of *H19* and miR-106b-5p, miR-19a-3p, FUS protein and PCBP1. In time, these observations could offer distinctive insights into the CML molecular mechanisms (Yang et al., 2020). Maternally expressed gene 3 (*MEG3*) was considerably reduced in imatinib-resistant-K562 cell lines and -CML patients. Furthermore, overexpression of *MEG3* has greatly decreased the *ABCG2*, *MDR1*, and *MRP1* expression, resistance to imatinib was overturned, cell apoptosis was increased and cell proliferation was reduced in imatinib-resistant K562 cells. On an important note, binding of *MEG3* to miR-21 was negatively associated in patients diagnosed with CML. Additionally, the phenotype of *MEG3*-overexpression was overturned by miR-21 mimics in imatinib-resistant K562 cell lines. Overall, *MEG3* might be involved in the resistance to imatinib in CML cell lines and potentially contributing to the resistance to imatinib by miR-21 regulation and successive multidrug resistant transportes expression, cell apoptosis, and proliferation (Zhou et al., 2017).

10.5 LncRNAs APPLICABLE IN THERAPEUTICS FOR AML AND ALL

Recently there has been much evidence demonstrating that a small number of lncRNAs play crucial physiological roles in the treatment of leukemia because biologically lncRNAs regulate different processes such as apoptosis, proliferation, differentiation, etc. Studies have shown that lncRNAs have well-understood roles like targets for leukemia treatment. By assessment of transcripts quantitatively through using extremely sensitive qPCR assay, for minimal residual disease detection, it is clear that lncRNA can be a suitable candidate potential biomarker. The importance of lncRNAs in clinical and biological lympohoid and myeloid leukemia has been compiled from many recent studies. Numerous lncRNAs act as oncogenes, but here we are focusing on gathering information about the vital use or function of lncRNAs as tumor suppressors. For example, the lncRNA HOXA cluster antisense RNA2 (*HOXA-AS2*), was known to show oncogenic attributes in various human malignancies but recently was found to diminish GC sensitivity in ALL through the HOXA3/EGFR/Ras/Raf/MEK/ERK pathway (Zhao et al., 2019). In silico analysis of miRNA-miRNA and lncRNA-miRNA interactions has shown that the lncRNA *JPX* might serve as a decoy for miR-106-5p, miR-17-5p and miR-9-5p keeping them away from TGF-βR mRNAs which promote activate TGF-βR-induced non-canonical ERK1/2 and PI3K/AKT pathways for the modulation of megakaryoblastic cells differentiation in AMKL (Dahariya et al., 2021). LncRNA In Non-Homologous End Joining Pathway 1 (*LINP1*) was reported to control progression of AML by making use of the HNF4α/AMPK/WNT5A signaling pathway (Shi et al., 2019). By performing RNA sequencing for extensive lncRNA expression profiling it was observed that the lncRNA *AC008753.6*, lncRNA *CES1P1* and lncRNA *RP11-342M1.7* can act as prediction factors for AML risk both independently or in combination (Yong, 2019). miR-335-3p dysregulation, guided by the lncRNAs *MALAT1* and Nuclear Paraspeckle Assembly Transcript 1 (*NEAT1*) were correlated with a poor prognostics in pediatric ALL (Pouyanrad, Rahgozar, & Ghodousi, 2019). In a similar fashion, imatinib resistance in CML can be regulated by lncRNA *MEG3* through miR-21 suppression (Zhou et al., 2017). To a great extent, these discoveries have provided in-depth understanding to leukemia pathogensis in high-risk groups, which can further assist clinicians to further study the potential use of lncRNAs in treatment. For more efficient lncRNA-based therapy, the field of genomics and biotechnology needs to develop improvised vehicles for delivery of lncRNA with low toxicity and greater stability. Additionally, siRNAs can be packaged in nanoparticle vectors, for efficient targeting, to target oncogenic lncRNAs. Although the field of lncRNAs is being widely studied, their role as a potential therapeutic agent and as a biomarket in leukemia is not yet fully explored in detail.

10.5.1 lncRNAs AS A BIOMARKER

Several types of cancer studies have reported the dysregulation of lncRNAs and their correlation with the tumor or oncogenic suppression activities. As per the recent literature, lncRNAs offer promising potential as biomarkers in the leukemia disease-specific disorders. Caudros et al. found that the lncRNA *AL133346.1* along with a neighboring protein-coding mRNA (CCN2) were extensively over-expressed in leukemia samples in comparison to normal BM samples (Cuadros et al., 2020). By the Kaplan-Meier method the expression of AL133346.1/CCN2 (previously known as RP11-69I8.3/CTGF), was shown increased in B-ALL samples and positively correlated with higher overall survival of pediatric B-cell ALL patients. These data suggest that high of both lncRNA *AL133346.1* and *CCN2* expression was accompanying with higher survival rate of pediatric B-cell ALL patients that suggesting the potential role of lncRNA *AL133346.1/CCN2* as an extremely sensitive and specific biomarker that could lead to a better prognosis for the patients. Chen et al. examined the expression and clinical significance of lncRNA *CCDC26* in patients with diagnosed and relapsed patients AML by analyzing the publicly available GEO database (Chen et al., 2019). They showed the correlation between the AML patients' prognosis and the expression levels of CCDC26, which in turn revealed that the AML oncogenesis was correlation with the CCDC26 upregulation and might act as a prognosis indicator in AML patients. Similarly, another study determined the upregulation of lncRNA *SNHG5* expression in AML patients compared to healthy individuals and the correlation in between the overall survival and the lncRNA *SNHG5* expression was determined by using Kaplan-Meier analysis, which showed that the plasma lncRNA *SNHG5* might work as an independent biomarker for AML patients (Sun et al., 2018). Wang et al. analyzed the expression levels of lncRNA *LINC00899* in serum, and BM samples secured from AML patients, and normal samples were analyzed through qRT-PCR. They first reported by Kaplan-Meier and Cox proportional hazards analysis, that the higher levels of serum lncRNA *LINC00899* expression could be a possible and viable non-invasive biomarker target for the early AML prognosis and clinical detection (Wang et al., 2018). Ma et al. group systematically evaluated TCGA datasets and interpreted the clinical data multivariate analysis, univariate analysis, Kaplan-Meier analysisn and x^2 test. The significant upregulation of lncRNA *LINC00265* is associated with favorable prognosis of AML and may serve as a potential clinical prognostic and diagnostic biomarker by inhibiting cell proliferation, migrationn and invasion via activating PI3K/AKT signaling (Ma et al., 2018). In the same way, Serum lncRNA FBXL19 antisense RNA 1 (*FBXL19-AS1*) expression was determined to be up-regulated in AML, which might be useful as a novel prognostic and diagnostic biomarker for AML (Sheng et al., 2021). In addition, Wang group studied the expression of lncRNA *CRNDE* in AML clinical data and its impact on apoptosis and proliferation of AML cell line U937. Furthermore, it was clearly evident from the results that the lncRNA *CRNDE* expression in U937 cell lines and AML patients was considerably higher and negatively associated with the classification and overall survival of AML patients. Knockdown study by using siRNA-CRNDE has demonstrated that the cell cycle and proliferation of U937 cell line was promoted by lncRNA *CRNDE* along with inhibition of apoptosis, which results in malignant progression in AML. The essential role of lncRNA-CRNDE in promoting the proliferation and inhibiting the cell-apoptosis presents as a biomarker for AML (Ma et al., 2018).

10.5.2 lncRNAs AS THERAPEUTIC TARGET

Several research studies have reported that the lncRNA might play a crucial role in the progression and pathogenesis of ALL and AML. It could be therapeutic targets for various AML and ALL because of their disease- and stage-restricted expression. The high expression of lncRNA *HOTTIP* accelerates the cell cycle progression and proliferation ability of AML cells by upregulation of DDA1 expression via sponging microRNA-608 (Zhuang et al., 2019). Ma et al. by employing RNA-seq data of a large group of AML patients has highlighted a group of lncRNAs in correlation to pathogenesis, which were abnormally expressed in acute promyelocytic leukemia (APL). Among them, lncRNA *CRNDE* is one of the evolutionarily conservative lncRNA was highly expressed and coordinating with PML/RARα to promote APL development. Further mechanistic and knockdown studies showed the importance of oncogenic lncRNA *CRNDE* serves oncogenic effects via miR-181/NOTCH2 axis regulation in the AML development (Ma et al., 2020). The expression and function of lncRNA *LINC00641* in AML cells was explored by using an interactive analysis tool such as luciferase reporter assay and RT-PCR assay for gene expression profiling (GEP). Results showed lncRNA *LINC00641* as a unique lncRNA associated with AML whose knockdown inhibits cell proliferation, growth, migration, and invasion of AML cells, and by modulating miR-378a/ZBTB20 axis promotes apoptosis in AML (Wang et al., 2019). Besides, the upregulation of lncRNA *UCA1* with IL6R facilitates cell proliferation and suppresses apoptosis by targeting miR-331-3p to activate JAK2/STAT3 pathway in multiple myeloma (Li et al., 2019). Similarly, in another study, Sun et al. showed the role of lncRNA *UCA1* in AML associated with the promotion of cell proliferation, migration, and invasion of human leukemia cells via sponging miR-126, which may provide the feasibility for the use of lncRNA-based therapy in AML treatment (Sun et al., 2018). As lncRNA *UCA1* plays an oncogenic role in AML, another study showed that targeting lncRNA *UCA1* suppresses glycolysis via miR-125a/HK2 pathway, thus playing a positive role in overcoming the chemoresistance in AML pediatric patients (Sun et al., 2018). In addition, ANRIL is a lncRNA that promotes malignant cell survival and cell glucose metabolism to accelerate AML progression via implying ANRIL-AdipoR1-AMPK/SIRT1 signaling pathway in regulating cell glucose metabolism and is a possible therapeutic agent and prognosis marker in AML treatment (Sun et al., 2018). Xing reported in his study the high expression of HOTAIR found in leukemic cell lines and primary AML blasts, which promote KIT expression by competitively binding miR-193a that revealed the important oncogenic role of HOTAIR, which can be potential target for therapeutic intervention in AML prognosis (Xing et al., 2015). Chen et al. detected the over-activation of lncRNA colon cancer-associated transcript-1 (*CCAT1*) in AML adult patients of French-American-British M4 and M5 subtypes. They determined that CCAT1 repressed differentiation and maturation of monocytes and promotion of HL-60 cell growth via sequestering tumor suppressive miR-155 and promote c-Myc expression. Therefore, the regulation of expression of lncRNA *CCAT1* might provide as a viable treatment option in AML therapy (Chen et al., 2016). Alike, overexpression of the lncRNA *PVT1* was associated with the proliferation of leukemia cells and accelerates the development of myeloid leukemia in APL. Moreover, PVT1 knockdown by RNA interference led to suppression of the c-Myc protein level, and cell proliferation was inhibited in APL (Zeng et al., 2015). PVT1 is potential candidate for oncogenic lncRNA which play crucial part in the progression and initiation of several malignancies through impacting cell cycle, proliferation and apoptosis (Colombo et al., 2015; Cui et al., 2016). In AML, the lncRNA *PVT1* increases. In addition, it is reported that this lncRNA positively regulates c-Myc in turn results in enhanced proliferation of myeloid cells (Zeng et al., 2015). One of the other studies showed inhibition of PVT1 by using antisense LNA GapmeRs technology might considerably induce necrosis and apoptosis in human acute erythroleukemia cells. As PVT1 plays a crucial role in the pathogenesis of AML types, may be useful for future therapeutic target against AML progression (Salehi & Sharifi, 2018). High expression of TUG1 confers adriamycin resistance in AML by suppressing the expression of miR-34a epigenetically via recruiting EZH2 (Wang et al., 2018). Emmrich et al. recognized lncRNAs *MONC* (*MIR99AHG*) and *MIR100HG* are highly expressed in acute megakaryoblastic leukemia (AMKL) blasts which plays an important role in the maintenance and self-renewal of AMKL that revealed the exceptional function of lncRNAs *MONC* and *MIR100HG* as oncogenes in the myeloid leukemia development (Emmrich et al., 2014). Using onco-lncRNA-inhibitors initiated for treatment of cancer which accompanied by chemotherapy drugs.

10.5.3 lncRNAs TARGET ONCOGENE AND ONCO-MIRNAS

Recently, much attention has been given to identifying the oncogenic lncRNAs and their potential functional involvement in the leukemogenesis.

The expression of lncRNA *linc00460* in serum was considerably upregulated in AML patients and found to be in the pathogenesis of AML. Knockdown study of lncRNA *linc00460* showed lncRNA *linc00460* may be a potential target for AML, which inhibited cell viability, while initiating apoptosis and cell cycle inhibition by sponging of miR-320b, which could result in the inhibition of PBX homeobox three expressions in AML (Zhaung et al., 2021). Chen et al. reported upregulated expression of HOTAIRM1, an lncRNA involved in the autophagy pathways regulation to accelerate myeloid granulocytic differentiation and has a vital role in the degradation of onco-protein PML-RARA in APL cells (Chen et al., 2017). LncRNA *IRAIN* is one of new imprinted lncRNA which originated from the IGF1R promoter that mediates enhancer looping between the IGF1R enhancer and promoter through modulating expression of IGF1R in cis fashion. Expression analyzes reveal that lncRNA *IRAIN* is greater in low-risk patients than the AML high-risk patients that suggests the participation of lncRNA *IRAIN* exhibit as a tumor suppressor in the malignant leukemia (Sun et al., 2014). In different study Mangiavacchi et al. identified lncRNA *linc-223* as a novel onco-suppressor in AML via binding an oncogenic microRNA (miR-125-5p) in leukemia and competing for its repressing activity on IRF4. The results suggest the occurrence of a specific crosstalk between linc-223 and IRF4 mRNA through competition for miR-125-5p binding is relevant for inhibition of AML cell-cycle and promote the lineage-specific differentiation/maturation of AML cells (Mangiavacchi et al., 2016). Zhao et al. explored the participation of lncRNA NEAT as onco-suppressor negatively regulate the expression of miR-23a-3p therefore promote structural maintenance of chromosome 1 alpha (SMC1A) expression in primary AML cells that suggest the sponging action of NEAT against miR-23a-3p, inhibit cell proliferation and enhanced the apoptosis in AML (Zhao et al., 2019). LncRNA *CCDC26* as stated before is associated with pediatric AML since its copy number is distorted in AML patients. On the other side lncRNA *CCDC26* can be tumor suppressor via acting as inhibitor against hematopoietic tyrosine kinase which is a well-known oncogene. Interestingly, the overexpression of lncRNA *CCDC26* can mask the function of KIT which is likely to become effective treatment of AML by hematopoietic tyrosine kinase-targeted therapy (Hirano et al., 2015). Yang et al. has demonstrated the transcriptional activation of lncRNA *MEG3* by Wilms' tumor 1 (WT1) in AML cells. Furthermore, dysregulation of WT1 gene through epigenetic mutations or alterations could impair leukemic cell differentiation and initiate cell proliferation (Yang et al., 2007). Lyu et al. through his findings reported that the downregulation of *MEG3* expression can inhibit AML and suppress leukemogenesis, both in p53-independent and -dependent process (Lyu et al., 2017). While this is very exciting for those of us who featuring lncRNA-based prognostic scoring systems to integrate into our disease models and treatment practices towards ultimate improvement in clinical outcomes for leukemic patients.

10.6 CONCLUSION

Even though, the pattern of lncRNAs expression is extensively studied during the disease conditions; the role of lncRNAs and the exact intricate mechanisms involved in disease conditions, particularly in cancer, is not well known. LncRNAs might regulate the maintenance of stemness, invasion, migration, invasion, apoptosis, and proliferation during the development of cancer. In addition, recent research reports show the involvement of lncRNAs in tumor metastasis and remodeling of the tumor microenvironment. LncRNAs can be tailor-made and used in several therapeutic applications, either by restoration or inhibition of lncRNAs, which in turn finely regulate the mechanisms of cancer cells in a highly cancer type-specific manner. In future, lncRNAs-based therapy can be employed as a primary healthcare approach.

KEYWORDS

- **ALL**
- **AML**
- **CLL**
- **CML**
- **leukemia**
- **lncRNA**

REFERENCES

Affinito, O., Pane, K., Smaldone, G., et al., (2020). lncRNAs-mRNAs co-expression network underlying childhood B-cell acute lymphoblastic leukaemia: A pilot study. *Cancers (Basel), 12*(9).

Arab, K., Park, Y. J., Lindroth, A. M., et al., (2014). Long noncoding RNA TARID directs demethylation and activation of the tumor suppressor TCF21 via GADD45A. *Molecular Cell, 55*(4), 604–614.

Bach, C., & Slany, R. K., (2009). Molecular pathology of mixed-lineage leukemia. *Future Oncology, 5*(8), 1271–1281.

Barcenas-Lopez, D. A., Nunez-Enriquez, J. C., Hidalgo-Miranda, A., et al., (2020). Transcriptome analysis identifies LINC00152 as a biomarker of early relapse and mortality in acute lymphoblastic leukemia. *Genes (Basel), 11*(3).

Bennett, J. M., Catovsky, D., Maria-Theregse, T. D., et al., (1976). Proposals for the classification of the acute leukemias French-American-British (FAB) Co-operative Group. *British Journal of Haematology, 33*(4), 451–458.

Bug, G., Markus, R., Barbara, W., et al., (2005). Clinical trial of valproic acid and all-trans retinoic acid in patients with poor-risk acute myeloid leukemia. *Cancer, 104*(12), 2717–2725.

Campo, E., Steven, H. S., Nancy, L. H., Stefano, P., Harald, S., & Elaine, S. J., (2011). The 2008 WHO classification of lymphoid neoplasms and beyond: Evolving concepts and practical applications. *Blood, 117*(19), 5019–5032.

Cao, L., Xiao, P. F., Tao, Y. F., et al., (2016). Microarray profiling of bone marrow long noncoding RNA expression in Chinese pediatric acute myeloid leukemia patients. *Oncol. Rep, 35*(2), 757–770.

Carabia, J., Cecilia, C., Pau, A., et al., (2017). Microenvironment regulates the expression of MIR-21 and tumor suppressor genes PTEN, PIAS3 and PDCD4 through ZAP-70 in chronic lymphocytic leukemia. *Scientific Reports, 7*(1), 1–11.

Chen, C., Peipei, W., Wenjian, M., et al., (2019). lncRNA-CCDC26, as a novel biomarker, predicts prognosis in acute myeloid leukemia. *Oncology Letters, 18*(3), 2203–2211.

Chen, L., Wei, W., Lixia, C., Zhijun, L., & Xing, W., (2016). Long noncoding RNA CCAT1 acts as a competing endogenous RNA to regulate cell growth and differentiation in acute myeloid leukemia. *Molecules and Cells, 39*(4), 330–336.

Chen, Zhen-Hua, Wen-Tao, W., Wei, H., et al., (2017). The lncRNA HOTAIRM1 regulates the degradation of PML-RARA oncoprotein and myeloid cell differentiation by enhancing the autophagy pathway. *Cell Death and Differentiation, 24*(2), 212–224.

Chuang, T. D., & Omid, K., (2018). Expression profiling of lncRNAs, miRNAs, and mRNAs and their differential expression in leiomyoma using next-generation RNA sequencing. *Reproductive Sciences, 25*(2), 246–255.

Ciccone, M., & Alessandra, F., (2014). Management of chronic lymphocytic leukemia in elderly patients. *International Journal of Hematologic Oncology, 3*(6), 433–443.

Colombo, T., Lorenzo, F., Giuseppe, M., & Paola, P., (2015). PVT1: A rising star among oncogenic long noncoding RNAs. *BioMed Research International, 2015*, 17–21.

Costa, F. F., (2005). Noncoding RNAs: New players in eukaryotic biology. *Gene, 357*(2), 83–94.

Cuadros, M., Daniel, J. G., Alvaro, A., et al., (2020). LncRNA-mRNA co-expression analysis identifies al133346.1/CCN2 as biomarkers in pediatric b-cell acute lymphoblastic leukemia. *Cancers, 12*(12), 1–15.

Cui, M., Lei, Y., Xiaoxia, R., Wenjing, Z., Quan, L., & Yupei, Z., (2016). Long noncoding RNA PVT1 and cancer. *Biochemical and Biophysical Research Communications, 471*(1), 10–14.

Dahariya, S., Paddibhatla, I., Kumar, S., Raghuwanshi, S., Pallepati, A., & Gutti, R. K., (2019). Long noncoding RNA: Classification, biogenesis and functions in blood cells. *Mol. Immunol., 112*, 82–92.

Davidovich, C., & Cech, T. R., (2015). The recruitment of chromatin modifiers by long noncoding RNAs: Lessons from PRC2. *RNA, 21*(12), 2007–2022.

De Clara, E., Gourvest, M., Ma, H., et al., (2017). Long noncoding RNA expression profile in cytogenetically normal acute myeloid leukemia identifies a distinct signature and a new biomarker in NPM1-mutated patients. *Haematologica, 102*(10), 1718–1726.

Delás, M. J., & Gregory, J. H., (2017). lncRNAs in development and disease: From functions to mechanisms. *Open Biology, 7*(7).

Delgado, J., Ferran, N., Dolors, C., & Elias, C., (2020). Chronic lymphocytic leukemia: From molecular pathogenesis to novel therapeutic strategies. *Haematologica, 105*(9), 2205–2217.

Dorantes-Acosta, E., & Rosana, P., (2012). Lineage switching in acute leukemias: A consequence of stem cell plasticity? *Bone Marrow Research, 2012*, 1–18.

Emmrich, S., Alexandra, S., Franziska, S., Veera, R. T., Dirk, R., & Jan-Henning, K., (2014). LincRNAs MONC and MIR100HG act as oncogenes in acute megakaryoblastic leukemia. *Molecular Cancer, 13*(1), 171.

Fabris, L., Jaroslav, J., & George, C., (2020). Noncoding RNAs as cancer hallmarks in chronic lymphocytic leukemia. *International Journal of Molecular Sciences, 21*(18), 1–18.

Feng, Y., Shen, Y., Chen, H., et al., (2018). Expression profile analysis of long noncoding RNA in acute myeloid leukemia by microarray and bioinformatics. *Cancer Sci., 109*(2), 340–353.

Fernando, T. R., Rodriguez-Malave, N. I., Waters, E. V., et al., (2015). LncRNA expression discriminates karyotype and predicts survival in B-lymphoblastic leukemia. *Mol. Cancer Res., 13*(5), 839–851.

Gallipoli, P., George, G., & Brain, H. J. P., (2015). Epigenetic regulators as promising therapeutic targets in acute myeloid leukemia. *Therapeutic Advances in Hematology, 6*(3), 103–119.

Garding, A., Nupur, B., Rainer, C., et al., (2013). Epigenetic upregulation of lncRNAs at 13q14.3 in leukemia is linked to the in Cis downregulation of a gene cluster that targets NF-kB. *PLoS Genetics, 9*(4), e1003373–e1003373.

Garitano-Trojaola, A., Jose-Eneriz, E. S., Ezponda, T., et al., (2018). Deregulation of linc-PINT in acute lymphoblastic leukemia is implicated in abnormal proliferation of leukemic cells. *Oncotarget, 9*(16), 12842–12852.

Garzon, R., Stefano, V., Dimitrios, P., et al., (2014). Expression and prognostic impact of lncRNAs in acute myeloid leukemia. *Proceedings of the National Academy of Sciences of the United States of America, 111*(52), 18679–18684.

Ghazavi, F., De Moerloose, B., Van, L. W., et al., (2016). Unique long noncoding RNA expression signature in ETV6/RUNX1-driven B-cell precursor acute lymphoblastic leukemia. *Oncotarget, 7*(45), 73769–73780.

Gioia, R., Simon, D., Manon, O., et al., (2017). LncRNAs downregulated in childhood acute lymphoblastic leukemia modulate apoptosis, cell migration, and DNA damage response. *Oncotarget, 8*(46), 80645–80650.

He, B., Hlavka-Zhang, J. C., Richard, B. L., & Duohui, J., (2019). Epigenetic landscape in leukemia and its impact on antileukemia therapeutics. *Germ Line Mutations Associated Leukemia*, 1–19.

Hirano, T., Ryoko, Y., Hironori, H., Yuka, H., Atsuhiko, I., & Takeshi, Y., (2015). Long noncoding RNA, CCDC26, controls myeloid leukemia cell growth through regulation of KIT expression. *Molecular Cancer, 14*, 90.

Hong, W., Rui, G., Andrew, R. H., Jifan, H., & Li, W., (2015). Effect of long noncoding RNA RUNXOR on the epigenetic regulation of RUNX1 in acute myelocytic leukemia. *Journal of Clinical Oncology, 151*(33), 10–17.

Ishii, N., Kouichi, O., Hiroshi, S., et al., (2006). Identification of a novel noncoding RNA, MIAT, that confers risk of myocardial infarction. *Journal of Human Genetics, 51*(12), 1087–1099.

James, A. R., Schroeder, M. P., Neumann, M., et al., (2019). Long noncoding RNAs defining major subtypes of B cell precursor acute lymphoblastic leukemia. *J. Hematol. Oncol., 12*(1), 8.

Jones, R., Susanne, W., John, H., & Aditi, K., (2020). A long intergenic noncoding RNA regulates nuclear localization of DNA methyl transferase-1. *bioRxiv, 44*.

Kang, Z. J., Yu, F. L., Ling, Z. X., et al., (2016). The Philadelphia chromosome in leukemogenesis. *Chinese Journal of Cancer, 35*(1), 1–15.

Kantarjian, H. M., & Jorge, C., (2006). New strategies in chronic myeloid leukemia. *International Journal of Hematology, 83*(4), 289–293.

Kazemzadeh, M., Reza, S., & Ayla, V. O., (2015). LncRNAs: Emerging players in gene regulation and disease pathogenesis. *Journal of Genetics, 94*(4), 771–784.

Klein, U., Marie, L., Marta, C., et al., (2010). The DLEU2/miR-15a/16-1 cluster controls B cell proliferation and its deletion leads to chronic lymphocytic leukemia. *Cancer Cell, 17*(1), 28–40.

Koch, U., & Freddy, R., (2011). Notch in T-ALL: New players in a complex disease. *Trends in Immunology, 32*(9), 434–442.

Kornienko, A. E., Philipp, M. G., Denise, P. B., & Florian, M. P., (2013). Gene regulation by the act of long noncoding RNA transcription. *BMC Biology, 11*, 59.

Lee, R. S., & Charles, W. M. R., (2013). Linking the SWI/SNF complex to prostate cancer. *Nature Genetics, 45*(11), 1268–1269.

Lei, L., Siyu, X., Dan, L., et al., (2018). Genome-wide characterization of lncRNAs in acute myeloid leukemia. *Briefings in Bioinformatics, 19*(4), 627–635.

Li, J. L., Liu, X. L., Guo, S. F., Yang, Y., Zhu, Y. L., & Li, J. Z., (2019). Long noncoding RNA UCA1 regulates proliferation and apoptosis in multiple myeloma by targeting MIR-331-3p/IL6R axis for the activation of JAK2/STAT3 pathway. *European Review for Medical and Pharmacological Sciences, 23*(21), 9238–9250.

Li, J., & Sun, C. K., (2018). Long noncoding RNA SNHG5 is up-regulated and serves as a potential prognostic biomarker in acute myeloid leukemia. *European Review for Medical and Pharmacological Sciences, 22*(11), 3342–3347.

Li, J., Zhenyu, X., & Changning, L., (2013). Long noncoding RNAs and complex human diseases.

Llinàs-Arias, P., & Manel, E., (2017). Epigenetic inactivation of tumor suppressor coding and noncoding genes in human cancer: An update. *Open Biology, 7*(9).

Long, Y., Xueyin, W., Daniel, T. Y., & Thomas, R. C., (2017). How do lncRNAs regulate transcription? *Science Advances, 3*(9), eaao2110–eaao2110.

Lu, Y., Yan, L., Xiao, C., et al., (2017). Long noncoding RNA HULC promotes cell proliferation by regulating PI3K/AKT signaling pathway in chronic myeloid leukemia. *Gene, 607*, 41–46.

Lyu, Y., Lou, J., Yang, Y., et al., (2017). Dysfunction of the WT1-MEG3 signaling promotes AML leukemogenesis via p53-dependent and -independent pathways. *Leukemia, 31*(12), 2543–2551.

Ma, L., Kuai, W. X., Sun, X. Z., Lu, X. C., & Yuan, Y. F., (2018). Long noncoding RNA LINC00265 predicts the prognosis of acute myeloid leukemia patients and functions as a promoter by activating pi3k-akt pathway. *European Review for Medical and Pharmacological Sciences, 22*(22), 7867–7876.

Ma, X., Wei, Z., Ming, Z., Shufen, L., Wen, J., & Kankan, W., (2020). Oncogenic role of lncRNA CRNDE in acute promyelocytic leukemia and NPM1-mutant acute myeloid leukemia. *Cell Death Discovery, 6*(1).

Mangiavacchi, A., Melissa, S., Silvia, M., et al., (2016). The miR-223 host noncoding transcript linc-223 induces IRF4 expression in acute myeloid leukemia by acting as a competing endogenous RNA. *Oncotarget, 7*(37), 60155–60168.

Mattick, J. S., & Igor, V. M., (2006). Non-coding RNA. *Human Molecular Genetics, 15*(1), R17–R29.

Medyouf, H., Samuel, G., Hongfang, W., et al., (2011). High-level IGF1R expression is required for leukemia-initiating cell activity in T-ALL and is supported by notch signaling. *Journal of Experimental Medicine, 208*(9), 1809–1822.

Mercer, T. R., Marcel, E. D., & John, S. M., (2009). Long noncoding RNAs: Insights into functions. *Nature Reviews Genetics, 10*, 155–159.

Merry, C. R., Megan, E. F., Jessica, N. S., et al., (2015). DNMT1-associated long noncoding RNAs regulate global gene expression and DNA methylation in colon cancer. *Human Molecular Genetics, 24*(21), 6240–6253.

Neff, T., Amit, U. S., Michael, J. K., et al., (2012). Polycomb repressive complex 2 is required for MLL-AF9 leukemia. *Proceedings of the National Academy of Sciences of the United States of America, 109*(13), 5028–5033.

Nitzan, M., Steiman-Shimony, A., Yael, A., Ofer, B., & Hanah, M., (2014). Interactions between distant ceRNAs in regulatory networks. *Biophysical Journal, 106*(10), 2254–2266.

Ohtake, S., (2007). Acute myeloid leukemia. *Cancer and Chemotherapy, 34*(13), 2175–2179.

Ouimet, M., Drouin, S., Lajoie, M., et al., (2017). A childhood acute lymphoblastic leukemia-specific lncRNA implicated in prednisolone resistance, cell proliferation, and migration. *Oncotarget, 8*(5), 7477–7488.

Pan, J. Q., Yan, Q. Z., Jing, H. W., Ping, X., & Wei, W., (2017). LncRNA co-expression network model for the prognostic analysis of acute myeloid leukemia. *International Journal of Molecular Medicine, 39*(3), 663–671.

Papaioannou, D., Andreas, P., Oliver, M. D., et al., (2019). The long noncoding RNA HOXB-AS3 regulates ribosomal RNA transcription in NPM1-mutated acute myeloid leukemia. *Nature Communications, 10*(1).

Pelayo, R., Rob, W., Scott, S. P., et al., (2005). Lymphoid progenitors and primary routes to becoming cells of the immune system. *Current Opinion in Immunology, 17*(2), 100–107.

Pouyanrad, S., Soheila, R., & Elaheh, S. G., (2019). Dysregulation of miR-335–3p, targeted by NEAT1 and MALAT1 long noncoding RNAs, is associated with poor prognosis in childhood acute lymphoblastic leukemia. *Gene, 692*(2018), 35–43.

Rinn, J. L., & Howard, Y. C., (2012). Genome regulation by long noncoding RNAs. *Annual Review of Biochemistry, 81*(1), 145–166.

Ronchetti, D., Manzoni, M., Agnelli, L., et al., (2016). LncRNA profiling in early-stage chronic lymphocytic leukemia identifies transcriptional fingerprints with relevance in clinical outcome. *Blood Cancer Journal, 6*(9).

Salehi, M., & Mohammadreza, S., (2018). Induction of apoptosis and necrosis in human acute erythroleukemia cells by inhibition of 47Tlong noncoding RNA PVT1. *Molecular Biology Research Communications, 7*(2), 89–96.

Sattari, A., Hasan, S., Farzaneh, M., et al., (2016). Upregulation of long noncoding RNA MIAT in aggressive form of chronic lymphocytic leukemias. *Oncotarget, 7*(34), 54174–54182.

Schwarzer, A., Stephan, E., Franziska, S., et al., (2017). The noncoding RNA landscape of human hematopoiesis and leukemia. *Nature Communications, 8*(1), 218.

Sheng, H., Jiajia, Z., Yan, M., Yuhua, Z., Yunpeng, D., & Renpeng, J., (2021). lncRNA FBXL19-AS1 is a diagnosis biomarker for paediatric patients with acute myeloid leukemia. *The Journal of Gene Medicine, 23*(3), e3317–e3317.

Shi, J., Rongqin, D., Yuqing, C., Honggang, G., Yinyin, H., & Yin, Z., (2019). LncRNA LINP1 regulates acute myeloid leukemia progression via HNF4α/AMPK/WNT5A signaling pathway. *Hematological Oncology, 37*(4), 474–482.

Spitale, R. C., Miao-Chih, T., & Howard, Y. C., (2011). RNA templating the epigenome: Long noncoding RNAs as molecular scaffolds. *Epigenetics, 6*(5), 539–543.

Subhash, S., Per, O. A., Subazini, T. K., Chandrasekhar, K., & Meena, K., (2016). Global DNA methylation profiling reveals new insights into epigenetically deregulated protein coding and long noncoding RNAs in CLL. *Clinical Epigenetics, 8*(1), 1–16.

Sun, C., Shuping, L., Guili, Z., Na, W., Huiyuan, S., & Caifu, L., (2017). CEBPA-mediated upregulation of the lncRNA PLIN2 promotes the development of chronic myelogenous leukemia via the GSK3 and Wnt/β-catenin signaling pathways. *American Journal of Cancer Research, 7*(5), 1054–1067.

Sun, J., Wei, L., Yunpeng, S., et al., (2014). A novel antisense long noncoding RNA within the IGF1R gene locus is imprinted in hematopoietic malignancies. *Nucleic Acids Research, 42*(15), 9588–9601.

Sun, L. Y., Xiao, J. L., Yu, M. S., et al., (2018). LncRNA ANRIL regulates AML development through modulating the glucose metabolism pathway of AdipoR1/AMPK/SIRT1. *Molecular Cancer, 17*(1), 1–6.

Sun, M. D., Zheng, Y. Q., Wang, L. P., Zhao, H. T., & Yang, S., (2018). Long noncoding RNA UCA1 promotes cell proliferation, migration and invasion of human leukemia cells via sponging miR-126. *European Review for Medical and Pharmacological Sciences, 22*(8), 2233–2245.

Terwilliger, T., & Abdul-Hay, M., (2017). Acute lymphoblastic leukemia: A comprehensive review and 2017 update. *Blood Cancer Journal, 7*(6), e577–e577.

Tian, X., Jie, T., Xinyi, T., Jie, M., & Shengjun, W., (2016). Long noncoding RNAs in the regulation of myeloid cells. *Journal of Hematology and Oncology, 9*(1), 99.

Trimarchi, T., Erhan, B., Panagiotis, N., et al., (2014). Genome-wide mapping and characterization of notch-regulated long noncoding RNAs in acute leukemia. *Cell, 158*(3), 593–606.

Tsiodras, S., George, S., Michael, J. K., & Dlmitrios, P. K., (2000). Infection and immunity in chronic lymphocytic leukemia. *Mayo Clinic Proceedings, 75*(10), 1039–1054.

Ulitsky, I., & David, P. B., (2013). LincRNAs: Genomics, evolution, and mechanisms. *Cell, 154*(1), 26–46.

Wang, F., Xiaoxue, T., Jie, Z., et al., (2018). A three-lncRNA signature for prognosis prediction of acute myeloid leukemia in patients. *Molecular Medicine Reports, 18*(2), 1473–1484.

Wang, G., Jingen, P., Lu, Z., Yajun, W., & Cheng, W., (2017). Long noncoding RNA CRNDE sponges miR-384 to promote proliferation and metastasis of pancreatic cancer cells through upregulating IRS1. *Cell Proliferation, 50*(6), e12389–e12389.

Wang, H., Qian, L., Shusen, T., et al., (2017). The role of long noncoding RNA HOTAIR in the acquired multidrug resistance to imatinib in chronic myeloid leukemia cells. *Hematology, 22*(4), 208–216.

Wang, H., Yan, W., Shangqian, X., Yizhi, L., & Zhi, X., (2017). Global and cell-type specific properties of lincRNAs with ribosome occupancy. *Nucleic Acids Research, 45*(5), 2786–2796.

Wang, J., Liu, Z. H., & Yu, L. J., (2019). Long noncoding RNA LINC00641 promotes cell growth and migration through modulating miR-378a/ZBTB20 axis in acute myeloid leukemia. *European Review for Medical and Pharmacological Sciences, 23*(17), 7498–7509.

Wang, X. Q. D., & Josée, D., (2017). Reciprocal regulation of chromatin state and architecture by HOTAIRM1 contributes to temporal collinear HOXA gene activation. *Nucleic Acids Research, 45*(3), 1091–1104.

Wang, X., Lalit, S., Neeraj, J., Tamer, K., Rohit, M., & Felipe, S., (2016). LncRNA MALAT1 promotes the development of mantle cell lymphoma by associating with ezh2. *Journal of Translational Medicine, 14*(1), 1–14.

Wang, Y., (2019). Comprehensive long noncoding RNA expression profiling by RNA sequencing reveals potential biomarkers for acute myeloid leukemia risk. *Cancer Biomarkers, 26*(1), 93–108.

Wang, Y., Li, Y., Song, H. Q., & Sun, G. W., (2018). Long noncoding RNA LINC00899 as a novel serum biomarker for diagnosis and prognosis prediction of acute myeloid leukemia. *European Review for Medical and Pharmacological Sciences, 22*(21), 7364–7370.

Wierzbicki, A. T., (2012). The role of long noncoding RNA in transcriptional gene silencing. *Current Opinion in Plant Biology, 15*(5), 517–522.

Wurm, A. A., & Cristina, P., (2019). Long noncoding RNAs as functional and structural chromatin modulators in acute myeloid leukemia. *Frontiers in Oncology, 9*, 1–9.

Wutz, A., Theodore, P. R., & Rudolf, J., (2002). Chromosomal silencing and localization are mediated by different domains of Xist RNA. *Nature Genetics, 30*(2), 167–174.

Xiao, Y., Changjie, J., Yiqiang, L., et al., (2017). lncRNA UCA1 contributes to imatinib resistance by acting as a ceRNA against miR-16 in chronic myeloid leukemia cells. *DNA and Cell Biology, 36*(1), 18–25.

Xing, C. Y., Xiao, Q. H., Fei, Y. X., et al., (2015). Long noncoding RNA HOTAIR modulates c-KIT expression through sponging miR-193a in acute myeloid leukemia. *FEBS Letters, 589*(15), 1981–1987.

Xu, J., Min, W., Yichen, S., Hongqian, Z., Yujie, W., & Jie, G., (2020). Identifying dysregulated lncRNA-associated ceRNA network biomarkers in CML based on dynamical network biomarkers. *BioMed Research International, 2020*.

Yadav, R. K., Gautam, D. K., Muj, C., Gajula, B. M. B., & Paddibhatla, I., (2021). Methotrexate negatively acts on inflammatory responses triggered in *Drosophila* larva with hyperactive JAK/STAT pathway. *Dev. Comp. Immunol., 123*, 104161.

Yang, F., Xinxian, D., Wenxiu, M., et al., (2015). The lncRNA Firre anchors the inactive X chromosome to the nucleolus by binding CTCF and maintains H3K27me3 methylation. *Genome Biology, 16*(1), 1–17.

Yang, J., Zhao, Y., Yumin, L., et al., (2020). The identification of long noncoding RNA H19 target and its function in chronic myeloid leukemia. *Molecular Therapy – Nucleic Acids, 19*, 1368–1378.

Yang, L., Han, Y., Saurez, S. F., & Minden, M. D., (2007). A tumor suppressor and oncogene: The WT1 story. *Leukemia, 21*(5), 868–876.

Yu, W., Gius, D., Onyango, P., et al., (2008). Epigenetic silencing of tumour suppressor gene p15 by its antisense RNA. *Nature, 451*(7175), 202–206.

Yuan, M., Shujun, W., Lijie, Y., et al., (2017). Long noncoding RNA profiling revealed differentially expressed lncRNAs associated with disease activity in PBMCs from patients with rheumatoid arthritis. *PLoS One, 12*(11), e0186795–e0186795.

Zan, H., & Paolo, C., (2015). Epigenetics of peripheral B-cell differentiation and the antibody response. *Frontiers in Immunology, 6*.

Zeng, C., Xibao, Y., Jing, L., Lijiang, Y., Shaohua, C., & Yangqiu, L., (2015). Overexpression of the long noncoding RNA PVT1 is correlated with leukemic cell proliferation in acute promyelocytic leukemia. *Journal of Hematology and Oncology, 8*(1), 126–126.

Zeng, H., Wu, H., Yan, M., Tang, L., Guo, X., & Zhao, X., (2020). Characterization of a 4 lncRNAs-based prognostic risk scoring system in adults with acute myeloid leukemia. *Leuk. Res., 88*, 106261.

Zeng, J., Li, Y., Wang, Y., et al., (2020). lncRNA 00312 attenuates cell proliferation and invasion and promotes apoptosis in renal cell carcinoma via miR-34a-5p/ASS1 axis. *Oxid. Med. Cell Longev., 2020*, 5737289.

Zhao, C., Shanshan, W., Yang, Z., et al., (2019). Long noncoding RNA NEAT1 modulates cell proliferation and apoptosis by regulating miR-23a-3p/SMC1A in acute myeloid leukemia. *Journal of Cellular Physiology, 234*(5), 6161–6172.

Zhaung, Q., Zhenlin, J., Xiangkuo, Z., Ting, J., & Lina, X., (2021). Long noncoding RNA LINC00460 serves as a potential biomarker and oncogene via regulation of the miR-320b/PBX3 axis in acute myeloid leukemia. *Molecular Medicine Reports, 23*(6), 1–12.

Zheng, J., Song, Y., Li, Z., Tang, A., Fei, Y., & He, W., (2019). The implication of lncRNA expression pattern and potential function of lncRNA RP4-576H24.2 in acute myeloid leukemia. *Cancer Med., 8*(17), 7143–7160.

Zhou, X., Ping, Y., Qi, L., & Zhiqiang, L., (2017). LncRNA MEG3 regulates imatinib resistance in chronic myeloid leukemia via suppressing microRNA-21. *Biomolecules and Therapeutics, 25*(5), 490–496.

Zhuang, M. F., Li, L. J., & Ma, J. B., (2019). LncRNA HOTTIP promotes proliferation and cell cycle progression of acute myeloid leukemia cells. *European Review for Medical and Pharmacological Sciences, 23*(7), 2908–2915.

CHAPTER 11

Minimal Residual Disease (MRD) as a Prognostic Marker in Acute Myeloid Leukemia

ANPREET CHOUDHARY,[1] REKHA SHARMA,[1] RAGHUNATH MANCHALA,[2] K. RAJENDER RAO,[3] SATTI VISHNUPRIYA,[4] SUVIR SINGH,[5] OBUL REDDY BANDAPALLI,[6] PRASHANT SURAVAJHALA,[7] RAGHUNADHARAO DIGUMARTHI,[8] and SUGUNAKAR VUREE[9]

[1]*School of Bioengineering and Biosciences, Lovely Professional University, Phagwara, Punjab, India*

[2]*Endocrinology and Metabolism Division, National Institute of Nutrition (NIN), Uppal Road, Hyderabad, Telangana, India*

[3]*Molecular Genetics, National Center for Laboratory Animal Sciences, National Institute of Nutrition, Hyderabad, Telangana, India*

[4]*Department of Genetics, Osmania University, Hyderabad, Telangana, India; #102, Apoorva Towers, Rd. No. 2, Banjara Hills, Hyderabad 500034, India*

[5]*Clinical Hematology and Bone Marrow Transplantation, Christian Medical College, Ludhiana, Punjab, India*

[6]*Division of Molecular Genetic Epidemiology, German Cancer Research Center (DKFZ), Im Neuenheimer Feld 580, Heidelberg, Germany*

[7]*Amrita School of Biotechnology, Amrita Vishwa Vidyapeetham, Clappana, Kerala, India*

[8]*Oncology Services, KIMS ICON, Vizag Steel General Hospital, Visakhapatnam, Andhra Pradesh, India*

[9]*MNR Foundation for Research and Innovation, MNR Medical College and Hospital, MNR University, Fasalwadi Village, Sangareddy (District), Telangana 502294, India*

Hematopoiesis: Biochemical, Cellular, Molecular, and Genomic Perspectives. Ravi Kumar Gutti (Ed.)

ABSTRACT

AML remains incurable four decades after its discovery. AML (t-AML) treatment remained unchanged without any improvements in outcomes. AML is characterized by cytogenetic abnormalities along with genetic and epigenetic changes. Molecular aberrated mutations, like FLT3-ITD and NPM1mut, are associated with high relapse rates. However, the persistence of leukemic cells in such patients can be characterized as minimal/measurable residual disease. The main reason for the development of the MRD technique is to quantify the leukemic cells in a patient's peripheral blood and bone marrow (BM). Detection of MRD using specific techniques could be used to tailor further treatment. There is numerous quantification technique to detect MRD, such as multiparameter flow cytometry (MPFC), Quantitative real-time polymerase reaction (RQ-PCR), digital-PCR, and next-generation sequencing (NGS). Several studies identify positive MRD test results during AML treatment that predict patients who will have an increased risk of leukemic relapse when compared with MRD-negative patients. The MRD method cannot detect patients who relapse after 18 months of clinical remission (CR) or after five years of CR. The MRD detection technique requires further rectification and standardization. In this chapter, the role of MRD and the Prognostic markers in AML are discussed, along with the detection methods.

11.1 INTRODUCTION

Assessment for Minimal Residual Disease (MRD) in acute myeloid leukemia (AML) has significantly expanded our ability to access chemosensitivity to treatment along with the duration of treatment responses (Gomez-Arteaga et al., 2018). It is usually accepted that the oncogenic change of hematopoietic stem cells (HSCs) or progenitors initiates leukemogenesis. Acute myeloid leukemia (AML) is a malignant disorder of hematopoietic stem cells which is characterized by an abnormal proliferation and differentiation arrest of myeloid progenitors or lineages and subsequent failure of bone marrow (BM) (Principe et al., 2016; Yi et al., 2020). AML occurs in people of all ages but predominantly in adults older than 65 years. Clinically, the comprehension of AML genetic alterations have overshadowed factors such as age, and gender. In fact, genetic attributes have been incorporated into AML prognostic models. Molecular evaluation at diagnosis has become a routine clinical practice (Vedula et al., 2017).

Currently, the therapy of patients with AML incorporates intensive induction treatments as well as autologous and allogeneic stem cell transplantation. However, while most patients enter remission, a relapse is the rule in about 80% of patients (Kern et al., 2011).

For more than 30 years, it has been perceived that there are residual undetected cells at the point of morphologic complete remission (CR) that are not eliminated by current regimens and hence additional post-remission treatment is perhaps needed, without which patients would relapse. Subsequently, it has become imperative to improve the capability to diagnose, quantify and measure residual disease in AML (Gomez-Arteaga et al., 2018). The objective of identifying is to be able to predict leukemic re-growth and overt relapse as a prognostic factor for available therapies and perhaps to help modify treatments early on to improve survival.

11.1.1 EPIDEMIOLOGY

AML is one of the most common types of leukemia in adults. Leukemia hits around 4% of males, whereas it is 3% in the case of females (Siegel et al., 2019). In 1990, AML represented 18.0% of total leukemia cases around the world. This extent increased to 23.1% in 2017. The age-standardized incidence rate (ASIR) of AML expanded from 1.35/100,000 to 1.54/100,000 during this period (Dong et al., 2020). In India, the incidence of AML is about 15% out of all hematological malignancies, respectively (Ahirwar et al., 2018). AML represents about 20% of acute leukemia (AL) in children and 80% of AL in adults. The incidence rate of AML progressively increases with age, and in adults beyond the age of 65 years, the incidence rate is around 30 times the incidence rate in children. According to the Surveillance, Epidemiology, and End Results (SEER) data set, the number of new cases of AML were 4.2 per 100,000 male and female for each year. The number of deaths were 2.8 per 100,000 male and female per year (Shoket et al., 2019).

11.2 ACUTE MYELOID LEUKEMIA (AML)

AML is a type of blood cancer; it is a heterogenous clonal disorder defined by clonal expansion of immature myeloid cells proliferate from myelogenous progenitor (blast) cells present in the BM and peripheral blood. For several decades, AML remained incurable. At present, approximately 30%-40%

under 60 years of age are cured (Döhner et al., 2015). AML blast cells are morphologically distinct from normal leukemic cells, these cells vary in size having inconsiderably larger than lymphocytes and monocytes. The nuclei of AML blast cells are generally larger in size and varying shape containing one or more nucleoli. They express cluster of differentiation markers like CD13, Cd33, and CD34 are found on both AML blast cells and normal premature myeloid cells (Campos et al., 1989). There are several other differentiation markers named as monocytic markers (e.g., CD4, CD14, CD11b), erythroid-specific markers (e.g., CD36, CD71) and megacaryocytes markers (e.g., CD41a, CD61). Some antigens such as terminal deoxynucleotidyl transferase (TdT), Human leukocyte antigen-D related (HLA-DR), CD19 and CD7 are confined to lymphocyte lineage (B and T-cells) and may co-express on AML blastic cells (Wolach et al., 2015).

11.2.1 CLASSIFICATION

In the mid-70s, the French-American-British classification system; classified AML into eight subtypes based on their morphological, immune-phenotypical and cytochemical characteristics (FAB M0 to M7) (Tables 11.1 and 11.2) (Bennett et al., 1976). Classification of AML does not know whether it is categories based on definite somatic genetic abnormal appearance that direct the AML sub-classes or whether on oncogenic mutations that result into different treatment-related outcomes (Lindsley et al., 2015).

11.2.2 CYTOGENETIC ABNORMALITIES IN AML

Cytogenetic abnormalities which are non-random such as deletion and translocation are seen in primary AML in nearly 52% of the adults (Byrd et al., 2002). There are certain recurrent chromosomal abnormalities involving translocation of t (8;21) (q22; q22), t (15;17) (q22; 12) and inversion of inv (16) (p13.1; q22), which are compatible with long duration of remission and patient survival. On the contrary, genetic anomalies present on chromosome number 5 and 7 (defined by greater than 3 chromosomal abnormalities) and 11q23 translocation are related to inferior response and short-lived overall survival (OS) (Döhner et al., 2015).

TABLE 11.1 In accordance with 2008 Revision by the World Health Organization (WHO), AML into Seven Related Sub-Classes

Subclass	Description	References
I.	AML with revert genetic abnormalities including gene rearrangement and gene mutations such as RUNX1, CBFB, PML-RARA, MLL and NPM1, CEBPA, respectively.	Vardiman et al. (2009)
II.	Multilineage myelodysplasia associated with alternations involve undesirable cytogenetic abnormalities, high level expression of Multi-Drug Resistant glycoproteins and produce unexpected outcomes in response to treatment.	Vardiman et al. (2009)
III.	Therapy-associated with acute myeloid leukemia (AML), myelodysplastic syndrome (MDS) and myeloproliferative neoplasm (MPN).	Vardiman et al. (2009)
IV.	Not otherwise specified (NOS) just as AML (matches with FAB M0-M7 classification system including acute erythroid leukemia/acute megacaryoblastic leukemia may further classified AML based on myelodysplasia-related alternations.	Vardiman et al. (2009)
V.	Granulocyte's sarcoma act as pathological diagnosis for the blast cells and extra medullary multiple myeloma that have an ability to destruct the normal structure of the tissue in that they are reside.	Vardiman et al. (2009)
VI.	AML and MDS are related to Down Syndrome because they share the identity with the down syndrome in a biological way.	Vardiman et al. (2009)
VII.	In myeloid-associated neoplasm, the tumor originated from the precursor of specific subsets of dendritic cells and dendritic plasmocytes cells, it is referred to as blastic plasmocytoid dendritic-cells neoplasm (BPDCN).	Vardiman et al. (2009)

TABLE 11.2 Classification of AML based on Causality

Group	Description	References
I.	Development of AML from antecedent myeloid neoplasm called secondary AML (S-AML).	Lindsley et al. (2015)
II.	AML patient exposed to leukemogenic-related therapy/ therapy-related to AML shows delayed obstacles.	Lindsley et al. (2015)
III.	Cause of De novo AML due to unidentified exposure or introductory stem cells disorder.	Lindsley et al. (2015)

11.2.3 CHARACTERIZATION OF AML ASSOCIATED MOLECULAR ABNORMALITIES

Molecular analysis of AML blast cells provides the information in related to heterogeneity in regard with occurrence and non-occurrence of genetic mutations, alteration of genes and expression of miRNA (Marcucci et al., 2011). As compared with other adult cancer, the AML leukemic cells have few numbers of mutations in its genomic material possessing, on an average of 13 mutations in the gene, 8 sets of the mutations are referred to as "random or passenger" mutation and 5 sets of mutation as "recurring or driver" mutation (Cancer Genome Atlas Research Network, 2013). There are some fundamental molecular abnormalities in association with identification and determination of the outcomes in therapy-related AML patients, which are described below in sequential and relative order of frequency mostly associated with genetic anomalies (Table 11.3).

Factor such as age and response to current therapies represents an additional status other than cytogenetic and selected molecular abnormalities for the determination and analysis of outcomes in AML patients (Dohner et al., 2010). In the year of 2010, working panel of European Leukemia Net (ELN) named as the International Working Panel recommended a standard classification system for describing chromosomal and molecular abnormalities that provide genetic knowledge on therapeutic results and its significant contrast of AML adult patients outcomes (Figure 11.1) (Dohner et al., 2010).

11.2.4 RISK STRATIFICATION AND PROGNOSIS IN AML

The new standard classification system proposed by ELN originated from 2008 revision by WHO classify the myeloid neoplasm and AML (Vardiman et al., 2009). According to ELN, four groups are recognized: AML patients are divided as Favorable, Intermediate 1, Intermediate 2 and adverse group (Table 11.4). Favorable risk group is characterized by APL (Acute premyelocytic leukemia) t (15, 17) (q22; q22) mutation, abnormally balanced mutation t (8;21) (q22; q22), inv (16) (p13.1q22), t (16;16) (p13.1q22), NPM1 mutated gene without FLT3-ITD mutation and CEBPA bi-allelic gene mutations. Intermediate 1 risk group involves NPM1 gene with FLT3-ITD mutated gene, normal NPM1 gene with or without gene FLT3-ITD.

TABLE 11.3 Repetitive Molecular Abnormalities Associated with Adult AML Patients: Gene Mutation, Type of Genetic Aberration, Relevant Frequency of Occurrence, Physiological Attributes, and Clinical Significance

Gene Mutation	Type of Genetic Aberration	Relevant Frequency of Occurrence	Physiological Features/Clinical Significance
Nucleophosmin 1 (NPM1)	Cytoplasmic mislocation in the gene at exon 12 (recurrent genetic mutation) (Haferlach et al., 2009)	1/3rd in adult AML patients (Haferlach et al., 2009) 20%-30% AML patient's predominately female patients (Schnittger et al., 2005; Falini et al., 2007)	Physiological features: NPM1 genetic mutation is associated with secondary chromosomal aberration/additional chromosomal aberration namely +8, +4 del(9q) (Haferlach et al., 2009). NPM1 associated with frequent gene mutation FLT3 (Boissel et al., 2010; Dohner et al., 2008; Marcucci et al., 2010; Mro´zek et al., 2007; Paschka et al., 2010; Schlenk et al., 2008; Schnittger et al., 2010; Wagner et al., 2010) Abnormal mRNA expression of mutated NPM1 gene as protein in cytoplasm instead of nucleus, regulating abnormal myeloid cell division and development (Falini et al., 2006, 2007; Cheng et al., 2010).
DNA Methyltransferase 3A (DNMT3A)	Non-synonymous missense mutation on arginine codon 882 (R882-DNMT3A) not affecting another genetic codon (non-R882-DNMT3A) (Marcucci et al., 2012)	Approximately 18%-22% in all cases of AML 34% cytogenetically normal-AML (CN-AML) patients (Ley et al., 2010)	Physiological feature: Affecting hematopoiesis and methylation process (Marcucci et al., 2012). Clinical significance: Frequent in CN-AML (cytogenetically normal-AML) patients DNMT3A mutation clinically significant on age-related factor of the patients, unfavorable prognosis factor in old patients (associated with R882-DNMT3A) and young patients (associated with non-R882-DNMT3A) showed lower overall survival (OS) rate as well as disease-free survival (DFS) rate (Marcucci et al., 2012)

TABLE 11.3 *(Continued)*

Gene Mutation	Type of Genetic Aberration	Relevant Frequency of Occurrence	Physiological Features/Clinical Significance
FMs-like tyrosine kinase 3 (FLT3)	In-frame internal tandem duplication (ITD) in the region of intracellular juxtamembrane (JM) which results into activation FLT3	20% in all cases of AML	Clinical significance: Mutation in FLT3-ITD is related with poor prognosis in AML patients (Ward et al., 2010) AML patients with non-ITD mutation on JM domain associated with FLT3 activation shows significantly adverse effect as compared to FLT3-ITD mutation in JM domain (Kayser et al., 2009)
	Secondary mutation in Tyrosine kinase domain (TKD) at FLT3 gene (Frohling et al., 2005; Kelly et al., 2002) Further activation of FLT3 achieved by mutation on another functional domain Tyrosine kinase domain (TKD) (Frohling et al., 2005)	30%-45% in all CN-AML patients (Fro¨ hling et al., 2005; Kelly et al., 2002)	
Isocitrate Dehydrogenase (IDH)	Arginine residual mutation on codon 132 of IDH1 and codon172 of IDH2 both type of mutations is found in brain tumor and AML (Marcucci et al., 2010; Paschka et al., 2010)	Combining frequency of these two mutations (IDH1 and IDH2) is approximately 15%-20% in all AML patients 25%-30% of CN-AML patients having either IDH1 mutation or IDH2 mutation (Paschka et al., 2010)	Phenotypic features: IDH mutation produces new class of mutated gene product which is a metabolizing enzyme Mutant IDH1 and IDH2 associated with loss of normal physiological enzymatic action, resulting into conversion of alpha-ketoglutarate into 2-hydroxyglutarate (oncogenic metabolite) (Ward et al., 2010) Patients having IDH1 and IDH2 R140 mutations associated with NPM1 genetic mutation shows adverse effect in AML patients with NPM1 mutation and without FLT3-ITD (Marcucci et al., 2010; Paschka et al., 2010; Boissel et al., 2010; Wagner et al., 2010) Clinical significance: Mutation R172 IDH2 results in shorter period of CR and inferior outcomes in AML patients (Marcucci et al., 2010; Paschka et al., 2010; Boissel et al., 2010)

TABLE 11.3 *(Continued)*

Gene Mutation	Type of Genetic Aberration	Relevant Frequency of Occurrence	Physiological Features/Clinical Significance
c-KIT	c-KIT mutation occurs at codon D816, KiT mutation for translocation t (8;21) (q22; q22) and inversion inv (16) (p13.1; q22) (Paschka et al., 2006)	Infrequent in <5% AML patient 22%-29% as CBF mutation, for example, AML with t (8;21) (q22; q22), inv (16) (p13.1q22) and genetically fused gene (RUNX1/RUNX1T1 and CFB/MYH11) (Paschka et al., 2006) c-KIT mutation is rare in another AML subtype (Paschka et al., 2008)	c-KIT mutation confers high rate of relapse-risk (RR), lower OS and DFS (Paschka et al., 2006). CBF-AML patient with c-KIT mutation have fewer OS rate than wild type KIT mutation t (8;21) (q22; q22) accept KIT mutation with inv (16) (p13.1q22) (Boissel et al., 2006) Mutated KIT related with inferior outcomes Mutated KIT also associated with high level of CBF-AML as compared to other AML (Bullinger et al., 2004; Valk et al., 2004)
CCAAT Enhancer Binding Protein alpha (CEBPA)	Two mutations or double mutations involving the combination of both N-terminal and bZIP gene (Biallelic mutation) (Fasan et al., 2014)	Occurrence of mutation 6%-10% in all AML patients and 15%-19% in CN-AML patients associated with del (9q) (Döhner et al., 2015; Mrózek et al., 2007)	Physiological features: Transcription (CEBPA) perform the function of gene expression in hematopoiesis process (Koschmieder et al., 2009) CEBPA mutation in AML patient is similar with that of CEBPA wild-type mutation in AML (Marcucci et al., 2011; Wouters et al., 2009) Clinical significance: Mutation in CEBPA associated with higher rate of CR and OS in 4%-5% of AML patients (Fasan et al., 2014)

TABLE 11.3 *(Continued)*

Gene Mutation	Type of Genetic Aberration	Relevant Frequency of Occurrence	Physiological Features/Clinical Significance
Mixed Lineage Leukemia (MLL)	Partial tandem duplication (PTD) and MLL gene translocation (Dohner et al., 2008; Mro´zek et al., 2007)	PTD-MLL occurrence in 5%-11% of CN-AML patients PTD-MLL is frequent in AML patient with trisomy of chromosome 11 (Dohner et al., 2008; Mrozek et al., 2007)	Physiological features: In chromosome 11q23 gene named as mixed lineage leukemia translate a protein that performs the function of chromatin modification (part of regulator and complex) with histone methyltransferase activity (Ernest et al., 2002) Translocation of AML gene results in ALL/AML with poor prognosis due to high mRNA expression of HOX gene (Meyers et al., 1993) Clinical significance: AML patients with MLL-PTD mutation show short-lived CR duration and relapse free survival MLL-PTD shows less significant prognostic effect in CN-AML patient, those effectively treated with Allo-HSCT or 4-cycle of immunotherapy (Dohner et al., 2008; Mrozek et al., 2007)
Runt-Related Transcription Factor (RUNX1)	RUNX1 also called AML1/CBF subunit alpha 2 (CBFA2) located on chromosome number 21, and it is intermittently translocated with gene named as – ETO/MTG8/RUNX1T1 present at chromosome 8q22 and forming the fused product in the form of fusion protein AML-ETO/t (8;21) (q22; q22) AML (Tang et al., 2009)	Out of 470 Chinese adult AML patient-13.2% of the patients were having RUNX1 mutation (Tang et al., 2009)	Physiological features: RUNX1 is an important transcription factor for normal hematopoiesis (Meyers et al., 1993)
	Mutation of RUNX1 have been related to undifferentiation morphology characteristics and chromosomal abnormalities particularly with trisomy on chromosome in 21 and 13 (Tang et al., 2009)	Mutation in RUNX1 is frequent in old CN-AML patient and less common in younger CN-AML patient (Gaidzik et al., 2011)	Clinical significance: Mutations in RUNX1 gene render the blasts resistant to standard induction therapy with a lower rate of OS in both older and younger AML patients (Mendler et al., 2012)

TABLE 11.3 *(Continued)*

Gene Mutation	Type of Genetic Aberration	Relevant Frequency of Occurrence	Physiological Features/Clinical Significance
Additional sex comb-like 1 (ASXL1)	Loss of functionality of gene ASXL1 (Metzeler et al., 2011)	Occurrence of mutation in 5%-11% in AML patients (Metzeler et al., 2011)	Physiological features: ASXL1 protein perform the function of epigenetics regulation of DNA and histone alternations (Gaidzik et al., 2012)
		This typeof mutation is more common approximately 16.2% in older patient greater than equal to 60 years old as compared to patient having age < 60 approximately 3.2% (Metzeler et al., 2011)	Clinical significance: Older AML patients with ASXL1 mutation along with t (8;21), NPM1 wild type gene, without FLT3-ITD shows lower rate CR and OS (Alpermann et al., 2015; Micol et al., 2014)
Ten-eleven translocation 2 (TET2)	Translocation of 10-11 oncogenes family member 2 causes the mutation and loss of functionality (Chou et al., 2011)	Occurrence of this type of mutation approximately 9%-23% in AML patients (Chou et al., 2011)	Physiological features: Normal TET1 gene performs the enzymatic function such as converting 5-methyl cytosine (5mC) into 5-hydroxymethyl cytosine (5hmC) in the genetic material (DNA), which plays a key role in DNA demethylation (Chou et al., 2011)
			Clinical significance: TET2 gene mutation represents an adverse factor for CR and OS (Metzeler et al., 2011)
WT1	Mutation in WT1 gene associated with small insertion in exon 1, 7 and non-sense mutation (mutation in wilm's tumor gene WT in leukemia) in exon 9 (Paschka et al., 2008)	This type of mutation occurs in 10%-13% in all CN-AML patients (Paschka et al., 2008; Virappane et al., 2008; Gaidzik et al., 2009)	Physiological features: Mutation of WT1 related to poorer outcomes (Paschka et al., 2008; Virappane et al., 2008; Gaidzik VI et al., 2009) WT1-SNPrs16754 found on mutational hotspot of WT1 gene with very poor outcomes in CN-AML patients (Damm et al., 2010)

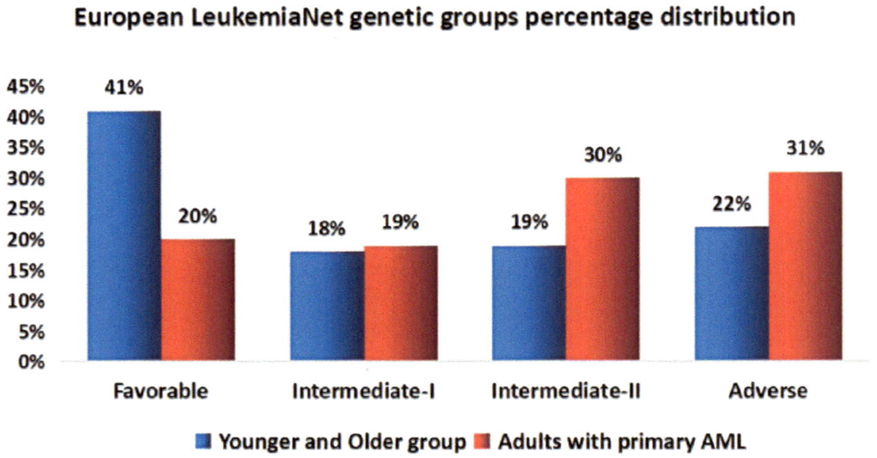

FIGURE 11.1 Graphical representation of European LeukemiaNet genetic groups percentage distribution of younger and older and adults with primary AML (Mrózek et al., 2012).

Source: Adapted from Mrózek et al. 2012

TABLE 11.4 Standard Classification System Proposed by ELN for AML Risk Group Stratification Based on Genetic Abnormalities

ELN Defined AML Risk Group	Cytogenetic Abnormalities	References
Favorable group	t (15, 17) (q22; q22) mutation; abnormally balanced mutation t (8;21) (q22; q22) inv (16) (p13.1q22); t (16;16) (p13.1q22); NPM1 mutated gene without FLT3-ITD mutation and CEBPA bi-allelic gene mutations CBFB-MHY11 associated with NPM1 mutation without FLT3-ITD represented as normal karyotype Normal karyotype mutation in CEBPA	Mrozek et al. (2012)
Intermediate 1 risk group	NPM1 gene with FLT3-ITD mutated gene; normal NPM1 gene with or without gene FLT3-ITD (normal karyotype)	Mrozek et al. (2012)
Intermediate 2 risk group	t(9;11), (p22; q23); MLLT3-KMT2A (chromosomal abnormality is not classified as favorable or adverse)	Mrozek et al. (2012)
Adverse risk group	aberrant karyotype inv (3) (q21; q26.2); t (3;3) (q21; q26.2); GATA-MECO (EVI1); DEK_NUP2w14; KMT2A rearranged genes del(5q), ab(17p) monosomal karyotype	Mrozek et al. (2012)

Group intermediate two involved t (9;11), (p22; q23), MLLT3-KMT2A. In this group, the chromosomal abnormality is not classified as favorable or adverse. Adverse risk group is categorized by subcomponents of aberrant karyotype that includes inv (3) (q21; q26.2), t (3;3) (q21; q26.2), GATA-MECO (EVI1), DEK_NUP2w14, KMT2A rearranged genes, del(5q), ab(17p) and along with monosomal karyotype, which are all related with improvised prognosis. The study also shows that older patient greater than 60 years of age is a self-sufficient factor for the prediction of AML-related outcomes in patients without consideration of ELN classification system (Dohner et al., 2010; Mrozek et al., 2012).

APL patients are classified based on their RR and initial WBCs count at the time of diagnosis. These APL patients are categorised into three levels:

1. **Low-Risk Group:** This includes WBCs count less than equal to 10×10^9/L and platelets count greater than 40×10^9/L.
2. **Intermediate Level Risk Group:** This includes WBC/platelets less than equal to 10×10^9/L or 40×10^9/L.
3. **High-Risk Group:** This includes WBCs count greater than 10×10^9/L therapy-related APL (t-APL) depend on risk stratification at the time of diagnosis and the compound named as arsenic trioxide (ATO) is used firstly in start of the treatment that shows the benefit to all risk associated group in APL (Sanz et al., 2000).

11.3 RECURRENT/RELAPSED DISEASE IN AML

Relapsed disease in AML patients is described by the reoccurrence of leukemic blast cells in bloodstream ≥ 5% present in BM or expansion of extramedullary disease. According to the International Working Group (IWG) criteria reappearance of the disease is associated with changes in new dysplastic cells or precancerous cells (Cheson et al., 2004). In some of the AML, blast cells ≥ 5% persist. These blast cells and these blast cells may show regenerative ability instead of neoplastic characteristics based on immunophenotyping technique.

However, in many AML patients who achieve CR one can detect with low-level disease by conventional cytogenetics, molecular genetic analysis, and fluorescence-based flow cytometry (Dohner et al., 2015).

Nearly 60% of AML patients achieve CR after induction therapy, but the majority of these patients will relapse (Dohner et al., 2015). AML cells become resistant to drugs used for induction therapy by undergoing specific molecular mechanism and transformed. The molecular basis for this induced resistance is poorly understood (Dohner et al., 2010, 2017).

As previously stated, that AML is referred to as heterogeneous disease, it is associated with a number of gene mutations which perform various functions, such as genes responsible for encoding myeloid transcription factor, anti-oncogenes, epigenetic modifier proteins, and gene encodes for spliceosomes.

Mutations of such normal functioning genes interferes with the normal hematopoiesis, more precisely, myelopoiesis. There are total of more than 60 genes that are frequently mutated and show high heterogeneity between the AML patients' cohort and within the AML patients (Cancer Genome Atlas Research, 2013; Metzeler et al., 2016; Papaemmanuil et al., 2016).

Molecular profiling (depending on different NGS and gene expression) in AML patients proposes that AML profile changes due to genetic and epigenetic alternation from the time of initial presentation to relapse (Dick et al., 2005; Short et al., 2018).

Whereas several studies have demonstrated individual patients acquire relapse-specific mutated gene, none of the gene mutation expressed to be frequently gained and some of the mutation rarely absent in the stage of the relapse. Gene mutation in DNMT3A, ASXL1, RUNX1 along with FLT3-ITD gene recurrently appear in AML relapse and showed the substandard outcomes, still those patients with these types of mutation effectively respond to induction therapy and attain CR (Cancer Genome Atlas Research, 2013; Metzeler et al., 2016; Papaemmanuil et al., 2016).

11.3.1 DISTINCT CLONAL EVOLUTION MODEL FOR AML

In the course of disease, an individual with AML may experience significant alternations in the genetic composition of the blasts that follow either a specific model of linear clonal evolution or a branching evolution. The linear evolution model for AML is defined as sequential single gene mutations with complete destruction of a dominant clone and development of secondary clone (sub-clone). In the absence of evolutionary pressure, AML cells undergo linear clonal evolution which enhances their survival.

In the branching clonal evolution model, some of the AML cells may drive by additional type of mutations that show resistance towards the treatment (linear clonal evolution). Whereas some of the population of leukemic cells may evolve through loss of mutation, which are sensitive to the therapy (branching clonal evolution). In linear clonal evolution, additional mutation occurs in the dominant clone at the time of diagnosis and at the stage of relapse (Figure 11.2). There are two types of cells, cells that reside in dominate clone and the non-essential cells have ability to gain new type of mutation that resist treatment and relapse disease. These new types of mutations are characterized by "driver mutations" responsible for clonal expansion and "passenger mutations" that priory exist in the cell and not affecting the leukemic cell division and expansion. New clone of cells evolves in a sequential manner in linear clonal expansion. This newly evolved clone shares common ancestry for specific genomic mutation background besides newly acquired mutations

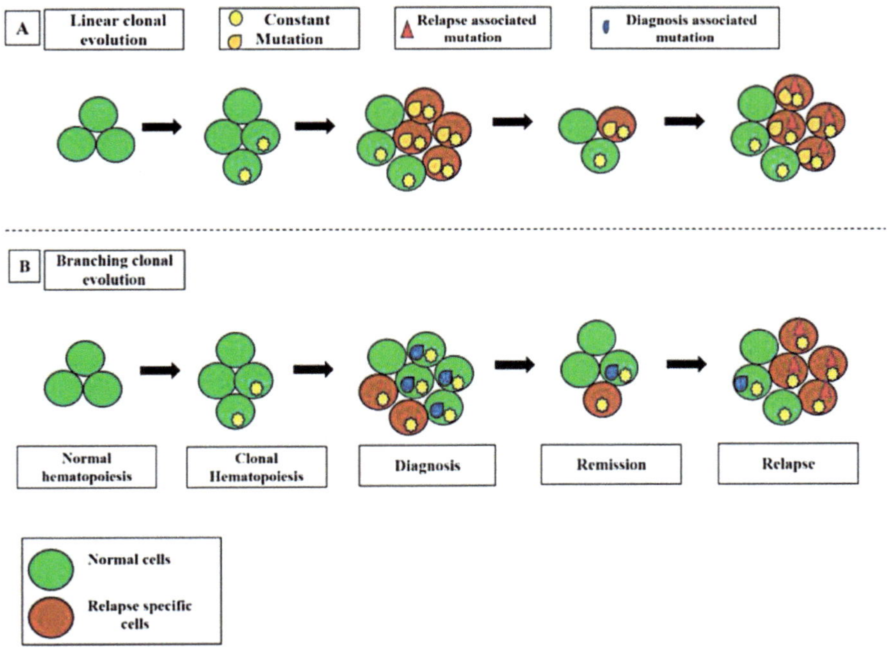

FIGURE 11.2 Clonal evolution model for AML diagnosis and AML relapse. Every single circle depicted as individual cell whereas bunch of identical cells represented as clone associated with similar type of mutation. Cells not carrying any symbol are termed as wild type meaning, not related to any other somatic aberration. Somatic molecular aberrations are depicted as colored symbols: light orange and orange (constant mutation), red color (relapse-associated mutation), and blue (diagnosis-associated mutation). Cells that carry red color symbols are relapse-specific cloned cells. (A) AML relapse related to linear clonal evolution delineates the sequentially acquired mutations, and relapse-associated clone develops originally from a primary clone of diagnosis. (B) Relapse related to linear clonal evolution delineates the origination of new clone called secondary clone and destruction of original clone, relapse associated with subclone developed during the diagnosis stage of AML.

along with driver and passenger mutation. Genetic mutation classifies based on clonal and sub-clonal mutation is obtained from variant allelic frequency (VAF) and gives the knowledge that how many cells mutated with specific variant. In case gene mutations are present at both point of time then relapse in the patient is associated with initial AML disease and evolve during the diseased period. In contrary, where the profiles of mutation at diagnosis and relapse are entirely distinct, then it shows that such relapse has not originated from the initial clone. It is then referred to as secondary AML. Corresponding mutation can be present in distinct and autonomous malignant tumor along with solid tumors, for example mutation in gene KRAS (hotspot present at G12, G13, and Q16) and BRAF (hotspot at V600). The clonal expression for the AML patients may exist at epigenetic extent and these epigenetic levels have different resistance to therapy (Li et al., 2016).

11.3.2 RELAPSED AML FOLLOWING INDUCTION THERAPY/ CHEMOTHERAPY

Many scientists have classified the genetic aberrations in AML by offering different and large number of AML genetic subclones which are showing definite clinical character like therapy-related response and relapse risk (Metzeler et al., 2016; Papaemmanuil et al., 2016; Welch et al., 2012). Distinction in mutational landscape and recurring mutational changes which a are associated with resistance against therapy can be recognize based on comparison between the diagnostic sample and relapse sample of AML. Possible mode of action of relapse in AML patients due to cytogenetic alternations, which were not present at diagnosis are found in approximately 30%-40% of patients (Garson et al., 1989; Kern et al., 2002; Kim et al., 2013; Testa et al., 1979). Patients who carry complex karyotype, have a lower rate of survival. Less than 10% of AML patients at relapse state carry reduced complex karyotype. Trisomy and deletion in chromosome 8, 12 and q arm of chromosome 9 at AML relapse, confer resistant against the treatment (Alpermann et al., 2015; Grimwade et al., 2016; Peniket et al., 2005). In contrary to, deletion in part of q arm of both the chromosomes 5 and 7 gained at relapse state and shows inferior outcomes (Papaemmanuil et al., 2016; Grimwade et al., 2010). With the advent of Next Generation Sequencing (NGS), researchers can identify single nucleotide polymorphisms at relapse. Identification as to whether the relapse occurred in a linear or non-linear evolution can be achieved through Whole Genome Sequencing (WGS) of leukemic cell samples at diagnosis and relapse (Table 11.5) (Ding et al., 2012).

TABLE 11.5 Study based on Classification of Clinically Relevant Sub-Type of AML

AML Subclass	Total Number of Cases	Number of Combined Samples	Technique of Sequencing	Clonal Evolution Model	Dx Related Mutations	Rel Related Mutation	References
AML associated with FLT3-ITD	N= 80	WES (n=13) (Dx+Rel+CR), GPS (n=37) (Dx+Rel+CR)	Initially with WES and Extented with GPS	Linear, branching evolution	NPM1, CEBPA, GATA2, SRCAP	SETD1A, ASXL1	Garg et al. (2015)
AML associated with inv (16) or t (8;21)	N= 13	(n=10) (Dx+Rel+CR), (n=3) (Dx+Rel)	WES	Linear, branching evolution	NRAS	Del GATA2	Sood et al. (2016)
AML associated with MLL-PTD	N= 85	WES (n=5) (Dx+Rel+CR), GPS (n=8) (Dx+Rel)	Initially with WES and extended with GPS	Not applicable	STAG2	FAT1	Sun et al. (2017)
CN-AML	N= 50	(n=50) (Dx+Rel+CR)	WES, verified by GPS	Linear, branching evolution	NPM1, FLT3-SNV, PTPN11, NRAS	FLT3-ITD, IDH1, WT1, KDM6A	Greif et al. (2018)
AML associated with NPM1	N= 104	(n=11) (Dx+Rel)	GPS	Not applicable	NPM1, PTPN11	RUNX1, TP53	Hollein et al. (2018)
AML associated with t (8;21)	N= 331	(n=19) (Dx+Rel+CR)	GPS	Not applicable	ASXL2	G2E3	Christen et al. (2019)
AML associated with RUNX1-RUNX1T1	N= 94	(n=17) (Dx+Rel)	GPS	Not applicable	ASXL1, ASXL2, NRAS	KIT	Hollein et al. (2019)
AML associated with NPM1	N= 129	WES (n=20) (Dx+Rel+CR), GPS (n=109) (Dx+Rel)	Initially with GPS, selected WES	Linear, branching evolution	NPM1, NRAS, FLT3-SNV	FLT3-ITD, MLL-PTD, RUNX1	Cocciardi et al. (2019)

The main focal point of these studies is a precise classification of clinically relevant sub-classes of AML. The number of genes analyzed ranged from 5 by targeted gene panel sequencing (GPS) to greater than 20,000 genes detected by the technique of Whole Exome Sequencing (WES). The combined sample at diagnosis and relapse (Dx+Rel) from AML patients were collected at the time of initial diagnosis and at relapse and are analyzed by GRS technique, it is approximately for 11 to 129 patients and by WES for 5 to 50 patients. Generally, WES technique is used to analyze a finite number of combined samples from AML patients, while the GPS technique is used to work with large cohorts for the determination of genetic mutational frequency. To identify the genetic mutational frequency at the individual level, it is acceptable to collect either diagnosis sample or relapse sample. Earlier, maximum studies worked on identification of mutational frequency instead of longitudinal characteristics of single genetic mutations. Research carried out by Garg et al. in 2015 based on WES method revealed of FLT3-ITD mutation is carried out by three combined samples which were collected at diagnosis, relapse, and CR (Garg et al., 2015).

11.3.3 CLASSIFICATION OF PATIENTS WITH HIGH RISK OF RELAPSE AFTER ALLO-HSC

11.3.3.1 AML-ASSOCIATED PARAMETER

Several AML-related factors utilized for the identification of pre-SCT setup associated higher risk of post-HSC relapse (Table 11.6) (Cornelissen et al., 2012; Middeke et al., 2012; van Gelder et al., 2013; Fang et al., 2011; Pfeiffer et al., 2013; Brunet et al., 2012). Consequently, the underlying classification system given by the ELN co-operative working group is also applicable for the Allo-HSCT context (Schmid et al., 2015). To provide long-term CR in AML patients, most of the clinicians adopt morphological CR (mCR) to allo-HSCT.

The advent of new techniques has improved the molecular diagnostic sensitivity and identification of minor concentration of leukemic cells. MRD detection in AML disease is focus point to use as a prognostic marker for risk stratification. Detection of MRD in patients foretells a high rate of relapse in context of HSCT.

Numerous researchers proposed that detection of MRD in pre-SCT is related with inferior results (Leung et al., 2012; Laane et al., 2006; Walter et al., 2011, 2015; Zhou et al., 2016). Recent research based on a retrospective cohort study of 359 adult AML patients treated with allo-HSCT showed three years of relapse-free survival (RFS) and overall survival (OS) in the

TABLE 11.6 AML-Relapse Associated Factor After Allo-HSCT in Patients

AML-Associated Parameter	Risk Factor	Extent of Relapse-Risk	References
Cytogenetic risk group	Favorable, intermediate 1, intermediate 2 and adverse risk group	Lower, intermediate, higher	Cornelissen et al. (2012); Middeke et al. (2012)
Cytogenetic aberrations and prognostic markers	NPM1 mutant, FLT3-WT	Lower	Pfeiffer et al. (2013); Brunet et al. (2012); Schmid et al. (2015)
	Monosomal karyotype	Higher	van Gelder et al. (2013); Fang et al. (2011)
	biallelic mutation	Lower	Bejar et al. (2014); Hamilton et al. (2015); Heuser et al. (2015)
	NPM1-WT	Intermediate	Tsirigotis et al. (2016)
	CEBPA-WT	Intermediate	Tsirigotis et al. (2016)
	Additional mutation	Intermediate	Tsirigotis et al. (2016)
CR condition	First complete remission (CR1)	Lower	Duval et al. (2010); Armistead et al. (2009); Craddock et al. (2011)
	secondar complete remission (CR2)	Higher	Tsirigotis et al. (2016)
MRD condition	MRD (+) at allo-HSCT	Higher	Leung et al. (2012); Laane et al. (2006); Walter et al. (2011, 2015); Zhou et al. (2016)

patient without MRD detected by MFC. AML patients with detected with positive MRD morphological complete remission (CRm) along with those associated to live leukemia cells, they possess proportionate relapse state of three years (either 65% or 67%) and OS (either 26% or 23%) (Araki et al., 2016). According to traditional classification system based on AML risk was due to AML heterogeneity in other related results. In a recent year, Bejar and colleagues (Bejar et al., 2014) identified gene mutations using massive parallel sequencing technique in 87 cases of MDS patients. From the same study, it was depicted that there were many recurrent mutations such as ASXL1, TP53, DNMT3A and RUNX1 generally occur. Genetic mutations associated with TP53, TET2 along with DNMT3A showing inferior OS after the treatment of allo-HSCT (Bejar et al., 2014), TET2 mutation in patient generally inferior OS for both disease (MDS and AML) (Hamilton et al., 2015). Another report describes the outcomes of 308 cases of MDS/

secondary AML patients gave the information of two mutations in 82% suffered patients (Heuser et al., 2015).

11.3.4 LATE RELAPSE ASSOCIATED WITH AML PATIENTS

Number of patients who are diagnosed recently with AML and after treatment of intensive induction therapy attain CR. Nearly 2/3rd of the patients relapse after first induction therapy, most of relapses occur in 18 months. Late relapses occur after five years of CR in 1%-3% of AML patients (Verma et al., 2010; Medeiros et al., 2007; Kantarjian et al., 1988). Majority of late relapses typically carrying normal karyotype (Medeiros et al., 2007). The question that arises in a late relapse, is whether it is true relapse or relapse due to previous treatment of intensive chemotherapy. Recent studies on WES using the mechanism for late relapse in AML patients is not understand properly till now. There is one study in a group of patients those who are in CR after intensive induction therapy and undergo relapse after five years of CR (n=15). By the utilization of WES technique, patient's BM samples were analyzed at diagnosis (n=10), CR (n=6) and initial relapse (n=10). Analysis showed there are 41 driver mutations out of which 11 mutations associated with primary AML, 17 mutations are AML relapse specific, and 13 mutations detected in samples that carry both primary mutations at diagnosis and relapse associated. It was reported that 12 mutations out of 13 were shared mutations including epigenetic modification and spliceosome genes (Yilmaz et al., 2019).

11.4 COMPLETE REMISSION (CR)

Complete remission (CR) in patients with AML is the essential endpoint for the assessment of induction treatment (De Greef et al., 2005). Many patients with AML will accomplish CR with contemporary treatments (Buccisano et al., 2017).

Around 2 out of 3 individuals with AML who gets the induction chemotherapy go into remission. This generally implies that the BM contains blast cells fewer than 5%, hence the blood cell count returns to the normal limits, and there is no symptom or sign of leukemia. The actual possibility of remission depends on an individual's particular prognostic factors, for example, their age and the presence of certain gene changes in the leukemia cells. For example, older people generally do not do as well as those younger than 60. They often have trouble tolerating intensive treatment and often have chromosome changes in leukemia cells that are linked to a poorer outlook. About half of these patients go into remission after initial treatment. In case, remission is achieved, patients usually get more chemotherapy to get rid of any leftover leukemia cells. Up to half of patients with consolidation goes into long-term remission which may be cured. However, this number is likewise influenced by prognostic factors. Utilizing an allogenic stem cell transplant as consolidation has a higher achievement rate, yet it also has a higher complication with the risk of death (Figure 11.3) (De Greef et al., 2005).

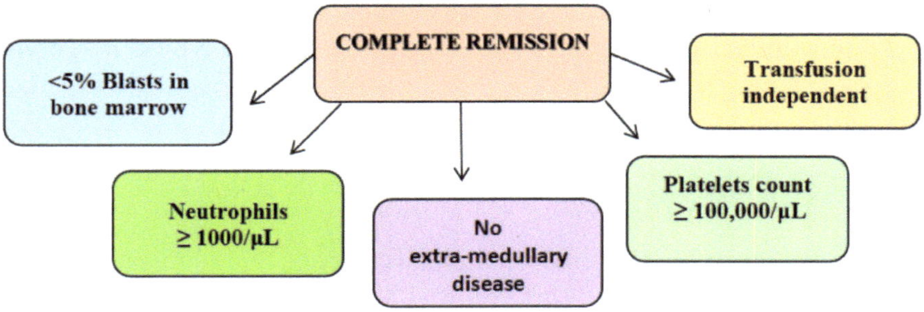

FIGURE 11.3 Criteria for achieving complete remission in AML.

11.4.1 MORPHOLOGIC COMPLETE REMISSION (CR)

A CR assignment needs that the patient accomplishes the morphologic leukemia-free state and have an outright neutrophil check of more than 1,000/L and platelets of ≥100,000/L. Hemoglobin concentration or hematocrit does not matter to the remission status, even though the patient should be autonomous of transfusion. Recent arrangement recommends that circulating blast might be recognized in small numbers and still be similar with a comparative disease-free survival (DFS) as patients without circulating blasts. Generally, nonetheless, persistent blasts in the peripheral blood correspond with repetitive AML and blast infiltration of the BM. Sometimes, an uncommon peripheral blood blast might be recognized during recovery; nonetheless, if the patient were in CR, the BM would have under 5% impact and no Auer rods. Flow cytometry may likewise be valuable to recognize leukemia and a recovering BM. There is no necessity for BM cellularity. The ramifications of dysplasia after treatment are not satisfactory. In any case, if the dysplasia is found, diligence proposes residual leukemia or a myelodysplasia in blast crisis (Cheson et al., 2003). Hence, a classification of patients who satisfy the morphologic criteria for CR could be considered (Cheson et al., 2003).

11.4.1.1 CYTOGENETIC COMPLETE REMISSION (CRC)

Most of patients who accomplish a CR are not cured. The utilization of a morphologically decided cut-off of 5% blast is arbitrary. Based on the ongoing cytogenetic information, a different class of CR is proposed to incorporate attaining an ordinary karyotype at CR. Current information states that patients with residual cytogenetic anomalies have a poorer prognosis than those in whom the chromosomal abnormality is not detectable. As such, it is critical to depict expressly how CRc is characterized; for example, the minimum number of metaphases needed to characterize a typical karyotype (Cheson et al., 2003).

11.4.1.2 MOLECULAR COMPLETE REMISSION (CRM)

Clinical research of MRD utilizing both molecular and multidimensional flow cytometric procedures have clearly shown that AML patients in morphologic and cytogenetic CR have detectable residual disease. Therefore, automated quantitative reverse transcriptase polymerase chain reaction (RT-PCR) techniques are quite sensitive in detecting residual disease in AML cases with a specific genetic markers such as – PML-RARα in t(15;17), AML1/ETO fusion in t(8;21), CBFβ-MYH11 fusion in inv(16). Some other molecular targets, such as WT-1, may also be assessed via quantitative RT-PCR approaches and might be an important target for MRD testing. Using these quantitative MRD approaches in patients during the clinical research studies, quantitative thresholds in the range of one leukemia cell in 1,000 to 10,000 cells may predict impending relapse compared with mCR. If CRm is used as a measure in AML clinical trials, it should then be defined precisely, including the markers examined and the sensitivity of the quantitative PCR assay (Cheson et al., 2003).

11.4.1.3 MORPHOLOGIC COMPLETE REMISSION (CR) WITH INCOMPLETE BLOOD COUNT RECOVERY (CRI)

After chemotherapy, a few patients satisfy the measures for CR despite residual neutropenia (1,000/L) or thrombocytopenia (100,000/L). For the three CR classes, extramedullary leukemia, for example, CNS or soft tissue involvement, should be absent. Without symptoms, CSF examination is not suggested except as a part of a clinical study (Cheson et al., 2003).

11.5 MINIMAL RESIDUAL DISEASE (MRD)

Minimal residual disease (MRD) is defined as the number of leukemic cells that remain in the body after the treatment. The number of remaining cells could be so small that they do not even cause any physical signs or symptoms and cannot be even detected via traditional methods, such as analyzing cells under a microscope or other quantitative methods, for example, RT-PCR and flow cytometry, etc. But because of the impact of MRD on the progression of the disease, the ability to detect residual leukemic cells must be at a high sensitivity. It is seen that, if the sensitivity increases to a maximum, then there is a higher chance of achieving a longer relapse free survival (RFS) and potentially achieve a cure. MRD is a time-point dependent variable. In AML, the prognostic significance of each time point is well reported (e.g., end of induction, end of consolidation, before transplant). A more accurate terminology proposed is "measurable" instead of "minimal" based on the sensitivity to its approach and thus MRD will reflect measurable residual disease (Figure 11.4) (Gomez-Arteaga et al., 2018).

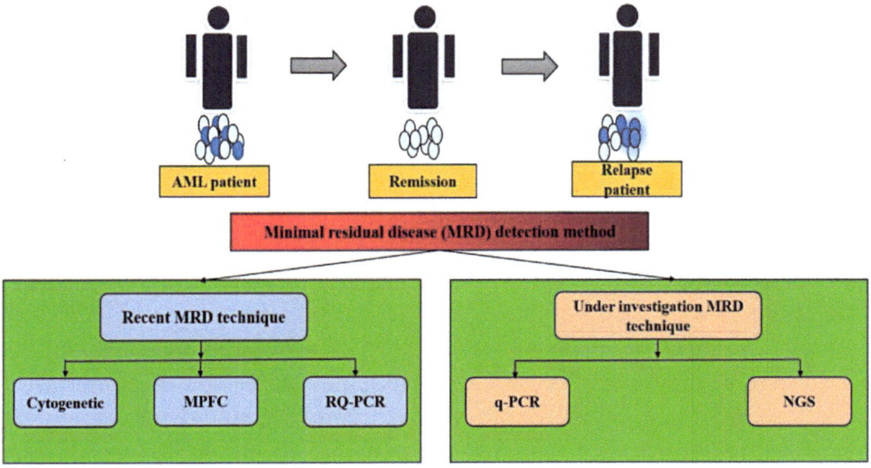

FIGURE 11.4 Minimal residual disease (MRD) detection method: recent MRD techniques include cytogenetic, MPFC and RQ-PCR, under investigation MRD technique: Q-PCR and NGS.

11.5.1 DETECTION METHODS FOR MRD

The approach to MRD should be highly specific and sensitive. There are two main clinical approaches to use MRD: (i) the first one is to use it as a prognostic marker; and (ii) the second one as a surrogate endpoint to define response to treatment and to define treatment allocation (Gomez-Arteaga et al., 2018). Successful treatment of AML depends on the achievement of mCR, defined as less than 5% leukemic cells. However, it is becoming clear that the morphology assessment does not pick up a minority of myeloid blasts that could survive after induction and consolidation chemotherapy. There are different methods for the detection of MRD, such as next-generation sequencing (NGS), polymerase chain reaction (PCR), and multiparameter flow cytometry (MPFC), all with varying levels of sensitivity (Principe et al., 2016). These tests require samples of BM cells (via aspiration) or peripheral blood cells through a venepuncture (Table 11.7) (Buccisano et al., 2017).

11.5.1.1 MULTIPARAMETER FLOW CYTOMETRY (MPFC)

MRD monitoring by MPFC is based on antigen expression patterns that specify the diverse lineages of normal hematopoietic cells (Buccisano et al., 2019; Cruz et al., 2017). AML blasts have an irregular antigen expression patterns that are not detectable on the surface of BM cells from healthy donors (Cruz et al., 2017). Therefore, MPFC uses multiple fluorescent labeled monoclonal antibodies to identify the markers expressed on cell surfaces, which further helps define leukemia phenotype as part of the standard diagnostic workup (Tomlinson et al., 2017). The different immuno-phenotypic patterns are called leukemia-associated immunophenotype (LAIP) and are usually present in about 80% to 100% of AML patients (Cruz et al., 2017). This detection technique has the ability to monitor small numbers of cell populations, and it also has been used to detect and monitor MRD. This Flow cytometry needs proper identification of immunophenotypes unique to the AML at diagnosis. These can include co-expression of markers, for example, CD34 and CD15, expression of lymphoid markers, overexpression of normal myeloid or stem cell antigens and absence of normal lineage markers (CD33) (Table 11.8) (Al-Mawali et al., 2009; Tomlinson et al., 2017).

AML blast cells have shown different antigenic expression on the outer surface of BM cells that are not identified in non-diseased persons (Kern et al., 2012). The different types of antigenic expression is known as leukemic cells associated immunophenotype (LAIP) (Vidriales et al., 2003). There

TABLE 11.7 Various Techniques Utilized for the Detection of MRD

Techniques Used for Detection of MRD	Sensitivity-Associated with Techniques	Advantages	Disadvantages
Multiparameter Flow Cytometry (MPFC)	Approximately 10^{-4} to 10^{-5}	Applicable for broad range, i.e., 90% patients. Comparatively give quick outcomes less than equal to one day. Showing immense precision using delineate LAIP. Enable to analyze phenotypic expression of cells associated with LSC. Enable to differentiate between alive and diseased cells. Easy to store the data related to cell characterization.	To some extent, this technique is showing challenges and instinctive clarification. MFC sensitive to utilization of immunoglobulin group. Shows the limitation related to compliance and standardization in all working laboratory. Leukemia cells related phenotype generally not stable over time, e.g., LAIP may be unable to detect sub-clone related to relapse.
Next-generation DNA sequencing (NSG)-based MRD assay	Approximately 10^{-3} to 10^{-5}	Ease to operate. Shows sensitivity (10^{-3} to 10^{-5}). Relevant according to define sub-classes.	Limitations related to standardization. Occurrence of any error which results in lower sensitivity of mutational sequence. Existence of few of the cytogenetic abnormalities in AML patients. Contaminated associated risk.
Reverse transcription quantitative-polymerase chain reaction (RT-qPCR)-based MRD assay	Approximately 10^{-3} to 10^{-5}	Applicable for broad range. This technique can be carried out by any certified laboratory having the ability to perform it sufficiently. Shows immense sensitivity, i.e., greater than MFC or its equal. Quality assurance is regularly done. Well standardized.	Turnaround time = days. Costly and time consuming. More expertise and limited threshold value are required.

TABLE 11.8 Different Types of LAIP (Leukemic Cells-Associated Immunophenotype) – Relevant Frequency of Occurrence and Associated Markers

Phenotype Associated with LAIP	Relevant Frequency of Occurrence	Associated Markers
Overexpression of antigen	20% to 30%	CD4, CD33, CD34, CD45, CD123, CD64
Absence of antigenic expression	20% to 30%	HLA-DR
Cross-lineage expression of antigen	30% to 40%	T lymphocytes associated markers CD2, CD3, CD5, CD7
		B lymphocytes associated marker CD19
		Natural killer cells marker CD56
Asynchronous antigen expression	60% to 70%	Immature myeloid-derived suppressor cells CD34, CD117
		Mature myelogenous cell markers CD11c, CD13, CD14, CD15, CD33, CD38 CD65

are different types of LAIP classified into three major categories based on relevant frequency of occurrence and associated markers:

i. **Overexpression of Antigen:** Expression of normal antigen expressed at high level on each leukemic cell (CD4, CD33, CD34, CD45, CD123, CD64).

ii. **Absence of Antigenic Expression:** Antigen expression is absent (HLA-DR).

iii. **Cross-Lineage Expression of Antigen:** T-lymphocytes associated markers CD2, CD3, CD5, CD7, B lymphocytes associated marker CD19 and Natural killer cells marker CD56 (Ommen et al., 2016; Kern et al., 2012; Principe et al., 2017). Expression of LAIP in AML patients shows heterogenous pattern that means distinct LAIP detected in individual patients, therefore single LAIP are observe in less than 5% of all patients. After the use of immunoglobulin associated panel that define one or more different type of LAIP in all AML patients (Kern et al., 2012).

11.5.1.2 REVERSE TRANSCRIPTASE-QUANTITATIVE PCR (RT-QPCR)

Real-time quantitative reverse transcription polymerase chain reaction (RT-QPCR) is the most used molecular technique in AML-MRD testing. It is a quantitative test of mRNA transcripts and are specifically expressed in AML (Zhou et al., 2017). RT-QPCR is profoundly reproducible between research areas. The turnaround time is short. The risk of contamination is low. Moreover, the ability to give a quantitative assessment of the residual transcripts and the standardization of gene identification to evade negative outcomes makes this technique the most favored for quantitative monitoring of MRD. The basic target of RT-QPCR are the fusion genes produced by chromosomal rearrangements (e.g., PML-RARA t[15;17], RUNX1-RUNX1T1 t[8;21], CBFB-MYH11 [inv{16} t{16;16}, t{11q23}], DEK-CAN[NUP214] t[6;9]), insertions or duplications (e.g., NPM1, FLT3-ITD, MLL-PTD), point mutations (e.g., CEBPA, IDH1/2, KIT, RAS, and so on) Among these objectives, CBF-AML and also NPM1 mutant AML have got a broad clinical approval (Grimwade et al., 2014; Ivey et al., 2016).

This test generally consists of a two-step reaction. The total RNA is extracted. It is initially converted into cDNA via reverse transcription; at that point, the overall amount of cDNA is estimated. The RT-QPCR reaction by utilizes oligonucleotides that only hybridize to the sequence of interest. Once measurement is done, the copy number of tested cDNA is determined by comparing with a standard curve. ABL1 is the most utilized gene, having the steadiest expression in normal and leukemic cells (Beillard et al., 2003). The result of RT-QPCR is generally communicated as the percentage of copy number of the tested gene transcript to the copy number of ABL1. In AML MRD detection method, the combination of PCR amplification and overexpression of tested transcripts in leukemia makes RT-QPCR systematically the most sensitive testing procedure. This method has been generally applied in the detection of AML with repetitive chromosomal translocations like NPM1 mutations, and overexpression of WT1 (Zhou et al., 2017).

11.5.1.3 NEXT GENERATION SEQUENCING (NGS)

As with the new fact of comprehensive genetic landscapes for AML, next generation sequencing (NGS) has been investigated in monitoring the response in AML after treatment (Metzeler et al., 2016; Papaemmanuil et al., 2016). The

NGS-based MRD detection is like PCR-based MRD detection method, except the genetic anomalies which are distinguished directly by DNA re-sequencing, with high specificity. In NGS MRD test, the DNA fragments of interest are captured and are further amplified by PCR, the PCR products sequenced in a parallel manner, and then are reassembled to get compared with reference sequences. Unlike to quantitative PCR, NGS does not need oligonucleotides that hybridize particularly to a specific sequence. The benefit of NGS in AML MRD testing is best shown in the identification of the mutant NPM1 exon 12 (Thol et al., 2012; Salipante et al., 2014). The test has a direct powerful reach and lower limit of identification compared to QPCR and can recognize all sub-types of NPM1 mutation without utilizing allele-specific oligonucleotide tests. Nonetheless, identification of mutated NPM1 likely represents a most ideal situation in NGS-based AML MRD testing given that all subtypes of NPM1 changes comprise of $\geq$ 4 base-pair insertion in exon 12. Hence, irregularities possibility generated by sequencing error is uncommon, and mutation in NPM1 is a leukemia-driver (Zhou et al., 2017).

Since AML patients may relapse with a different mutant clone, and in such case, the marker being monitored was not informative to predict relapse. With this technique, there is no need for patient-specific assays as practically all mutations are detected (Thol et al., 2012). Overall, NGS-based methods can detect sub-clinical disease in AML samples that do not qualify for the established leukemia-specific RQ-PCR assays. Likewise, NGS gives the ability to identify treatment emergent leukemia. In addition, NGS gives exact data about leukemic tumor burden, a significant boundary to measure treatment response (Grimwade et al., 2014).

11.6 PROGNOSTIC MARKERS

Over the last two decades, it has been shown that there were not any advances in the treatment of AML due to its heterogeneity (Byrd et al., 2002; Grimwade et al., 2010). Researchers identified a number of essential prognostic markers in studies such as RUNX1-RUNX1T1, WT1, NPM1, FLT3-ITD/TKD, and IDH1/IDH2 included all such molecular abnormalities. These markers play a significant role in the assessment of MRD in AML patients and help enroll into specific clinical trials for AML patients (Vardiman et al., 2009). Prognostic marker is generally dependent upon therapy related to AML (t-AML). Adult AML patients, with poor prognostic markers may after post-remission treatment with high dose of chemotherapy drug such as cytarabine, benefit from allo-HSCT. Some prognostic markers that predict poor outcome include high EVI1 gene expression along with ITD mutational alternation in the FLT3 gene. Such features predict poor outcome even with allo-HSCT. Some micromolecules and monoclonal antibody (mAb) that target the receptor CXCR4 (chemokines) could be tried, as high expression of this is predictive of poor prognosis (Koreth et al., 2009).

Recent prognostic marker panel guidance proposed by the ELN co-operative group classifies them as favorable markers, such as, core markers like CD33, CD34, CD45, CD117 along with HLA-DR, mandatory marker like CD13, CD14, CD15, CD11b, CD38 (Schuurhuis et al., 2018), CD123 (Angelini et al., 2015, Roug et al., 2018), CD371 (Roug et al., 2018, Larsen et al., 2012), CD45RA (Kersten et al., 2016, Zeijlemaker et al., 2016) along with CD99 (Angelini et al., 2015, Strehl et al., 2013) and optional marker like CD56. A marker such as CD11a is generally present on normal leukocyte cells and CD34+ expressed precursor cells in BM (Dworzak et al., 1998, Kansas et al., 1990). Boztug and colleagues described the deficiency related to marker such as CD11a as being associated with AML-M7, Down syndrome-related AML and myeloproliferative neoplasm (MPN), that make it optimal parameter for MRD assessment in all such AML sub-classes (Table 11.6).

Advances in technologies involving GEP (gene expression profiling) or mass spectrometry that identify cell surface and its related novel CD markers for detection of MRD by flow cytometry (Coustan-Smith et al., 2018; Mirkowska et al., 2013). Coustan-Smith and his colleagues were reported that distinct marker pattern in all AML patients and immature blood cells by utilizing genome-wide analysis of gene expression. It was also shown that 22 prognostic markers can differentiate the normal cells and leukemic cells and expression associated with these markers showed stability during chemotherapy and at relapse (Table 11.9) (Coustan-Smith et al., 2018).

Using NGS, FLT3-ITD (Bibault et al., 2015) along with IDH1/IDH2 (Debarri et al., 2015) CEBPA and NPM1 can also be identified (Dohner et al., 2017). The WT1 (wilm's tumor gene) marker is not used in monitoring of MRD in AML patients as it is less sensitive and specific (Schuurhuis et al., 2018). Recommendations by ELN co-operative working party do not support the use of Fms-like tyrosine kinase internal tandem duplication (FLT3-ITD), FLT3-TKD, NRAS, KRAS, IDH1, IDH2, MLL-PTD and EVI1 expression as individual marker for assessing MRD. These mutations are more frequently loss or gain but they show prognostic significance in MRD detection by RT-QPCR (Schuurhuis et al., 2018).

TABLE 11.9 Technique Used for Assessment of MRD in AML Patients with Utilization of Prognostic Markers

MRD Assessment Techniques	Key Prognostic Markers
Molecular flow cytometry (MFC)	Core markers CD33, CD34, CD45, CD117 and HLA-DR
	mandatory marker like CD13, CD14, CD15, CD11b, CD38 (Schuurhuis et al., 2018), CD123 (Angelini et al., 2015; Roug et al., 2018)
	CD371 (Roug et al., 2018; Larsen et al., 2012)
	CD45RA (Kersten et al., 2016; Zeijlemaker et al., 2016) along with CD99 (Angelini et al., 2015; Strehl et al., 2013)
	Optional marker named as CD56.
Next generation DNA sequencing (NGS)	FLT3-ITD (Bibault et al., 2015) along with IDH1/IDH2 (Debarri et al., 2015)
	CEBPA
	NPM1 (Dohner et al., 2017).
Reverse transcription quantitative-polymerase chain reaction (RT-qPCR)	PML-RARA1
	CBFB-MYH11
	RUNX1-RUNX1T1
	WT1
	FLT3-ITD
	FLT3-TKD
	NRAS
	KRAS
	IDH1/IDH2
	MLL-PTD and EVI1 (Schuurhuis et al., 2018)

11.7 CHALLENGES ASSOCIATED WITH MRD DETECTION IN AML

11.7.1 EXPANSION OF OPTIMAL PARAMETER IN MRD DETECTION

In the present day, optimization in molecular assessment of MRD in AML should be accomplished through the consideration of numerous techniques and rational aspects like patient's age, regimen target, selection of favorable source of sample either (BM or PB), preferential prognostic markers, monitoring associated sensitivity defined by mRNA expression level of leukemic transcriptional factor regulated by the gene along with AML relapse disease associated kinetic process. Ivey and colleagues reported detection of the MRD using NPM1 mutant transcript at the time of quantification of PB sample and reported the prognostic results. Therefore, BM sampling provide important substrate for analyzing an PM in AML patients (Grimwade et al., 2014).

11.7.2 IMMUNOPHENOTYPIC PATTERN (LAIP) ACCURACY

In the monitoring of MRD in AML patients, the phenotypic aberration in the LAIP may also be present on some healthy cells that may negate its accuracy. Rossi et al. reported using a 6-color assay that indicated the CD15+ and CD17+ markers cells could be analyzed in BM sample of patients (Rossi et al., 2013). In the GIMEMA (AML treatment according to Gruppo Italiano Malattie Ematologiche dell'Adulto) clinical trial that analyzed LAIP in 500 adult patients with de novo AML, in the accuracy was 91% using an eight-color assaying technique (unpublished data).

11.7.3 MAJOR SHIFTS IN IMMUNOPHENOTYPE

The alternation of antigen expression, such as deletion and incrementation on leukemic cells of paired sample in AML patients at relapse, demonstrates the selectivity of LAIP changes. Most of the antigen expression, such as CD11b, CD14, and CD15 as markers are not found, although frequency of CD34 and CD117 markers is higher (Langebrake et al., 2005; Baer et al., 2001; Voskova et al., 2004; Van der Velden et al., 2010). Furthermore, these type of immunophenotypic alternations/shift possibly relate to cytogenetic shifts. 7 AML patients with altered mutational pattern at relapse as compared to the time of diagnosis, could be identified by flow cytometry. In effect, some of the mutations were identified in lower frequency at the time of relapse in earliest populations of CD34 positive and CD38 negative leukemic cells (Bachas et al., 2012). In agreement with this, Angelini et al. (2015) compared mutations such as FLT3 and NPM1, using 9 color MPFC. Some fraction of CD123, CD99, CD25, expression of CD34 positive type of phenotype greater than equal to 11.7%, with sensitivity greater than 90% containing CD34, CD123, CD99, and CD25 clones and identified in 3 AML patients who were having other mutations such as normal FLT3 and NPM1 mutation.

11.7.4 SELECTION OF SAMPLE SOURCE (EITHER PB OR BM) FOR MRD ASSESSMENT BY MPFC

PB is the most suitable source as compared to sample collected from BM because BM sampling is quite painful for the patients. Moreover, peripheral blood related MRD shows high specificity as normal progenitor cells are typically absent in peripheral blood. After Zeijmaker and his colleagues identified the important relation between peripheral blood and BM, PB-based MRD assessment is more precise than BM (Zeijlemaker et al., 2016).

11.8 CONCLUSION

Patients with AML who achieve a CR with the presence of MRD after treatment are particularly a challenging subset of patients. MRD is an important prognostic factor in AML. Even though advances in prognostic risk stratifications have streamlined post-induction therapy for AML, overall long-term survival remains poor. As per the new ELN recommendation establishing complete remission without measurable residual disease (CRMRD) as a separate category of treatment response assessments represents an achievement in incorporating MRD into clinical practice. The detection of MRD immunophenotyping by MPFC is a good option for monitoring MRD in most AML patients. Overall, there is an urgent need to employ data mining and precision-based therapies using machine learning integration (Iddamalgoba et al., 2016) for rapid development of companion diagnostics in Leukemia research (Joergensen et al., 2019). Nevertheless, it is still a matter of concern what could be the best therapy to use for MRD in AML.

With the availability of newer agents for treating myeloma, it has become possible to achieve more profound responses leading to better long-term survival. With an increasing number of patients achieving responses deeper than CR, more sensitive techniques must assess the depth of response and potentially tailor therapy. Measurable residual disease (MRD) is a valuable tool to help with the same and allows a sensitivity of up to 1 in 10^6 cells, depending on the technique (Romano et al., 2019). Current methods of choice for sensitive MRD detection include multiparameter next-generation flow cytometry and next-generation sequencing. Guidelines for next-generation flow cytometry assessment in myeloma are provided by the Euro-flow consortium and are widely applicable in laboratories using eight color instruments (Roshal et al., 2017). The advantages of NGF include broad applicability to more than 99% of patients, with a rapid turnaround and no need for a pre-treatment sample. NGS is applicable in about 90% of patients and provides a higher sensitivity at the cost of a longer turnaround time (Bai et al., 2018). Several studies have reported an association between the depth of MRD and clinical outcomes, and it has emerged as one of the most vital prognostic markers in myeloma (Kostopoulos et al., 2020). Even among patients with CR, MRD provides prognostic differentiation (Kunacheewa et al., 2020). It is recommended that MRD be performed at the time of disease reassessment, preferably from a BM sample. Further evaluation of MRD with molecularly targeted therapies in myeloma will allow us to achieve better disease control and identify patients for escalation or de-escalation of treatment.

ACKNOWLEDGMENTS

Supported by grants from the Indian Council of Medical Research (ICMR), Government of India (ICMR ltr_12th batch (3/1/3/PDF (12)/2015-HRD_ICMR centenary Post-Doctoral Fellowship) and University on Grants Commission – Major Research Project, New Delhi.

KEYWORDS

- **acute myeloid leukemia**
- **complete remission**
- **digital-PCR**
- **FMs-like tyrosine kinase 3**
- **measurable residual disease**
- **multiparameter flow cytometry**
- **next-generation sequencing**
- **nucleophosmin 1**
- **quantitative real-time polymerase reaction**
- **relapse**
- **therapy-related AML**

REFERENCES

Ahirwar, R., Nigam, R. K., & Parmar, D., (2018). A study of leukemias profile in central India. *Tropical Journal of Pathology and Microbiology, 4*(2), 183–185. doi: https://doi.org/10.17511/jopm.2018.i02.12.

Al-Mawali, A., Gillis, D., & Lewis, I., (2009). The role of multiparameter flow cytometry for detection of minimal residual disease in acute myeloid leukemia. *American Journal of Clinical Pathology, 131*(1), 16–20. doi: https://doi.org/10.1309/AJCP5TSD3DZXFLCX.

Alpermann, T., Haferlach, C., Eder, C., et al., (2015). AML with gain of chromosome 8 as the sole chromosomal abnormality (+8sole) is associated with a specific molecular mutation pattern including ASXL1 mutations in 46.8% of the patients. *Leukemia Research, 39*(3), 265–272. doi: https://doi.org/10.1016/j.leukres.2014.11.026.

Angelini, D. F., Ottone, T., Guerrera, G., et al., (2015). A leukemia-associated CD34/CD123/CD25/CD99+ immunophenotype identifies FLT3–mutated clones in acute myeloid leukemia. *Clinical Cancer Research, 21*(17), 3977–3985. doi: https://doi.org/10.1158/1078-0432.CCR-14-3186.

Araki, D., Wood, B. L., Othus, M., et al., (2015). Allogeneic hematopoietic cell transplantation for acute myeloid leukemia: Is it time to move toward a minimal residual disease-based definition of complete remission? *Blood, 126*(23), 2571. doi: https://doi.org/10.1200/JCO.2015.63.3826.

Armistead, P. M., De Lima, M., Pierce, S., et al., (2009). Quantifying the survival benefit for allogeneic hematopoietic stem cell transplantation in relapsed acute myelogenous leukemia. *Biology of Blood Marrow Transplantion, 15*(11), 1431–1438. doi: https://doi.org/10.1016/j.bbmt.2009.07.008.

Bachas, C., Schuurhuis, G. J., Assaraf, Y. G., et al., (2012). The role of minor subpopulations within the leukemic blast compartment of AML patients at initial diagnosis in the development of relapse. *Leukemia, 26*, 1313–1320. doi: https://doi.org/10.1038/leu.2011.383.

Baer, M. R., Stewart, C. C., Dodge, R. K., et al., (2001). High frequency of immunophenotype changes in acute myeloid leukemia at relapse: Implications for residual disease detection (Cancer and Leukemia Group B Study, 8361). *Blood, 97*, 3574–3580. doi: https://doi.org/10.1182/blood.v97.11.3574.

Bai, Y., Orfao, A., & Chim, C. S., (2018). Molecular detection of minimal residual disease in multiple myeloma. *British Journal of Haematology, 181*(1), 11–26. doi: https://doi.org/10.1111/bjh.15075.

Beillard, E., Pallisgaard, N., Van, D. V. V. H. J., et al., (2003). Evaluation of candidate control genes for diagnosis and residual disease detection in leukemic patients using "real-time" quantitative reverse-transcriptase polymerase chain reaction (RQ-PCR) – a Europe against cancer program. *Leukemia, 17*(12), 2474–2486. doi: https://doi.org/10.1038/sj.leu.2403136.

Bejar, R., Stevenson, K. E., Caughey, B., et al., (2014). Somatic mutations predict poor outcome in patients with myelodysplastic syndrome after hematopoietic stem-cell transplantation. *Journal of Clinical Oncology, 32*(25), 2691–2698. doi: https://doi.org/10.1200/JCO.2013.52.3381.

Bennett, J. M., Catovsky, D., Daniel, M. T., et al., (1976). Proposals for the classification of the acute leukaemias French-American-British (FAB) co-operative group. *British Journal of Haematology, 33*(4), 451–454. doi: https://doi.org/10.1111/j.1365-2141.1976.tb03563.x.

Bibault, J. E., Figeac, M., Hélevaut, N., et al., (2015). Next-generation sequencing of FLT3 internal tandem duplications for minimal residual disease monitoring in acute myeloid leukemia. *Oncotarget, 6*(26), 22812–22821. doi: https://doi.org/10.18632/oncotarget.4333.

Boissel, N., Leroy, H., Brethon, B., et al., (2006). Incidence and prognostic impact of c-Kit, FLT3, and RAS gene mutations in core binding factor acute myeloid leukemia (CBF-AML). *Leukemia, 20*(6), 965–970. doi: https://doi.org/10.1038/sj.leu.2404188.

Boissel, N., Nibourel, O., Renneville, A., et al., (2010). Prognostic impact of isocitrate dehydrogenase enzyme isoforms 1 and 2 mutations in acute myeloid leukemia: A study by the acute leukemia French association group. *Journal of Clinical Oncology, 28*(23), 3717–3723. doi: https://doi.org/10.1200/JCO.2010.28.2285.

Boztug, H., Schumich, A., Pötschger, U., et al., (2013). Blast cell deficiency of CD11a as a marker of acute megakaryoblastic leukemia and transient myeloproliferative disease in children with and without down syndrome. *Cytometry Part B: Clinical Cytometry, 84*(6), 370–378. doi: https://doi.org/10.1002/cyto.b.21082.

Brunet, S., Labopin, M., Esteve, J., et al., (2012). Impact of FLT3 internal tandem duplication on the outcome of related and unrelated hematopoietic transplantation for adult acute myeloid leukemia in first remission: A retrospective analysis. Journal of *Clinical Oncology, 30*(7), 735–741. doi: https://doi.org/10.1200/JCO.2011.36.9868.

Buccisano, F., Maurillo, L., Schuurhuis, G. J., et al., (2019). The emerging role of measurable residual disease detection in AML in morphologic remission. *Seminars in Hematology, 56*(2), 125–130.

Buccisano, F., Walter, R. B., et al., (2017). Should patients with acute myeloid leukemia and measurable residual disease be transplanted in first complete remission? *Current Opinion in Hematology, 24*(2), 132–133. doi: https://doi.org/.

Buckley, S. A., Wood, B. L., Othus, M., et al., (2017). Minimal residual disease prior to allogeneic hematopoietic cell transplantation in acute myeloid leukemia: A meta-analysis. *Haematologica, 102*(5), 865. doi: https://doi.org/.

Bullinger, L., Döhner, K., Bair, E., et al., (2004). Use of gene-expression profiling to identify prognostic subclasses in adult acute myeloid leukemia. *New England Journal of Medicine, 350*(16), 1605–1616. doi: https://doi.org/.

Byrd, J. C., Mro´zek, K., Dodge, R. K., et al., (2002). Pretreatment cytogenetic abnormalities are predictive of induction success, cumulative incidence of relapse, and overall survival in adult patients with de novo acute myeloid leukemia: Results from cancer and leukemia group B (CALGB 8461). *Blood, 100*(13), 4325–4336. doi: https://doi.org/.

Campos, L., Guyotat, D., Archimbaud, E., et al., (1989). Surface marker expression in adult acute myeloid leukaemia: Correlations with initial characteristics, morphology and response to therapy. *British Journal of Haematology, 72*(2), 161–166. doi: https://doi.org/10.1111/j.1365-2141.1992.tb08168.x.

Cheng, K., Sportoletti, P., Ito, K., et al., (2010). The cytoplasmic NPM mutant induces myeloproliferation in a transgenic mouse model. *Blood, 115*(16), 3341–3345. doi: https://doi.org/10.1182/blood-2009-03-208587.

Cheson, B. D., Bennett, J. M., Kopecky, K. J., et al., (2003). Revised recommendations of the international working group for diagnosis, standardization of response criteria, treatment outcomes, and reporting standards for therapeutic trials in acute myeloid leukemia. *Journal of Clinical Oncology, 21*(24), 4642–4629. doi: https://doi.org/10.1200/JCO.2003.04.036.

Chou, W. C., Chou, S. C., Liu, C. Y., et al., (2011). TET2 mutation is an unfavorable prognostic factor in acute myeloid leukemia patients with intermediate-risk cytogenetics. *Blood, The*

Journal of the American Society of Hematology, 118(14), 3803–3810. doi: https://doi.org/10.1182/blood-2011-02-339747.

Christen, F., Hoyer, K., Yoshida, K., et al., (2019). Genomic landscape and clonal evolution of acute myeloid leukemia with t(8;21): An international study on 331 patients. *Blood, The Journal of the American Society of Hematology, 133*(10), 1140–1151. doi: https://doi.org/10.1182/blood-2018-05-852822.

Cornelissen, J. J., Breems, D., Van, P. W. L., et al., (2012). Comparative analysis of the value of allogeneic hematopoietic stem-cell transplantation in acute myeloid leukemia with monosomal karyotype versus other cytogenetic risk categories. *Journal of Clinical Oncology, 30*(17), 2140–2146. doi: https://doi.org/10.1200/JCO.2011.39.6499.

Cornelissen, J. J., Gratwohl, A., Schlenk, R. F., Sierra, J., Bornhäuser, M., & Juliusson, G., (2012). The European LeukemiaNet AML working party consensus statement on allogeneic HSCT for patients with AML in remission: An integrated-risk adapted approach. *Nat. Rev. Clin. Oncol., 9,* 579–590. doi: https://doi.org/10.1038/nrclinonc.2012.150.

Coustan-Smith, E., Song, G., Shurtleff, S., et al., (2018). Universal monitoring of minimal residual disease in acute myeloid leukemia. *JCI Insight, 3*(9). doi: https://doi.org/10.1172/jci.insight.98561.

Craddock, C., Labopin, M., Pillai, S., Finke, J., Bunjes, D., & Greinix, H., (2011). Factors predicting outcome after unrelated donor stem cell transplantation in primary refractory acute myeloid leukaemia. *Leukemia, 25*, 808–813. doi: 10.1038/leu.2011.13.

Cruz, N. M., Mencia-Trinchant, N., Hassane, D. C., & Guzman, M. L., (2017). Minimal residual disease in acute myelogenous leukemia. *International Journal of Laboratory Hematology, 39*, 53–56. doi: https://doi.org/10.1111/ijlh.12670.

Damm, F., Heuser, M., Morgan, M., et al., (2010). Single nucleotide polymorphism in the mutational hotspot of WT1 predicts a favorable outcome in patients with cytogenetically normal acute myeloid leukemia. *Journal of Clinical Oncology, 28*(4). doi: https://doi.org/10.1200/JCO.2009.23.0342.

De Greef, G. E., Van, P. W. L. J., Boogaerts, M., Huijgens, P. C., Verdonck, L. F., & Vellenga, E., (2005). Criteria for defining a complete remission in acute myeloid leukaemia revisited. An analysis of patients treated in HOVON-SAKK co-operative group studies. *British Journal of Haematology, 128*(2), 184, 185. doi: https://doi.org/10.1111/j.1365-2141.2004.05285.x.

Del, P. M. I., (2016). minimal residual disease in acute myeloid leukemia of adults: Determination, prognostic impact and clinical applications. *Mediterranean Journal of Hematology and Infectious Diseases, 8*(1), 2016052. doi: https://doi.org/10.4084/MJHID.2016.052.

Dick, J. E., & Lapidot, T., (2005). Biology of normal and acute myeloid leukemia stem cells. *International Journal of Hematology, 82*(5), 391, 392. doi: https://doi.org/.

Dohner, H., Estey, E. H., Amadori, S., et al., (2009). Diagnosis and management of acute myeloid leukemia in adults: Recommendations from an international expert panel, on behalf of the European LeukemiaNet. *Blood, 115*(3), 453–474. doi: https://doi.org/10.1182/blood-2009-07-235358.

Döhner, H., Estey, E., Grimwade, D., et al., (2017). Diagnosis and management of AML in adults: 2017 ELN recommendations from an international expert panel. *Blood, 129*(4), 424–447. doi: https://doi.org/10.1182/blood-2016-08-733196.

Döhner, H., Weisdorf, D. J., & Bloomfield, C. D., (2015). Acute myeloid leukemia. *New England Journal of Medicine, 373*(12), 1136–1152. doi: https://doi.org/10.1056/NEJMra1406184.

Dohner, K., & Dohner, H., (2008). Molecular characterization of acute myeloid leukemia. *Haematologica, 93*(7), 976–9782. doi: https://doi.org/10.3324/haematol.13345.

Dong, Y., Shi, O., Zeng, Q., Lu, X., Wang, W., Li, Y., & Wang, Q., (2020). Leukemia incidence trends at the global, regional, and national level between 1990 and 2017. *Experimental Hematology and Oncology, 9*(1). doi: https://doi.org/10.1186/s40164-020-00170-6.

Duval, M., Klein, J. P., He, W., et al., (2010). Hematopoietic stem-cell transplantation for acute leukemia in relapse or primary induction failure. *Journal of Clinical Oncology, 28*, 3730–3738. doi: https://doi.org/10.1200/JCO.2010.28.8852.

Dworzak, M. N., Fritsch, G., Fleischer, C., Printz, D., Froschl, G., & Buchinger, P., (1998). Comparative phenotype mapping of normal vs. malignant pediatric B-lymphopoiesis unveils leukemia-associated aberrations. *Experimental Hematology, 26*(4), 305–313.

Ernst, P., Wang, J., & Korsmeyer, S. J., (2002). The role of MLL in hematopoiesis and leukemia. *Current Opinion in Hematology, 9*(4), 282–285. doi: https://doi.org/10.1097/00062752-200207000-00004.

Falini, B., Bolli, N., Shan, J., et al., (2006). Both carboxy-terminus NES motif and mutated tryptophan(s) are crucial for aberrant nuclear export of nucleophosmin leukemic mutants in NPMc + AML. *Blood, 107*(11), 4514–4523. doi: https://doi.org/10.1182/blood-2005-11-4745.

Falini, B., Nicoletti, I., Martelli, M. F., & Mecucci, C., (2007). Acute myeloid leukemia carrying cytoplasmic/mutated nucleophosmin (NPMc + AML): Biologic and clinical features. *Blood, 109*, 874–880. doi: https://doi.org/10.1182/blood-2006-07-012252.

Fang, M., Storer, B., Estey, E., Othus, M., Zhang, L., & Sandmaier, B. M., (2011). Outcome of patients with acute myeloid leukemia with monosomal karyotype who undergo hematopoietic cell transplantation. *Blood, 118*, 1490–1494. doi: 10.1182/blood-2011-02-339721.

Fasan, A., Haferlach, C., Alpermann, T., et al., (2013). The role of different genetic subtypes of CEBPA mutated AML. *Leukemia, 28*(4). doi: https://doi.org/10.1038/leu.2013.273.

Frohling, S., Scholl, C., Gilliland, D. G., & Levine, R. L., (2005). Genetics of myeloid malignancies: Pathogenetic and clinical implications. *Journal of Clincal. Oncology, 23*, 6284–6289. doi: https://doi.org/.

Gaidzik, V. I., Bullinger, L., Schlenk, R. F., et al., (2011). RUNX1 mutations in acute myeloid leukemia: Results from a comprehensive genetic and clinical analysis from the AML study group. *Journal of Clinical Oncology, 29*(10), 1364–1372. doi: https://doi.org/10.1200/JCO.2010.30.7926.

Gaidzik, V. I., Paschka, P., Späth, D., et al., (2012). TET2 mutations in acute myeloid leukemia (AML): Results from a comprehensive genetic and clinical analysis of the AML study group. *Journal of Clinical Oncology, 30*(12), 1350–1355. doi: https://doi.org/10.1200/jco.2011.39.2886.

Gaidzik, V. I., Schlenk, R. F., Moschny, S., et al., (2009). Prognostic impact of WT1 mutations in cytogenetically normal acute myeloid leukemia: A study of the German-Austrian AML study group. *Blood, 113*. doi: https://doi.org/10.1182/blood-2008-10-183392.

Garg, M., Nagata, Y., Kanojia, D., et al., (2015). Profiling of somatic mutations in acute myeloid leukemia with FLT3-ITD at diagnosis and relapse. *Blood, 126*(22), 2491–2501. doi: https://doi.org/10.1182/blood-2015-05-646240.

Garson, O. M., Hagemeijer, A., Sakurai, M., et al., (1989). Cytogenetic studies of 103 patients with acute myelogenous leukemia in relapse. *Cancer Genetics and Cytogenetics, 40*(2). doi: https://doi.org/10.1016/0165-4608(89)90024-1.

Gomez-Arteaga, A., & Guzman, M. L., (2018). Minimal residual disease in acute myeloid leukemia. *Biological Mechanisms of Minimal Residual Disease and Systemic Cancer, 1100*, 111–125. doi: https://doi.org/10.1007/978-3-319-97746-1_7.

Greif, P. A., Hartmann, L., Vosberg, S., et al., (2018). Evolution of cytogenetically normal acute myeloid leukemia during therapy and relapse: An exome sequencing study of 50 patients. *Clinical Cancer Research, 24*(7), 1716–1726. doi: https://doi.org/10.1158/1078-0432.CCR-17-2344.

Grimwade, D., & Freeman, S. D., (2014). Defining minimal residual disease in acute myeloid leukemia: Which platforms are ready for "prime time"? *Blood, 124*(23), 3345–3347. doi: https://doi.org/10.1182/blood-2014-05-577593..

Grimwade, D., Hills, R. K., Moorman, A. V., et al., (2010). Refinement of cytogenetic classification in acute myeloid leukemia: Determination of prognostic significance of rare recurring chromosomal abnormalities among, 5876 younger adult patients treated in the United Kingdom medical research council trials. *Blood, 116*(3), 354–365. doi: https://doi.org/10.1182/blood-2009-11-254441.

Grimwade, D., Ivey, A., & Huntly, B. J. P., (2015). Molecular landscape of acute myeloid leukemia in younger adults and its clinical relevance. *Blood, 127*(1), 29–35. doi: https://doi.org/10.1182/blood-2015-07-604496.

Haferlach, C., Mecucci, C., Schnittger, S., et al., (2009). AML with mutated NPM1 carrying a normal or aberrant karyotype show overlapping biologic, pathologic, immunophenotypic, and prognostic features. *Blood, 114*(14), 3024–3032. doi: https://doi.org/10.1182/blood-2009-01-197871.

Hamilton, B., Majhail, N., Hirsch, C., et al., (2015). Prognostic impact of molecular mutations in acute myeloid leukemia (AML) and myelodysplastic syndromes (MDS) on allogeneic hematopoietic cell transplant (HCT) outcomes: Adverse impact of TET2 mutations. *Blood, 126*(23), 740. doi: https://doi.org/10.1182/blood.V126.23.740.740.

Heuser, M., Koenecke, C., Gabdoulline, R., et al., (2015). *Molecular Predictors of Outcome in Patients with MDS and AML Following MDS After Allogeneic Hematopoietic Stem Cell Transplantation.* American Society of Hematology 57th Annual Meeting and Exposition: Orlando, FL, USA. doi: https://doi.org/10.1007/s00277-017-3027-5.

Hollein, A., Meggendorfer, M., & Dicker, F., (2018). NPM1 mutated AML can relapse with wild-type NPM1: Persistent clonal hematopoiesis can drive relapse. *Blood Advances, 2*(22), 3118–3123. doi: https://doi.org/10.1182/bloodadvances.2018023432.

Höllein, A., Nadarajah, N., Meggendorfer, M., et al., (2019). Molecular characterization of AML with RUNX1-RUNX1T1 at diagnosis and relapse reveals net loss of co-mutations. *HemaSphere, 3*(1), 178. doi: https://doi.org/10.1097/HS9.0000000000000178.

Houria, D., Delphine, L., Christophe, R., et al., (*2015*). IDH1/2 but not DNMT3A mutations are suitable targets for minimal residual disease monitoring in acute myeloid leukemia patients: A study by the acute Leukemia French Association. *Oncotarget., 6*(39), 42345–42353. doi: https://doi.org/10.18632/oncotarget.5645.

Ivey, A., Hills, R. K., Simpson, M. A., et al., (2016). Assessment of minimal residual disease in standard-risk AML. *New England Journal of Medicine, 374*(5). doi: https://doi.org/10.1056/NEJMoa1507471.

Kansas, G. S., Muirhead, M. J., & Dailey, M. O., (1990). Expression of the CD11/CD18, leukocyte adhesion molecule 1, and CD44 adhesion molecules during normal myeloid and erythroid differentiation in humans. *Blood, The Journal of the American Society of Hematology, 76*, 2483–2494.

Kantarjian, H. M., Keating, M. J., Walters, R. S., McCredie, K. B., & Freireich, E. J., (1988). The characteristics and outcome of patients with late relapse acute myelogenous leukemia. *Journal of Clinical Oncology, 6*, 232–238. doi: 10.1200/jco.1988.6.2.232.

Kayser, S., Schlenk, R. F., Londono, M. C., et al., (2009). Insertion of *FLT3* internal tandem duplication in the tyrosine kinase domain-1 is associated with resistance to chemotherapy and inferior outcome. *Blood, The Journal of the American Society of Hematology, 114*, 2386–2392.

Kelly, L. M., Liu, Q., Kutok, J. L., Williams, I. R., Boulton, C. L., & Gilliland, D. G., (2002). FLT3 internal tandem duplication mutations associated with human acute myeloid leukemias induce myeloproliferative disease in a murine bone marrow transplant model. *Blood, The Journal of the American Society of Hematology, 99*, 310–318.

Kern, W., Bacher, U., Haferlach, C., Schnittger, S., & Haferlach, T., (2010). The role of multiparameter flow cytometry for disease monitoring in AML. *Best Practice and Research Clinical Haematology, 23*(3), 379–390.

Kern, W., Haferlach, T., Schnittger, S., Ludwig, W., Hiddemann, W., & Schoch, C., (2002). Karyotype instability between diagnosis and relapse in 117 patients with acute myeloid leukemia: Implications for resistance against therapy. *Leukemia, 16*(10), 2084–2091.

Kern, W., Schnittger, S., Schoch, C., & Haferlach, T., (2011). Monitoring of minimal residual disease in acute myeloid leukemia. *Atlas of Genetics and Cytogenetics in Oncology and Haematology,* (4), 352–354. doi: https://doi.org/10.4267/2042/38145.

Kersten, B., Valkering, M., Wouters, R., et al., (2016). CD45RA, a specific marker for leukaemia stem cell sub-populations in acute myeloid leukaemia. *British Journal of Haematology, 173*(2), 219–235.

Kim, Y., Jang, J., Hyun, S. Y., et al., (2013). Karyotypic change between diagnosis and relapse as a predictor of salvage therapy outcome in AML patients. *Blood Research, 48*(1), 24–30.

King-Underwood, L., Renshaw, J., & Pritchard-Jones, K., (1996). Mutations in the Wilms' tumor gene WT1 in leukemias. *Blood, 87*(6), 2171–2179.

Koreth, J., Schlenk, R., Kopecky, K. J., et al., (2009). Allogeneic stem cell transplantation for acute myeloid leukemia in first complete remission. *JAMA, 301*(22), 2349–2361.

Koschmieder, S., Halmos, B., Levantini, E., & Tenen, D. G., (2009). Dysregulation of the C/EBP_ differentiation pathway in human cancer. *Journal of Clinical Oncology, 27*(4), 619. doi: 10.1200/JCO.2008.17.9812.

Kostopoulos, I. V., Ntanasis-Stathopoulos, I., Gavriatopoulou, M., Tsitsilonis, O. E., & Terpos, E., (2020). Minimal residual disease in multiple myeloma: Current landscape and future applications with immunotherapeutic approaches. *Frontiers in Oncology, 10*, 860.

Kunacheewa, C., Lin, P., & Manasanch, E. E., (2020). Minimal residual disease in multiple myeloma 2019. *Advances in Cell and Gene Therapy, 3*(1), e71.

Laane, E., Derolf, A. R., Björklund, E., et al., (2006). The effect of allogeneic stem cell transplantation on outcome in younger acute myeloid leukemia patients with minimal residual disease detected by flow cytometry at the end of post-remission chemotherapy. *Haematologica, 91*, 833–836. https://doi.org/10.3324/%25x.

Langebrake, C., Brinkmann, I., Teigler-Schlegel, A., et al., (2005). Immunophenotypic differences between diagnosis and relapse in childhood AML: Implications for MRD monitoring. *Cytometry Part B: Clinical Cytometry: The Journal of the International Society for Analytical Cytology, 63B*, 1–9. http://dx.doi.org/10.1002/cyto.b.20037 .

Larsen, H. Ø., Roug, A. S., Just, T., Brown, G. D., & Hokland, P., (2011). Expression of the hMICL in acute myeloid leukemia-a highly reliable disease marker at diagnosis and during follow-up. *Cytometry Part B: Clinical Cytometry, 82B*(1), 3–8.

Leung, W., Pui, C. H., Coustan-Smith, E., Yang, J., Pei, D., & Gan, K., (2012). Detectable minimal residual disease before hematopoietic cell transplantation is prognostic but does not preclude cure for children with very-high-risk leukemia. *Blood, The Journal of the American Society of Hematology, 120*, 468–472. doi: 10.1182/blood-2012-02-409813.

Ley, T. J., Ding, L., Walter, M. J., et al., (2010). DNMT3A mutations in acute myeloid leukemia. New *England Journal of Medicine, 363*(25), 2424–2433.

Li, S., Garrett-Bakelman, F. E., Chung, S. S., et al., (2016). Distinct evolution and dynamics of epigenetic and genetic heterogeneity in acute myeloid leukemia. *Nature Medicine, 22*(7), 792–799.

Lindsley, R. C., Mar, B. G., Mazzola, E., et al., (2014). Acute myeloid leukemia ontogeny is defined by distinct somatic mutations. *Blood, The Journal of the American Society of Hematology, 125*(9), 1367–1376.

Marcucci, G., Haferlach, T., & Döhner, H., (2011). Molecular genetics of adult acute myeloid leukemia: Prognostic and therapeutic implications. *Journal of Clinical Oncology, 29*(5), 475–486.

Marcucci, G., Maharry, K., Wu, Y. Z., et al., (2010). IDH1 and IDH2 gene mutations identify novel molecular subsets within de novo cytogenetically normal acute myeloid leukemia: A cancer and leukemia group B study. *Journal of Clinical Oncology, 28*(14), 2348.

Marcucci, G., Metzeler, K. H., Schwind, S., et al., (2012). Age-related prognostic impact of different types of DNMT3A mutations in adults with primary cytogenetically normal acute myeloid leukemia. *Journal of Clinical Oncology, 30*(7), 742.

Medeiros, B. C., Minden, M. D., Schuh, A. C., et al., (2007). Characteristics and outcomes of acute myelogenous leukemia patients with very late relapse (>5 years). *Leukemia and Lymphoma, 48*, 65–71. doi: 10.1080/10428190601043252.

Mendler, J. H., Maharry, K., Radmacher, M. D., et al., (2012). RUNX1 mutations are associated with poor outcome in younger and older patients with cytogenetically normal acute myeloid Leukemia and with distinct gene and microRNA expression signatures. *Journal of Clinical Oncology, 30*(25), 3109.

Metzeler, K. H., Becker, H., Maharry, K., et al., (2011). ASXL1 mutations identify a high-risk subgroup of older patients with primary cytogenetically normal AML within the ELN favorable genetic category. *Blood, The Journal of the American Society of Hematology, 118*(26), 6920–6929.

Metzeler, K. H., Herold, T., Rothenberg-Thurley, M., Amler, S., Sauerland, M. C., & Gorlich, D., (2016). Spectrum and prognostic relevance of driver gene mutations in acute myeloid leukemia. *Blood, The Journal of the American Society of Hematology, 128*(5), 686–698.

Metzeler, K. H., Maharry, K., Radmacher, M. D., et al., (2011). TET2 mutations improve the new European leukemiaNet risk classification of acute myeloid leukemia: A cancer and leukemia group B study. *Journal of Clinical Oncology, 29*(10), 1337.

Meyers, S., Downing, J. R., & Hiebert, S. W., (1993). Identification of AML-1 and the (8;21) translocation protein (AML-1/ETO) as sequence-specific DNA-binding proteins: The runt homology domain is required for DNA binding and protein-protein interactions. *Molecular and Cellular Biology, 13*(10), 6336–6345. doi: https://doi.org/10.1128/mcb.13.10.6336.

Micol, J. B., Duployez, N., Boissel, N., et al., (2014). Frequent ASXL2 mutations in acute myeloid leukemia patients with t(8;21)/RUNX1-RUNX1T1 chromosomal translocations. *Blood, the Journal of the American Society of Hematology, 124*(9), 1445–1449.

Middeke, J. M., Beelen, D., Stadler, M., Göhring, G., Schlegelberger, B., & Baurmann, H., (2012). Outcome of high-risk acute myeloid leukemia after allogeneic hematopoietic cell transplantation: Negative impact of Abnl(17p) and -5/5q-. *Blood, The Journal of the American Society of Hematology, 120*, 2521–2528.

Mirkowska, P., Hofmann, A., Sedek, L., et al., (2013). Leukemia surfaceome analysis reveals new disease-associated features. *Blood, The Journal of the American Society of Hematology, 121*(25), e149–159.

Mrózek, K., Marcucci, G., Nicolet, D., et al., (2012). Prognostic significance of the European leukemiaNet standardized system for reporting cytogenetic and molecular alterations in adults with acute myeloid leukemia. *Journal of Clinical Oncology, 30*(36), 4515.

Mrozek, K., Marcucci, G., Paschka, P., Whitman, S. P., & Bloomfield, C. D., (2007). Clinical relevance of mutations and gene-expression changes in adult acute myeloid leukemia with normal cytogenetics: Are we ready for a prognostically prioritized molecular classification? *Blood, 109*(2), 431–448.

Ommen, H. B., (2016). Monitoring minimal residual disease in acute myeloid leukaemia: A review of the current evolving strategies. *Therapeutic Advances in Hematology, 7*(1), 3–16.

Ouyang, J., Goswami, M., Tang, G., et al., (2015). The clinical significance of negative flow cytometry immunophenotypic results in a morphologically scored positive bone marrow in patients following treatment for acute myeloid leukemia. *American Journal of Hematology, 90*(6), 504–510.

Papaemmanuil, E., Gerstung, M., Bullinger, L., et al., (2016). Genomic classification and prognosis in acute myeloid leukemia. *New England Journal of Medicine, 374*(23), 2209–2221.

Paschka, P., (2008). Core binding factor acute myeloid leukemia. *In Seminars in Onclogy, 35*(4), 412, 413. doi: 10.1053/j.seminoncol.2008.04.011.

Paschka, P., Marcucci, G., Ruppert, A. S., et al., (2006). Adverse prognostic significance of KIT mutations in adult acute myeloid leukemia with inv(16) and t(8;21): A cancer and leukemia group B study. *Journal of Clinical Oncology, 24*(24), 3904–3911.

Paschka, P., Marcucci, G., Ruppert, A. S., et al., (2008). Wilms' tumor 1 gene mutations independently predict poor outcome in adults with cytogenetically normal acute myeloid leukemia: A cancer and leukemia group B study. *Journal of Clinical Oncology, 26*(28), 4595.

Paschka, P., Schlenk, R. F., Gaidzik, V. I., et al., (2010). IDH1 and IDH2 mutations are frequent genetic alterations in acute myeloid leukemia and confer adverse prognosis in cytogenetically normal acute myeloid leukemia with NPM1 mutation without FLT3 internal tandem duplication. *Journal of Clinical Oncology, 28*(22), 3636–3643.

Pedersen-Bjergaard, J., Andersen, M. T., & Andersen, M. K., (2007). Genetic pathways in the pathogenesis of therapy-related myelodysplasia and acute myeloid leukemia. *ASH Education Program Book, 1*, 392–397. doi: 10.1182/asheducation-2007.1.392.

Peniket, A., Wainscoat, J., Side, L., et al., (2005). Del (9q) AML: Clinical and cytological characteristics and prognostic implications. *British Journal of Haematology, 129*(2), 210–220.

Pfeiffer, T., Schleuning, M., Mayer, J., et al., (2013). Influence of molecular subgroups on outcome of acute myeloid leukemia with normal karyotype in 141 patients undergoing salvage allogeneic stem cell transplantation in primary induction failure or beyond first relapse. *Haematologica, 98*, 518. doi: 10.3324/haematol.2012.070235.

Romano, A., Palumbo, G. A., Parrinello, N. L., Conticello, C., Martello, M., & Terragna, C., (2019). Minimal residual disease assessment within the bone marrow of multiple myeloma: A review of caveats, clinical significance and future perspectives. *Frontiers in Oncology, 9*(699).

Roshal, M., Flores-Montero, J. A., Gao, Q., et al., (2017). MRD detection in multiple myeloma: Comparison between MSKCC 10-color single-tube and EuroFlow 8-color 2-tube methods. *Blood Advances, 1*(12), 728–732.

Rossi, G., Nomdedéu, G. J. F., Fontana, A., Minervini, M. M., GarcíaDabrio, M. C., & Cascavilla, N., (2013). CD117-CD15 in acute myeloid leukemia: No role as LAIP in the study of minimal residual disease. *European Journal of Haematology, 90*(2), 171–174. http://dx.doi.org/10.1111/ejh.12042 PMid:23167809.

Roug, A. S., Larsen, H. Ø., Nederby, L., et al., (2013). hMICL and CD123 in combination with a CD45/CD34/CD117 backbone - a universal marker combination for the detection of minimal residual disease in acute myeloid leukaemia. *British Journal of Haematology, 164*(2), 212–222.

Salipante, S. J., Fromm, J. R., Shendure, J., Wood, B. L., & Wu, D., (2014). Detection of minimal residual disease in NPM1-mutated acute myeloid leukemia by next-generation sequencing. *Modern Pathology, 27*(11), 1438–1446. doi: https://doi.org/ 10.1038/modpathol.2014.57.

Sanz, M. A., Lo Coco, F., Martin, G., et al., (2000). Definition of relapse risk and role of nonanthracycline drugs for consolidation in patients with acute promyelocytic leukemia: A joint study of the PETHEMA and GIMEMA cooperative groups. *Blood, The Journal of the American Society of Hematology, 96*(4), 1247–1253.

Schlenk, R. F., Döhner, K., Krauter, J., et al., (2008). Mutations and treatment outcome in cytogenetically normal acute myeloid leukemia. *New England Journal of Medicine, 358*(18), 1909–1918.

Schmid, C., Labopin, M., Socie, G., et al., (2015). Outcome and risk factor analysis of molecular subgroups in cytogenetically normal AML treated by allogeneic transplantation. *Blood, 126*, 2062–2069. doi: 10.1182/blood-2015-06-651562.

Schnittger, S., Haferlach, C., Ulke, M., Alpermann, T., Kern, W., & Haferlach, T., (2010). IDH1 mutations are detected in 6.6% of 1414 AML patients and are associated with intermediate risk karyotype and unfavorable prognosis in adults younger than 60 years and unmutated NPM1 status. *Blood, 116*(25), 5486–5496. doi: https://doi.org/10.1182/blood-2010-02-267955.

Schnittger, S., Schoch, C., Kern, W., et al., (2005). Nucleophosmin gene mutations are predictors of favorable prognosis in acute myelogenous leukemia with a normal karyotype. *Blood, 106*, 3733–3739.

Schuurhuis, G. J., Heuser, M., Freeman, S., et al., (2018). Minimal/measurable residual disease in AML: Consensus document from ELN MRD working party. *Blood, The Journal of the American Society of Hematology, 131*(12), 1275–1291.

Shallis, R. M., Wang, R., Davidoff, A., Ma, X., & Zeidan, A. M., (2019). Epidemiology of acute myeloid leukemia: Recent progress and enduring challenges. *Blood Reviews, 36*, 70–87. doi: https://doi.org/10.1016/j.blre.2019.04.005.

Shoket, N., Muzamil, J., Zargar, T. B., et al., (2019). Clinical profile of acute myeloid leukemia in North India and utility of nontransplant measures in its management. *Indian Journal of Medical and Paediatric Oncology, 40*(5), S44–53.

Short, N. J., Rytting, M. E., & Cortes, J. E., (2018). Acute myeloid leukaemia. *The Lancet, 392*(10147), 593–606.

Siegel, R. L., & Miller, K. D., (2019). Cancer statistics. *CA: A Cancer Journal for Clinicians, 69*(1), 7–34. doi: https://doi.org/10.3322/caac.21551.

Strehl, S., König, M., Boztug, H., et al., (2012). All-trans retinoic acid and arsenic trioxide resistance of acute promyelocytic leukemia with the variant STAT5B-RARA fusion gene. *Leukemia, 27*(7), 1606–1610.

Sun, Q. Y., Ding, L. W., Tan, K. T., et al., (2016). Ordering of mutations in acute myeloid leukemia with partial tandem duplication of MLL (MLL-PTD). *Leukemia, 31*(1), 1–10. doi:10.1038/leu.2016.160.

Takahashi, K., Roh, W., Zhang, J., et al., (2015). Clonal evolution of acute myeloid leukemia relapsed after 19 years of remission. *American Journal of Hematology, 90*, E134–135. doi: 10.1002/ajh.24009.

Tang, J. L., Hou, H. A., & Chen, C. Y., (2009). AML1/RUNX1 mutations in 470 adult patients with de novo acute myeloid leukemia: Prognostic implication and interaction with other gene alterations. *Blood, 114*, 5353–5357.

Testa, J. R., Mintz, U., Rowley, J. D., Vardiman, J. W., & Golomb, H. M., (1979). Evolution of karyotypes in acute nonlymphocytic leukemia. *Cancer Research, 39*(9), 3619–3627.

Thol, F., Kölking, B., Damm, F., et al., (2012). Next-generation sequencing for minimal residual disease monitoring in acute myeloid leukemia patients with FLT3-ITD or NPM1 mutations. *Genes, Chromosomes and Cancer, 51*(7), 689–695. doi: https://doi.org/10.1002/gcc.21955.

Tomlinson, B., & Lazarus, H. M., (2017). Enhancing acute myeloid leukemia therapy – monitoring response using residual disease testing as a guide to therapeutic decision-making. *Expert Review of Hematology, 10*(6), 563–574. doi: https://doi.org/10.1080/17474086.2017.1326811.

Tsirigotis, P., Byrne, M., Schmid, C., et al., (2016). Relapse of AML after hematopoietic stem cell transplantation: Methods of monitoring and preventive strategies. A review from the ALWP of the EBMT. *Bone Marrow Transplantation, 51*(11), 1431–1438.

Valk, P. J. M., Verhaak, R. G. W., Beijen, M. A., et al., (2004). Prognostically useful gene-expression profiles in acute myeloid leukemia. *New England Journal of Medicine, 350*(16), 1617–1628.

Van, D. V. V. H. J., SLUIJS-Geling, A. V. D., Gibson, B. E. S., et al., (2010). Clinical significance of flow cytometric minimal residual disease detection in pediatric acute myeloid leukemia patients treated according to the DCOG ANLL97/MRC AML12 protocol. *Leukemia, 24*, 1599–1606. http://dx.doi.org/10.1038/leu.2010.153.

Van, G. M., De Wreede, L. C., Schetelig, J., et al., (2013). Monosomal karyotype predicts poor survival after allogeneic stem cell transplantation in chromosome 7 abnormal myelodysplastic syndrome and secondary acute myeloid leukemia. *Leukemia, 27*, 879–888. doi: 10.1038/leu.2012.297.

Vardiman, J. W., Thiele, J., Arber, D. A., et al., (2009). The 2008 revision of the World Health Organization (WHO) classification of myeloid neoplasms and acute leukemia: Rationale and important changes. *Blood, The Journal of the American Society of Hematology, 114*(5), 937–951.

Vedula, R. S., & Lindsley, R. C., (2017). Measurement of residual disease in acute myeloid leukemia. *Current Hematologic Malignancy Reports, 12*(6), 574–581. doi: 10.1007/s11899-017-0428-4.

Verma, D., Kantarjian, H., Faderl, S., et al., (2010). Late relapses in acute myeloid leukemia: Analysis of characteristics and outcome. *Leukemia and Lymphoma, 51*, 778–782. doi: 10.3109/10428191003661852.

Vidriales, M. B., San-Miguel, J. F., Orfao, A., Coustan-Smith, E., & Campana, D., (2003). Minimal residual disease monitoring by flow cytometry. *Best Practice and Research Clinical Haematology, 16*(4), 599–503.

Virappane, P., Gale, R., Hills, R., et al., (2008). Mutation of the Wilms' tumor 1 gene is a poor prognostic factor associated with chemotherapy resistance in normal karyotype acute myeloid leukemia: The United Kingdom medical research council adult leukaemia working party. *Journal of Clinical Oncology, 26*(33), 5429–5435.

Voskova, D., Schoch, C., Schnittger, S., Hiddemann, W., Haferlach, T., & Kern, W., (2004). Stability of leukemia-associated aberrant immunophenotypes in patients with acute myeloid leukemia between diagnosis and relapse: Comparison with cytomorphologic, cytogenetic, and molecular genetic findings. *Cytometry Part B: Clinical Cytometry, 62B*, 25–38.

Wagner, K., Damm, F., Göhring, G., et al., (2010). Impact of IDH1 R132 mutations and an IDH1 single nucleotide polymorphism in cytogenetically normal acute myeloid leukemia: SNP rs11554137 is an adverse prognostic factor. *Journal of Clinical Oncology, 28*(14), 2356–2364.

Walter, R. B., Gooley, T. A., Wood, B. L., et al., (2011). Impact of pretransplantation minimal residual disease, as detected by multiparametric flowcytometry, on outcome of myeloablative hematopoietic cell transplantation for acute myeloid leukemia. *Journal of Clinical Oncology, 29*, 1190–1197. doi: 10.1200/JCO.2010.31.8121.

Walter, R. B., Gyurkocza, B., Storer, B. E., et al., (2015). Comparison of minimal residual disease as outcome predictor for AML patients in first complete remission undergoing myeloablative or nonmyeloablative allogeneic hematopoietic cell transplantation. *Leukemia, 29*, 137–144. doi: 10.1038/leu.2014.173.

Ward, P. S., Patel, J., & Wise, D. R., (2010). The common feature of leukemia-associated *IDH1* and *IDH2* mutations is a neomorphic enzyme activity converting alpha-ketoglutarate to 2-hydroxyglutarate. *Cancer Cell, 17*, 225–234.

Welch, J. S., Ley, T. J., Link, D. C., et al., (2012). The origin and evolution of mutations in acute myeloid leukemia. *Cell, 150*(2), 264–278.

Wolach, O., & Stone, R. M., (2015). How I treat mixed-phenotype acute leukemia. *Blood, the Journal of the American Society of Hematology, 125*(16), 2477–2485.

Wouters, B. J., Lowenberg, B., Erpelinck-Verschueren, C. A. J., Van, P. W. L. J., Valk, P. J. M., & Delwel, R., (2009). Double CEBPA mutations, but not single CEBPA mutations, define a subgroup of acute myeloid leukemia with a distinctive gene expression profile that is uniquely associated with a favorable outcome. *Blood, The Journal of the American Society of Hematology, 113*(13), 3088–3091. doi: https://doi.org/10.1182/blood-2008-09-179895.

Yi, M., Li, A., Zhou, L., Chu, Q., Song, Y., & Wu, K., (2020). The global burden and attributable risk factor analysis of acute myeloid leukemia in 195 countries and territories from 1990 to 2017: Estimates based on the global burden of disease study 2017. *Journal of Hematology and Oncology, 13*(1), 1–6. doi: https://doi.org/10.1186/s13045-020-00908-z.

Yilmaz, M., Wang, F., Loghavi, S., et al., (2019). Late relapse in acute myeloid leukemia (AML): Clonal evolution or therapy-related leukemia? *Blood Cancer Journal, 9*(2), 1–6. doi: 10.1038/s41408-019-0170-3.

Zeijlemaker, W., Kelder, A., Oussoren-Brockhoff, Y. J. M., et al., (2015). A simple one-tube assay for immunophenotypical quantification of leukemic stem cells in acute myeloid leukemia. *Leukemia, 30*(2), 439–446.

Zeijlemaker, W., Kelder, A., Oussoren-Brockhoff, Y. J., et al., (2016). Peripheral blood minimal residual disease may replace bone marrow minimal residual disease as an immunophenotypic biomarker for impending relapse in acute myeloid leukemia. *Leukemia, 30*(3), 708–715. http://dx.doi.org/10.1038/leu.2015.255.

Zhou, Y., & Wood, B. L., (2017). Methods of detection of measurable residual disease in AML. *Current Hematologic Malignancy Reports, 12*(6), 557–567. doi: https://doi.org/10.1007/s11899-017-0419-5. 2013.

Zhou, Y., Othus, M., Araki, D., et al., (2016). Pre- and post-transplant quantification of measurable ('minimal') residual disease via multiparameter flow cytometry in adult acute myeloid leukemia. *Leukemia, 30*(7), 1457–1464. doi: 10.1038/leu.2016.46.

Index

D

E

L

M

N

O

P

Q

R

Z